ENCYCLOPAEDIA OF CHEMISTRY

Amit Biswas

ANMOL PUBLICATIONS PVT. LTD.
NEW DELHI - 110 002 (INDIA)

ANMOL PUBLICATIONS PVT. LTD.
H.O.: 4374/4B, Ansari Road, Darya Ganj,
New Delhi-110 002 (India)
Ph.: 23278000, 23261597

B.O.: No. 1015, Ist Main Road, BSK IIIrd Stage
IIIrd Phase, IIIrd Block
Bangalore - 560 085 (India)
Visit us at: www.anmolpublications.com

Encyclopaedia of Chemistry

First Published, 2007

PRINTED IN INDIA

Printed at Mehra Offset Press, Delhi.

Preface

Encyclopaedia of Chemistry covers all aspects of chemistry, including Agrochemistry, Analytical chemistry, Applied chemistry, Biochemistry, Biological Chemistry, Biotechnology, Chemical engineering, Combinatorial Chemistry, Cosmochemistry, Environmental Chemistry, Femtochemistry, Food Chemistry, General chemistry, Geochemistry, Inorganic chemistry , Makromolecular chemistry, Mechanochemistry, Medicinal chemistry, Nanochemistry, Organic chemistry, Physical chemistry, Electrochemistry, Solid State Chemistry and Theoretical chemistry.

This volume is ideal for the chemistry professional and student as well as general readers. It provides a comprehensive overview of chemistry with up-to-date entries covering all related disciplines.

This encyclopaedia integrates all of the diverse approaches that have gone into the making of the modern field of Chemistry. It is unique among available reference works in that it covers the most recent topics of research. Encyclopaedia of Chemistry is intended for all those interested in this field. It will be of immense value to students of all kinds, especially students of different fields of chemistry.

It will also be a major source of reference for professionals and students, who require an authoritative and up-to-date guide to assist them in their work. In preparation of this book, the author has freely consulted large number of books and journals so no authenticity is claimed. Author is especially thankful to Anmol Publications Pvt. Ltd., New Delhi for shaping this book in its final form. Suggestions for further improvement of this book are not only welcome but also greatly appreciated.

Author

A

Abatement To lower, depreciate, or reduce. Often refers to emission reductions in air quality control.

Abiotic This is a term used to describe anything which is characterized by the absence of life or incompatible with life. In toxicology and ecotoxicology it refers to physical (e.g. heat, sunlight) or chemical processes (e.g. hydrolysis) that are capable of modifying chemical structures.

Abiotic transformation Any process in which a chemical in the environment is modified by non-biological mechanisms.

Ablation The weathering of a glacier by surface melting, or rock weathering by hydraulic erosion.

Absolute configuration The spatial arrangement of the atoms of a chiral molecular entity (or group) and its stereochemical description.

Absolute error The difference between the approximate and exact value in any calculation.

Absolute zero of temperature The Absolute Zero of Temperature is the lowest temperature on the thermodynamic scale of temperature and is -273°C.

Absorbance Absorbance (A) is defined as $-\log(1-\alpha) = \log(1/\tau)$, where α is the absorptance and τ the transmittance of a medium through which a light beam passes.

Absorbed dose (In radiation) The energy imparted to matter in a suitably small element of volume by ionizing radiation divided by the mass of that element of volume (ISO, 1972). The SI unit for absorbed dose is joule per kilogram (J kgW-1) and its special name is gray (Gy).

Absorbed dose For any ionising radiation, absorbed dose (D) is the mean energy imparted to an element of irradiated matter divided by the mass of that element.

Absorbed dose (in toxicology) The amount of a chemical absorbed into the body or into organs and tissues of interest.

Absorptance Absorptance (α) is ratio of the radiant or luminous flux in a given spectral interval absorbed in a medium to that of the incident radiation. Also called absorption factor.

Absorption Absorption is the process whereby a gas is taken up and dispersed in the bulk of a solid or a liquid, or where a liquid is taken up and dispersed in the body of a solid.

Absorption (in colloid and surface chemistry) A process whereby, when two phases are brought into contact, a given component is transferred from one phase to the other . Experimental differentiation of absorption and adsorption may be difficult, and sometimes the two processes occur simultaneously; in such cases the term “sorption” is used.

Absorption (In radiation) A phenomenon in which radiation transfers to matter which it traverses some or all of its energy.

Absorption coefficient Absorption coefficient (a) is the relative decrease in the intensity of a collimated beam of electromagnetic radiation, as a result of absorption by a medium, during traversal of an infinitesimal layer of the medium, divided by the length traversed.

Absorption coefficient, molar Molar absorption coefficient (ε) is absorption coefficient divided by amount-of-substance concentration of the absorbing material in the sample solution (ε = a/c). The SI unit is m^2mol^{-1}. Also called extinction coefficient, but usually in units of $dm^3cm^{-1}mol^{-1}$.

Absorption cross section A measurement of an atom or molecule's ability to absorb light at a specified wavelength, measured in square cm/particle.

Absorption spectra Absorption Spectra result from the interaction of electromagnetic radiation with matter, where the energies that are removed from the continuous spectrum of light by the absorption medium show up as black lines or bands. The absorption spectrum of a substance is obtained by passing a beam of light through the substance in a spectrometer and examining the light that emerges. When a substance is capable of emitting a spectrum, the lines of the emission spectrum are in exactly the same positions as the lines and bands in the absorption spectrum.

Abstraction A chemical reaction or transformation, the main feature of which is the bimolecular removal of an atom (neutral or charged) from a molecular entity. For example: CH_3COCH_3 + $(i\text{-}C_3H_7)_2N^-$ $(CH_3COCH_2)^-$ + $(i\text{-}C_3H_7)_2NH$

(proton abstraction from acetone)

CH_4 + $Cl^\cdot$ $H_3C^\cdot$ + HCl

(hydrogen abstraction from methane).

Abstraction reaction A reaction that takes any atom away from another chemical species. Classical examples in atmospheric chemistry are the gas phase removal of hydrogen from methane by hydroxyl radical or the following solution phase reaction: $HSO_3^- + H_2O_2 \longrightarrow HSO_4^- + H_2O$.

Abundance of substances The Abundance of Substances is the ratio of the total mass of a specified element in the earth's crust to the total mass of the earth's crust. It is often expressed as a percentage.

Abzyme See catalytic antibody.

Acceleration Measure of how fast velocity is changing, so we can think of it as the change in velocity over time. The most common use of acceleration is acceleration due to gravity which can also appear as the gravitational constant (9.8 m/s^2).

Acceptable daily intake (Pesticide residues) The acceptable daily intake of a chemical is the daily intake which, during an entire life time, appears to be without appreciable risk to the health of the consumer on the basis of all the known facts at the time when a toxicological assessment is carried out. It is expressed in milligrams of the chemical per kilogram of body weight.

Acceptable daily intake for man (ADI) (Food additives) The acceptable daily intake (ADI) for man, expressed on a body weight basis (mg/kg body weight) is the amount of a food additive that can be taken daily in the diet, even over a lifetime, without risk. It is allocated only to substances for which the available data include either the results of adequate short-term and long-term toxicological investigations, or satisfactory information on the biochemistry and metabolic fate of the compound, or both.

Acceptable daily intake not specified (Pesticide residues) An ADI without an explicit indication of the upper limit of intake may be assigned to substances of very low toxicity, especially those that are food constituents or that may be considered as foods or normal metabolites in man. This expression was adopted as a more suitable expression than "ADI not limited" which was previously used. An additive having an "ADI not specified" must meet the criteria of good manufacturing practices, for example, it should have proved technological efficacy and be used at the minimum level of technological efficacy, it should not conceal inferior food quality or adulteration, and it should not create a nutritional imbalance. The above expression means that, on the basis of available data (chemical, biochemical, and toxicological), the total daily intake of the substance arising from its use or uses at levels necessary to achieve the desired effect and from its acceptable background in food, does not represent a hazard to health. For this reason, and for reasons stated in the individual evaluations, the establishment of an acceptable daily intake expressed in mg/kg body weight is not deemed necessary.

Acceptable residue Acceptable residues in human food that have been established for antibiotics found in foods.

Acceptable risk This concept relates to the probability of suffering disease or injury that will be tolerated by an individual, group or society. Acceptability of risk depends on the scientific data, social, economic and political factors, and on the perceived benefits arising from the a chemical or process.

Acceptor number (AN) A quantitative measure, devised by Gutmann (1976), of Lewis acidity.

Accretion The addition of air particles to hydrated drops (snow, rain, sleet, etc.) by coagulation as the drops fall through the sky.

Accumulation Successive additions of a substance to a target organism, or organ, or to part of the environment, resulting in an increasing quantity or concentration of the substance in the organism, organ, or environment.

Accumulation mode particles These are particles that are formed in the atmosphere due to both the chemical and physical processes that take place with the interactions of primary gaseous emissions. The primary gaseous emissions are injected into the atmosphere by combustion processes such as from a car or from a coal burning plant. .

Accuracy In measurement indicates that a set of measurements are close to the true answer, though they are not necessarily precise.

Acetaldehyde (C_2H_4O) This is a fairly simple aldehyde (second in the analogous series after formaldehyde) that is found in the atmosphere as a result of emissions from the manufacture of acetic acid, plastics, raw materials, and as a product in some polluted air oxidation reactions.

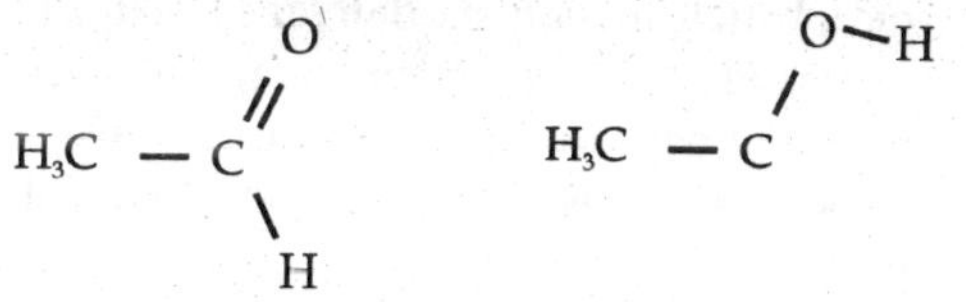

Acetate (CH_3COO^-, $C2H_3O_2^-$) acetate ion. 1. an ion formed by removing the acidic hydrogen of acetic acid, $HC_2H_3O_2$.

2. a compound derived by replacing the acidic hydrogen in acetic acid. 3. A fiber made of cellulose acetate.

Acetic acid ($C_2H_4O_2$) Is a carbonyl compound that is emitted into the troposphere by both natural and anthropogenic processes. In the troposphere, acetic acid is present in the gas phase and is highly water soluble. Since acetic acid is highly water soluble, it is found in high concentration as acidic precipitation, such as in fog water and cloud droplets in urban areas.

Acetone (C_3H_6O) Is a carbonyl compound that is found in the atmosphere as a reactive gas. Acetone is considered to be a volatile organic compound (VOC), which is emitted into the atmosphere by industrial processes. Acetone has been linked to the formation of ozone in the troposphere due to the fact that it is a source of free radicals.

Achiral A molecule that's superimposable on its mir ror image. Achiral molecules do not rotate plane-polarized light.

Acid Acid is a type of compound that contains hydrogen and dissociates in water to produce positive hydrogen ions. The reaction, for an acid HA is commonly written: $HA \rightleftarrows H^+ + A^-$

In fact, the hydrogen ion (the proton) is solvated, and the complete reaction is:

$HA + H_2O \rightleftarrows H_3O^+ + A^-$

This definition of acids comes from the Arrhenius theory. Such acids tend to be corrosive substances with a sharp taste, which turn litmus red and give colour changes with other indicators. They are referred to as protonic acids and are classified into strong acids, which are almost completely dissociated in water (e.g. sulphuric acid and hydrochloric acid), and weak acids, which are only partially dissociated (e.g. acetic acid and hydrogen sulphide). The strength of an acid depends on the extent to to which it dissociates, and is measured by its dissociation constant.

In the Lowry-Brønsted theory of acids and bases (1923), the definition was extended to one in which an acid is a proton donor (a Brønsted acid), and a base is a proton acceptor (a Brønsted base). An important feature of the Lowry-Brønsted concept is that when an acid gives up a proton, a conjugate base is formed that is capable of accepting a proton.

Acid $\rightleftarrows$ base + H^+

Similarly, every base produces its conjugate acid as a result of accepting a proton.

Base + $H^+ \rightleftarrows$ acid

For example, acetate ion is the conjugate base of acetic acid, and ammonium ion is the conjugate acid of ammonia.

As the acid of a conjugate acid/base pair becomes weaker, its conjugate base becomes stronger and vice versa.

A further extension of the idea of acids and bases was made in the Lewis theory. In this, a G. N. Lewis acid is a compound or atom that can accept a pair of electrons and a Lewis base is one that can donate an electron pair. This definition encompasses "traditional" acid-base reactions, but it also includes reactions that do not involve ions.

Acid pulse (dry deposition) deposit of powder-like substance over the ground surface; especially effecting plant leaves; that when contacted by water has a very low pH.

Acid rain Acid Rain results from the burning of fossil fuels containing sulphur, when sulphur dioxide is produced during the combustion process.

Although the nitrogen in the air is normally uncreative, nitrogen oxides are also produced during the high temperature combustion process.

Sulphur dioxide and nitrogen oxides, NO_x, are toxic acidic gases which readily react with the water in the atmosphere to form a mixture of sulphuric acid, nitric acid and nitrous acid. The dilute solutions of these acids which result give rain water a far greater acidity than normal (i.e. a lower pH) and this is known as acid rain.

Acid-base indicators Acid-Base Indicators are used when an acid is titrated against a base in Volumetric Analysis, to give a visual indication of the end-point of the titration. An acid-base indicator usually changes colour over a range of about 2 pH units.

Each indicators change colours at a different and unique pH.

Acid-base titration An Acid-Base Titration is the analytical technique in Volumetric Analysis , where an acid of known concentration is used to neutralise a known volume of a base, and the observed volume of the acid required is used to determine the unknown concentration of the base.

An acid-base indicator is used to determine the end-point of the titration.

Acidic Describes a solution with a high concentration of H^+ ions.

Acidification This process happens when compounds like ammonia, nitrogen oxides and sulphur dioxides are converted in a chemical reaction into acidic substances. Most of the compounds are a direct result of air pollution.

Ý $2FeS_2 + 6H_2O + 7O_2 = 4SO_42- + 8H^+ + 2Fe(OH)_2$.

Acidity 1. Of a compound: For Brønsted acids it means the tendency of a compound to act as a hydron donor. It can be quantitatively expressed by the acid dissociation constant of the compound in water or some other specified medium. For Lewis acids it relates to the association constants of Lewis adducts and -adducts.

2. Of a medium: The use of the term is mainly restricted to a medium containing Brønsted acids, where it means the tendency of the medium to hydronate a specific reference base. It is quantitatively expressed by the appropriate acidity function.

Acidity constant The equilibrium constant for splitting off a hydron from a Brønsted acid.

Acidity function Any function that measures the thermodynamic hydron-donating or -accepting ability of a solvent system, or a closely related thermodynamic property, such as the tendency of the lyate ion of the solvent system to form Lewis adducts. (The term "basicity function" is not in common use in connection with basic solutions.) Acidity functions are not unique properties of the solvent system alone, but depend on the solute (or family of closely related solutes) with respect to which the thermodynamic tendency is measured.

Commonly used acidity functions refer to concentrated acidic or basic solutions. Acidity functions are usually established over a range of composition of such a system by UV/VIS spectrophotometric or NMR measurements of the degree of hydronation (protonation or Lewis adduct formation) for the members of a series of structurally similar indicator bases (or acids) of different strength: the best known of these functions is the Hammett acidity function H_O (for uncharged indicator bases that are primary aromatic amines).

Acids and bases When the oxides of non-metals reacted with water they formed acids. For example, when sulphur dioxide is dissolved in water, an acidic solution of Sulphurous Acid results.

$SO_2 + H_2O \Rightarrow H_2SO_3$.

Similarly, the oxides of metals react with water to form bases. For example, when sodium oxide reacts with water, sodium hydroxide (i.e. caustic soda) is formed.

$Na2O + H_2O \Rightarrow 2\ NaOH$

Aconitase Trivial name for citrate (isocitrate)hydro-lyase (aconitate hydratase), EC4.2.1.3, which catalyzes the interconversion of citrate, cis-aconitate ((Z)-prop-1-ene-1,2,3-tricarboxylate) and isocitrate. The active enzyme contains a catalytic (4Fe-4S) cluster.

Actinides The actinides (actinons or actinoids) are the fourteen elements from thorium to lawrencium inclusive, which follow actinium in the periodic table. The position of actinium is somewhat equivocal and, although not itself an actinide, it is often included with them for comparative purpose. The series includes the following elements: thorium (Th), protactinium (Pa), uranium (U), neptunium (Np), plutonium (Pu), amercium (Am), curium (Cm), berkelium (Bk), californium (Cf), einsteinium (Es), fermium (Fm), mendelevium (Md), nobelium (No) and lawrencium (Lr). Every known isotope of the actinide elements is radioactive. Traces of Pa, Np and Pu are consequently found, but only Th and U occur naturally to any useful extent.

Action level (i) The level of a pollutant at which specified emergency countermeasures, such as the seizure and destruction of contaminated materials, evacuation of the local population or closing down the sources of pollution, are to be taken (ii) the level at which some kind of preventive action (not necessarily of an emergency nature) is to be taken; (iii) a level of exposure of workers to airborne harmful substances in workrooms to be determined by the competent authority; it is distinctly below the exposure limit and consequently such exposures below the action level do not usually necessitate application of all the preventive measures, especially of a medical nature, foreseen for exposures exceeding the action level. This level may lie between a third and a half of the exposure limit.

Activated complex An activated complex, often characterized by the superscript , is defined as that assembly of atoms which corresponds to an arbitrary infinitesimally small region at or near the col (saddle point) of a potential energy surface. See also transition state.

Activation energy In general, activation energy is the energy that must be added to a system in order for a process to occur, even though the process may already be thermodynamically possible. In chemical kinetics, the activation energy is the height of the potential barrier separating the products and reactants. It determines the temperature dependence of the reaction rate.

Active center The location in an enzyme where the specific reaction takes place.

Active site See active center.

Activity Activity (a) is a thermodynamic function used in place of concentration in equilibrium constants for reactions involving nonideal gases and solutions. For the species i activity is defined as

a = f × c

where a_i is the activity of the species i, c_i is its molar concentration, and f_i is a dimensionless quantity called the activity coefficient.

Activity coefficient Activity coefficient (γ or f) is a fractional number which when multiplied by the molar concentration of a substance in solution yields the chemical activity. This term gives an idea of how much interaction exists between molecules at higher concentration.

In solutions of very low ionic strength, when m is less than 0.01, the Debye-Hückel limiting law can be used to calculate approximate activity coefficients

$\log \gamma_i = -0.5.Z_i^2.\sqrt{\mu}$ where γ_i = activity coefficient of the species i, z_i = charge on the species i and μ = ionic strength of the solution.

Acute effects Effects that occur rapidly following exposure and are of short duration (WHO, 1979).

Acute toxicity test An experimental animal study in which the adverse effects occur in a short time (from l-7 days) following the administration of a single or multiple doses of a chemical. The most frequently used acute toxicity test involves determination of the median lethal dose (LD50) of the compound. The LD50 has been defined as "a statistically derived expression of a single administered dose of a material that can be expected to kill 50% of the animals" (WHO, 1978a).

Acute toxicity The adverse effects occurring within a short time of administration of a single dose or multiple doses given within 24 hours .

Acylaction reaction This reaction invol ves the introduction of an acyl group (RCO—) into a compound.

Method

An Alkyl halide is reacted with an alcohol or a carboxylic acid anhydride e.g.

RCOCl + R'OH ⇒ RCOOR' + HCl

The introduction of an acetyl group (CH3CO—) is Acetylation, a process used for protecting —OH groups in Organic synthesis.

Addition (alpha-addition) A chemical reaction resulting in a single reaction product from two or three reacting chemical species, with formation of two new chemical bonds to the same atom in one of the reactant molecular entities. The synonymous term 1/1/addition is also used. For example: Cl_2C: + CH_3OH Cl_2CHOCH_3

(This particular example can also be viewed as an insertion reaction.) In inorganic chemistry such -addition reactions, generally to a metallic central atom, are known as "oxidative additions".

Addition compound Complex compound.An addition compound contains two or more simpler compounds that can be packed in a definite ratio into a crystal. A dot is used to separate the compounds in the formula. For example, $ZnSO_4 \cdot 7\ H_2O$ is an addition compound of zinc sulfate and water. This represents a compound, and not a mixture, because there is a definite 1:7 ratio of zinc sulfate to water in the compound. Hydrates are a common type of addition compound.

Additive effect An additive effect is the overall consequence which is the result of two chemicals acting together and which is the simple sum of the effects of the chemicals acting independently.

Additivity principle The hypothesis that each of several structural features of a molecular entity makes a separate and additive contribution to a property of the substance concerned. More specifically, it is the hypothesis that each of the several substituent groups in a parent molecule makes a separate and additive contribution to the standard Gibbs energy change (or Gibbs energy of activation) corresponding to a particular equilibrium (or rate of reaction).

Adduct (pi-adduct) An adduct formed by electron-pair donation from a pi orbital into a sigma orbital, or from a sigma orbital into a pi orbital, or from a pi orbital into a pi orbital.

Adduct (sigma adduct) The product formed by the attachment of an electrophilic or nucleophilic entering group or of a radical to a ring carbon of an aromatic species so that a new sigma bond is formed and the original conjugation is disrupted.

Adduct A new chemical species AB, each molecular entity of which is formed by direct combination of two separate molecular entities A and B in such a way that there is change in connectivity, but no loss, of atoms within the moieties A and B. Stoichiometries other than 1:1 are also possible, e.g. a bis-adduct (2:1). An "intramolecular adduct" can be formed when A and B are groups contained within the same molecular entity.

This is a general term which, whenever appropriate, should be used in preference to the less explicit term complex. It is also used specifically for products of an addition reaction.

Adenosine 5'-triphosphate (ATP) Key nucleotide in energy-dependent cellular reactions, in combination with Mg(II). The reaction: ATP + water $\longrightarrow$ ADP + phosphate is used to supply the necessary energy.

Adiabatic lapse rate The rate of decrease of temperature with increasing altitude in the atmosphere. If heat is neither gained nor lost from the air parcel under consideration, then the lapse rate is said to be adiabatic.

Adiabatic process Adiabatic process is a thermodynamic process in which no heat enters or leaves the system. In general, an adiabatic change involves a fall or rise in temperature of the system.

Adolph strecker (1822-1871) Born in Darmstadt, the son of an archivist, and studied chemistry at Giessen under his fellow Darmstadter Justus von Liebig (1803-1873). He then became Professor of Chemistry at the University of Kristiania (Oslo), moving to Tübingen in 1860 and to Würzburg ten years later. Strecker synthesised alanine from acetaldehyde via its condensation product with ammonia and hydrogen cyanide in 1850. Twenty-five years later, Emil Erlenmeyer (1825-1909) developed this reaction into a general synthesis of amino-acids, and showed that the intermediate was probably the á-aminonitrile. It remains very popular today, nearly one and a half centuries after its initial appearance. Strecker also synthesised taurine and organometallic compounds of mercury, antimony and tin. His name is also associated with the Strecker degradation (1862) of á-amino-acids, and the Strecker sulphite alkylation (1868), a result of his work on alizarin.

Adrenodoxin A (2Fe-2S) ferredoxin involved in electron transfer from $NADPH^+$ (via a reductase) to cytochrome P-450 in the adrenal gland.

Adsorption Adsorption is the surface phenomenon involving the attachment of a gas or liquid to the surface of a solid. This surface property of solids is exploited in Chromatography for the separation of mixtures.

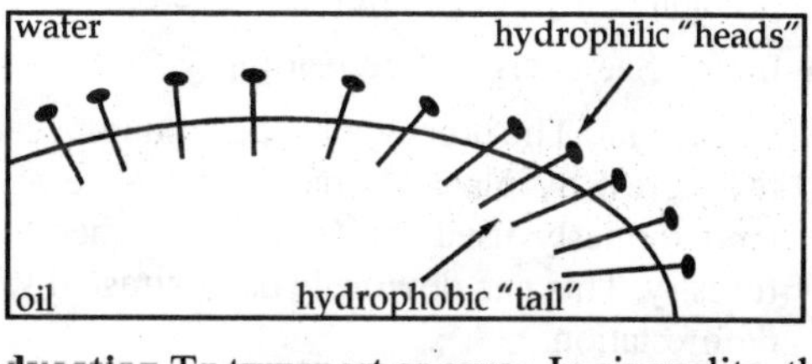

Advection To transport or carry. In air quality, the rate at which particulate matter is transported.

Adverse effect This is abnormal, undesirable or harmful effect to an organism, indicated by some result such as mortality, altered food consumption, altered body and organ weights,altered enzyme levels or visible (pathological) change. An effect may be classed as adverse effect if it causes functional or anatomical damage, causes irreversible changes or increases the susceptibility of the organism to other chemical or biological stress. A non-adverse effect will usually be reversed when exposure to the chemical ceases.

Aerobe An organism that needs dioxygen for respiration and thus for growth.

Aerobic Oxygen-dependent or requiring oxygen.

Aerobic respiration Organisms utilize oxygen to break down components, derive energy, and generate needed biomolecules. Carbohydrates are cycled into water and carbon dioxide.

Aerodynamic diameter of a particle The diameter of a spherical particle of unit density that has the same settling velocity in air as the particle in question

Aerosol Aerosols are colloidal dispersions of liquid or solid particles in a gas, as in a mist or smoke. The commonly used aerosol sprays contain an inert propellant liquefied under pressure. The pressure of the gas causes the mixture to be released as a fine spray (aerosol) or foam (aerogel) when a valve is opened.

Aerosol particles One of the components of an atmospheric air parcel, comprised of minute solids particles part of which is almost certainly water.

Aerosol scavengers Cloud droplets which attract an aerosol into snow, rain or some other water precipitate by coagulation and accretion

A-factor See energy of activation.

Afforestation The process or act of changing land into forest by planting trees, seeding, etc. on land formerly used for something other than forestry. This can obviously be contrasted with deforestation.

Agonist Any molecule that improves the activity of a different molecule; e.g., a hormone, which acts as an agonist when it binds to its receptor, thus triggering a biochemical response.

Agostic The term designates structures in which a hydrogen atom is bonded to both a carbon atom and a metal atom. The term is also used to characterize the interaction between a CH bond and an unsaturated metal centre, and to describe similar bonding of a transition metal with Si-H compounds. The expression "-hydrido-bridged" is also used to describe the bridging hydrogen.

Air parcel A volume of air, the component makeup and concentrations of which remain relatively static, that tends to travel around the earth changing geographic location and position above earth as an intact unit.

Air pollution Air pollution is the degradation of the quality of the atmosphere by chemical and particulate contamination. Acid Rain is a manifestation of the deterioration of the quality of the atmosphere as a result of the combustion of fossil fuels.

Air quality assessment A prescribed level of atmospheric pollution allowed for a certain compound during a specific time in a specific geographical area. Standards are set by some regulating body, office or agency.

Albedo Refers to reflection power or the portion of solar radiation that reaches a body (earth) surface and is immediately reflected back into the atmosphere. The uncertainty in the reflective power of clouds unfortunately increases the complexity of determining clouds' effects on modeling global warming.

Albert hofmann (1906) Born in Baden, Switzerland. Worked for Sandoz. Famous for his work on the ergot alkaloids, including the accidental discovery of the pyschotomimetic effects of LSD in 1943 and the total synthesis of ergotamine in 1961. Despite experiencing the first 'acid trip', Albert Hofmann is still alive and well at the age of 92.

Albumin A type of protein, especially aprotein of blood plasma which transports various substances, including metal ions, drugs and xenobiotics.

Alchemy Alchemy was the so-called black art of renaissance Europe which sought ways of converting everyday substances into gold.

Alcohols aliphatic The aliphatic alcohols are a series of homologous series organic compounds containing one or more hydroxyl groups (-OH) attached to an Alkyl Radical.

The aliphatic alcohols can be regarded as derivatives of alkanes in which one or more hydrogen atoms have been replaced by hydroxyl groups (-OH). The general formula of saturated aliphatic alcohols is CnH2n+1OH, where n=1,2,3, etc. The saturated carbon chain is often designated by the symbol R, so that ROH can represent any alcohol in the homologous series.

Methanol and ethanol are the first two members of the series. Compounds of this type with one hydroxyl group per molecule are known as monohydric alcohols.

Alcohols definition Alcohols are the homologous series of organic compounds of general formula, ROH, where R is any alkyl group or substituted alkyl group. An alcohol may be a primary alcohol, secondary alcohol, or tertiary alcohol, depending on whether the carbon atom attached to the hydroxyl group, -OH, is combined with an alkyl group attached to one, two or three carbon atoms, respectively.

Alcohols dihydric Ethylene glycol, CH2OHCH2OH, is the most important dihydric alcohol and approximately 75% of that produced is used as an anti-freeze agent.

Alcohols properties chemical The chemical properties of any given aliphatic alcohol depends on the nature of the alkyl group in the molecule and on the properties of the hydroxyl group.

Alcohols react with organic acids to form Esters. The reaction proceeds slowly but the rate of esterification is increased by the presence of hydrogen ions, which act as a catalyst in the reaction. Sulphuric Acid in addition to acting as a source of hydrogen ions also helps to increase the yield of ester by absorbing the water as it is formed in the reaction.

Alcohols are very weak acids, intermediate in strength between acetylene and water. They undergo substitution with strongly electropositive metals such as sodium.

Alcohols react with phosphorus pentachloride, when the hydroxyl group is replaced by a chlorine atom. Thus, when the hydroxyl group is replaced by a chlorine atom from phosphorus pentachloride fumes of hydrogen chloride are evolved and this is used in testing for the presence of the hydroxyl group in a compound.

Aldehyde A molecule containing a terminal carbonyl (CHO) group. Also a functional group.

Aldehydes Aldehydes are broad class of organic compounds having the generic formula RCHO, and characterized by unsaturated carbonyl group (C=O). They are formed from alcohols by either dehydrogenation or oxidation. Their chemical derivation is indicated by the name al(cohol) + dehyd(rogenation).

Aleksandr erminingeldovich arbuzov (1877-1968) Born in Arbuzov-Baran, near Kazan, and studied chemistry at the University of Kazan under Aleksandr Saytzev (or Zajcev). After a brief spell at the Institute of Forestry and Agriculture in Novo-Aleksandria, he spent the rest of his career at the University of Kazan. He was a major figure in the development of organophosphorus chemistry, notably the esters and thioesters of organophosphoric acids. His name is associated with the Arbuzov reaction (also called the Michaelis-Arbuzov rearrangement) which is used to prepare phosphonates from orthophosphates. He was also interested in the history of chemistry, perhaps not surprisingly given the association of Kazan with Butlerov and Saytzev.

Aleksandr porfirevið borodin (1833-1887) Born in St Petersburg, the illegitimate son of a prince, a curious link with another pioneer of condensation chemistry, Edward Frankland, who was also illegitimate. He studied chemistry under Nikolai Nikolaevið Zinin (1812-1880) and succeeded him as Professor of Chemistry at the St Petersburg Academy of Medicine and Surgery in 1864. He prepared methyl bromide from silver acetate in 1861, but another eighty years elapsed before Heinz and Cläre Hundiecker converted Borodin's synthesis into a general method, the Hunsdiecker or Hunsdiecker-Borodin reaction (1942). Borodin deserves to have more reactions named after him, as he developed general methods of condensing aldehydes and fluoridating organic compounds, a synthesis of fatty acids, and the well-known estimation of urea with hypobromite. He also pioneered the medical and chemical education of women, until the government stopped the teaching of women medical students in 1887. Borodin composed music as a hobby and kept a piano outside his laboratory.

Alicyclic compounds Alicyclic compounds are ring compounds containing single and/or multiple carbon to carbon bonds. Alicyclic compounds do not contain the conjugated double bonds which are characteristic of aromatic compounds.

Aliphatic compound An aliphatic compound is an organic compound which consists of carbon and hydrogen in a structure which consists of open unbranched chains of carbon atoms,

open branched chains of carbon atoms, or

Closed chain of carbon atoms (i.e. ring or cyclic structures where the compounds have chemical properties that resemble those in the alkanes, alkenes, or alkynes).

Alkaline earth An oxide of an alkaline earth metal, which produces an alkaline solution in reaction with water.

Alkali earth metal (Alkaline-earth metals) Elements in the second column (from the left) of the periodic table all fall into this series. These elements are in general white, differing by shades of color or casts; they are malleable, extrudable and machinable. These elements may be made into rods, wire or plate. Also, these elements are less reactive than the alkali metals and have higher melting points and boiling points.

Alkali metals Alkali metal is a term that refers to six elements: lithium (Li), sodium (Na), potassium (K), rubidium (Rb), caesium (Cs), and francium (Fr). These elements make up group 1 of the periodic table of elements. They all form singly charged positive ions, and are extremely reactive. They react violently with water, forming hydroxides and releasing hydrogen gas and heat. Caesium and francium are the most reactive and lithium is the least.

Alkaloids Alkaloids are basic nitrogen organic compounds (mostly heterocyclic) derived from plants and having diverse pharmacological properties. Alkaloids include morphine, cocaine, atropine, quinine, and caffeine, most of which are used in medicine as analgesics or anaesthetics. Some alkaloids are poisonous, e.g. strychnine and coniine, and colchicine inhibit cell division.

Alkane A molecule containing only C-H and C-C single bonds.

Alkanes Alkanes (paraffins) are acyclic branched or unbranched hydrocarbons having the general formula C_nH_{2n+2}, and therefore consisting entirely of hydrogen atoms and saturated carbon atoms. In systematic chemical nomenclature alkane names end in the suffix -ane. They form a homologous series (the alkane series) methane (CH_4), ethane (C_2H_6), propane (C_3H_8), butane (C_4H_{10}), etc. The lower members of the series are gases; the high-molecular mass alkanes are waxy solid. Generaly the alkanes are fairly unreactive. They form haloalkanes with halogens when irradiated with ultraviolet radiation. Alkanes are present in natural gas and petroleum.

Alkene A molecule containing one or more carbon-carbon double bonds. Also a functional group.

Alkenes The alkenes, (i.e. the olefins), are the second homologous series of hydrocarbons, where there is at least one double-bond between the carbon atoms in the molecules.

The alkenes are said to be unsaturated because of the existence of a multiple bond in the molecule. The general structure of the alkene series of hydrocarbons is C_nH_{2n}. The first member of the ethene series is ethene, (previously called ethylene). In the case of higher members of the alkene series, the double bond may be between the terminal carbon atoms of the chain, or may be between internal carbon atoms in the chain.

Ethene C_2H_4 $CH_2{=}CH_2$

Propene C_3H_6 $CH_3CH{=}CH_2$

Butene C_4H_8 $CH_3CH_2CH{=}CH_2$

Pentene C_5H_{10} $CH_3CH_2CH_2CH{=}CH_2$

Heptene C_6H_{12} $CH_3CH_2CH_2CH_2CH{=}CH_2$

Hexene C_7H_{14} C H $_3$ C H $_2$ C H $_2$ $CH_2CH2_2CH{=}CH_2$

Octene C_8H_{16} $CH_3CH_2CH_2CH_2CH_2CH_2CH{=}CH_2$

Nonene C_9H_{18} $CH_3CH_2CH_2CH_2CH_2CH_2CH_2CH{=}CH_2$

Decene $C_{10}H_{20}$ $CH_3CH_2CH_2CH_2CH_2CH_2CH_2CH_2CH{=}CH_2$

Ethene, propene and butene are planar compounds. The first bond on the unsaturated carbon atoms are sp^2 hybrids, and these bonds are arranged as far apart in space as possible (i.e. at 120 degree in the same plane). These are s bonds (sigma bonds) and are formed by the end-on overlap of sp^2 hydride orbitals of the carbon atoms. The second bond that makes up the double bond of the unsaturated carbon atoms in alkenes is a p-bond (pi-bond), formed by the side-on overlap of the p-orbitals of the carbon atoms. p-bond (pi-bonds) are much more reactive than the s bonds (sigma bonds).

Alkyl radical An alkyl group or an alkyl radical, is structurally similar to an alkane group, where a hydrogen atom is missing. Alkyl radicals are named by replacing the ending "-ane" from the name of the alkane, with the "-yl" suffix.

Alkyne A molecule containing one or more carbon-carbon triple bonds. Also a functional group.

Alkynes Alkynes (acetylenes) are acyclic branched or unbranched hydrocarbons having a one or more triple carbon-carbon bond. In systematic chemical nomenclature alkine names end in the suffix -yne. The general formula is $C_nH_{(2n+2)-4x}$ were x is the number of triple bonds. Alkynes that have only one triple bond form a homologous series: ethyne (acetylene), CHα CH, propyne, CH_3CHα CH, etc. Like alkenes, alkynes undergo addition reaction.

Allergen This descriptor may be used to any substance which produces an allergic reaction.

Allergy A broad term applied to disease symptoms following exposure to a previously encountered substance (allergen), often one which would otherwise be classified as harmless. Essentially it is a malfunction of the immune system..

Allobar A form of an element that has isotopic abundances that are different from the naturally occuring form. For example, "depleted" uranium has had most of the uranium-235 removed, and is an allobar of natural uranium.

Allosteric effector Specific small molecules that bind to aprotein at a site other than a catalytic site and modulate (by activation or inhibition) the biological activity.

Allosteric enzyme An enzymethat contains a region, separate from the region that binds the substrate for catalysis, where a small, regulatory molecule binds and affects that catalytic activity. This effector molecule may be structurally unrelated to the substrate, or may be a second molecule of substrate. If the catalytic activity is enhanced by binding, the effector is called an activator; if it is diminished, the effector iscalled an inhibitor.

Allotrope Allotropy; allotropic; allotropism.Some elements occur in several distinct forms called allotropes. Allotropes have different chemical and physical properties. For example, graphite and diamond are allotropes of carbon.

Allotropy Allotropy is the occurrence of an element in two or more crystalline forms.

Alloy An alloy is a solid solution of one metal in another. It is an example of a mixture, as no chemical bonding exists between the constituent elements in the alloy.

Common examples of alloys include

Brass (an alloy of copper and zinc, and sometimes other metals, known since Roman Times and widely used in industry, for ornament and decoration),

Bronze (an alloy of copper and Tin used for tools, weapons, machine parts and marine hardware),

Stainless steel (an alloy of iron, chromium and nickel used for cutlery and industrial components where corrosion resistance is required),

Duralium (an alloy of aluminium, copper, magnesium, manganese and silicon used in aircraft construction) and

Solder (an alloy of lead and tin used in making electrical connections).

Allylic carbon An sp^3 carbon adjacent to a double bond.

Allylic substitution reaction A substitution reaction occurring at position 1/ of an allylic system, the double bond being between positions 2/ and 3/. The incoming group may be attached to the same atom 1/ as the leaving group, or the incoming group becomes attached at the relative position 3/, with movement of the double bond from 2/3 to 1/2. For example:

$CH_3CH{=}CHCH_2Br$ $CH_3CH{=}CHCH_2OAc$
or $CH_3CH{=}CHCH_2Br$ $CH_3CH(OAc)CH{=}CH_2$
(written as a transformation).

Alpha particle Alpha particle is a helium nucleus emitted spontaneously from radioactive elements both natural and manufactured. Its energy is in range 4-8 MeV and is dissipated in a very short path, i.e. a few centimetres of air or less than 0.005 mm of aluminium. As a helium nucleus consists of two protons and two neutrons bound together as a stable entity the loss of an alpha particle involves a decrease in nucleon number of 4 and decrease of 2 in the atomic number, e.g. ${}^{226}_{88}Ra \rightarrow {}^{4}_{2}He + {}^{222}_{86}Rn$

A stream of alpha particles is known as an alpha ray or alpha-radiation.

Alpha; α, β(beta) These stereodescriptors are used in a number of different ways. 1. Relative stereodescriptors used in carbohydrate nomenclature to describe the configuration at the anomeric carbon by relating it to the anomeric reference atom. For simple cases the anomeric reference atom is the same as the configurational reference atom. Thus in α?D-glucopyranose the reference atom is C-5 and the OH at C-1 is on the same side as the OH at C-5 in the Fischer projection.

2. Relative stereodescriptors used by Chemical Abstracts Service to describe the configuration of a cyclic molecule (including suitable polycyclic systems) with several stereogenic centres whereby the α side of the reference plane is the side on which the substituent with CIP priority lies at the lowest numbered stereogenic centre.

Alternant A conjugated system of pi electrons is termed alternant if its atoms can be divided into two sets so that no atom of one set is directly linked to any other atom of the same set.

Example of aternate π system Example of non-alternate π system

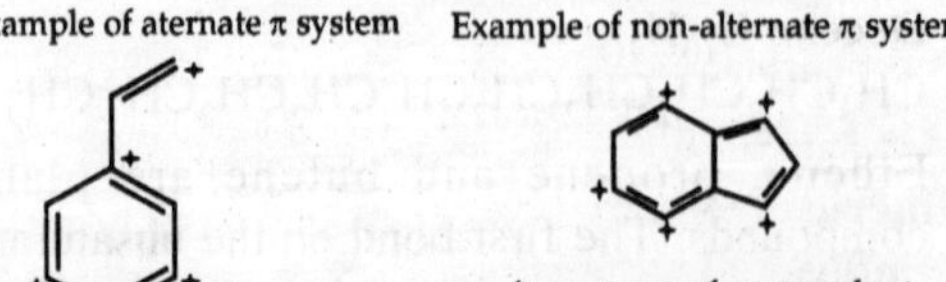

(two atoms of unstaned set are directly linked)

Ambident A description applied to a chemical species whose molecular entities each possess two alternative and strongly interacting distinguishable reactive centres, to either of which a bond may be made in a reaction: the centres must be connected in such a way that reaction at either site stops or greatly retards subsequent attack at the second site. The term is most commonly applied to conjugated nucleophiles, for example the enolate ion (which may react with electrophiles either at the -carbon atom or at oxygen) or -pyridones, and also to the vicinally ambident cyanide ion, cyanate ion, thiocyanate ion, sulfinate ion, nitrite ion, and unsymmetrical hydrazines. Ambident electrophiles are exemplified by carboxylic esters $RC(O)OCR_3$ which react with nucleophiles either at the carbonyl carbon or the alkoxy carbon.

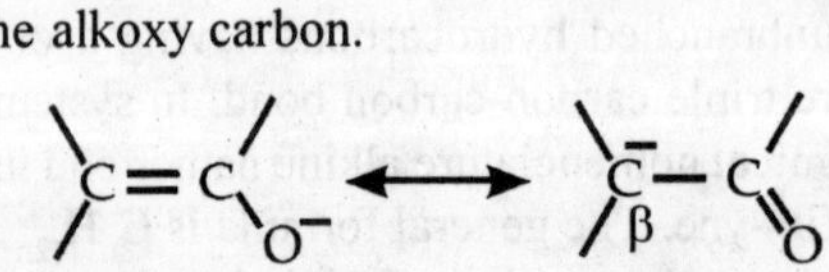

Molecular entities, such as dianions of dicarboxylic acids, containing two non-interacting (or feebly interacting) reactive centres, are not generally considered to be ambident and are better described as "bifunctional".

The Latin root of the word implies two reactive centres, but the term has in the past also incorrectly been applied to chemical species with more than two reactive centres. For such species the existing term "polydent" (or, better, "multident") is more appropriate.

Ambidentate Ligands, such as $(NCS)^-$, that can bond to a central atom through either of two or more donor atoms are termed ambidentate.

Ambo A prefix used to indicate that a molecule with two (or more) chiral elements is present as a mixture of the two racemic diastereoisomers in unspecified proportions. For example, the dipeptide formed from L-alanine and DL-leucine is L-alanyl-ambo-leucine.

American chemical society ACS A large and influential professional society for professionals and students in chemistry and related fields.

Amicyanin An electron transfer protein containing a type 1 copper site, isolated from certain bacteria.

Amide A molecule containing a carbonyl group attached to a nitrogen ($-CONR_2$). Also a functional group.

Amides Amides are organic compounds containing the amide functional group (i.e. $-C{=}O.NH_2-$).

Amides can be regarded as either 1. a carboxylic acid derivatives in which the hydroxyl group, -OH, of the carboxyl acid group is replaced by an amino functional group, $-NH_2$, or

2. as an ammonia derivatives in which a hydrogen atom is replaced by an acyl functional group, -C=O-

Amides are volatile solids.

Amine A molecule containing an isolated nitrogen = (NR_3). Also a functional group.

Amino acids Amino acids are compounds containing both a carboxylic acid group (-COOH) and an amino group ($-NH_2$). The most important are the a-amino acids, in which the $-NH_2$ group in attached to the C atom adjacent to the -COOH group. In the β-amino acids, there is an intervening carbon atom.

Ammonia The molecular formula for ammonia is NH_3. Ammonia is one of the most important inorganic nitrogen compounds in atmospheric water droplets. It reacts with strong acids and is one of the only known basic, gas phase atmospheric components. Atmospheric ammonia can also enhance the nucleation rate and the production of new particles in the atmosphere. These new particles can be activated to become condensation nuclei and then, through various processes, grow to a particle size of 0.05 micrometer or larger which can then be effective as cloud condensation nuclei. This process can, therefore, affect the global radiation budget. The major sources of ammonia are decaying natural organic matter, livestock wastes, fertilizers, and industrial activity.

Ammonium ion (NH4+) Ammonium.NH4+ is a cation formed by neutralization of ammonia, which acts as a weak base.

Ampere Ampere (A) is the SI base unit of electric current.

The ampere is that constant current which, if maintained in two straight parallel conductors of infinite length, of negligible circular cross-section, and placed 1 metre apart in vacuum, would produce between these conductors a force equal to 2×10^{-7} newton per metre of length.

Amphiphilic A compound containing a large organic cation or anion which possesses a long unbranched hydrocarbon chain, e.g. $H_3C(CH_2)_nCO_2^-M^+H_3C(CH_2)_nSO_3^-M^+H_3C(CH_2)_nN(CH_3)_3^+X^-$ $(n > 7)$.

The existence of distinct polar (hydrophilic) and non polar (hydrophobic) regions in the molecule promotes the formation of micelles in dilute aqueous solution.

Amphiprotic (solvent) Self-ionizing solvent possessing both characteristics of Brønsted acids and bases, for example H_2O and CH_3OH, in contrast to aprotic solvent.

Amphoteric A chemical species that behaves both as an acid and as a base is called amphoteric. This property depends upon the medium in which the species is investigated: H_2SO_4 is an acid when studied in water, but becomes amphoteric in superacids.

Amphoteric substance An amphoteric substance is one which has both acidic and basic properties and which can behave as a weak base or a weak acid under different experimental conditions.

Anabolism The processes of metabolism that result in the synthesis of cellular components from precursors of low molecular weight.

Anaerobe An organism that does not need dioxygen for growth. Many anaerobes are even sensitive to dioxygen. Obligate (strict) anaerobes grow only in the absence of dioxygen. Facultative anaerobes can grow either in the presence or in the absence of dioxygen.

Anaerobic respiration Living or acting in the absence of oxygen. Cellular respiration in the absence of oxygen.

Analog A drug whose structure is related to that of another drug but whose chemical and biological properties may be quite different.

Analysis Analysis involves the determination of the composition of a sample.

Qualitative Analysis involves determining the nature of a pure unknown compound or the compounds present in a mixture.

Quantitative Analysis involves measuring the proportions of known components in a mixture, and the chemical techniques include volumetric analysis and gravimetric analysis.

Instrumental Analysis include several physical techniques including spectroscopic techniques, mass spectrometry, polarography, nuclear magnetic resonance, etc.

Analytic study A hypothesis-testing method of investigating the association between a given disease or health state or other dependent variable and possible causative factors. In an analytic study, individuals in the study population may be classified according to absence or presence (or future development) of specific disease and according to "attributes" that may influence disease occurrence. Attributes may include age, race, sex, other disease(s), genetic, biochemical, and physiological characteristics, economic status, occupation, residence, and various aspects of the environment or personal behaviour. Three types of analytic study are cross-sectional (prevalence), cohort (prospective), and case control (retrospective) .

Anation Replacement of the ligand water by an anion in a coordination entity.

Anchimeric assistance See neighbouring group participation.

Anemia Condition in which there is a reduction in the number of red blood cells or amount of hemoglobin per unit volume of blood below the reference interval for a similar individual of the species under consideration, often causing pallor and fatigue.

Angle strain Strain due to a departure in bond angle from "normal" values. The term is often used in the context of non-aromatic cyclic compounds in which the internal angles differ from the regular tetrahedral angle of 109° 28'; in this sense angle strain is also known as Baeyer strain.

Angstrom (Å) is a unit of length equal to 10^{-10} m. The Ångström is defined in terms of the wavelength of the emission spectra's red line of an atom of cadmium (6438.4696 ⊕).

Angular momentum quantum number See azimuthal quantum number

Anhydrous Anhydrous compound; anhydride.A compound with all water removed, especially water of hydration. For example, strongly heating copper(II) sulfate pentahydrate (CuSO4·5H2O) produces anhydrous copper(II) sulfate (CuSO4).

Anion A negatively charged atom or molecule.

Anion exchange An anionic resin has negative ions built into its structure and therefore exchanges positive ions. In anion exchange, the side groups are ionized basic groups, such as (-NH_2, -NRH, -NR_2, -NR_3^+) to which anions OH^- are attached. The exchange reaction is one in which different anions in the solution displace the OH^- from the solid.

Anion hydrolysis Anion hydrolysis involves the reaction of the anion of a salt with water to give excess hydroxyl ions in solution.

Anionotropic rearrangement (or anionotropy) A rearrangement in which the migrating group moves with its electron pair from one atom to another.

Anisotropy Anisotropy is the property of molecules and materials to exhibit variations in physical properties along different molecular axes of the substance.

Annulation A transformation involving fusion of a new ring to a molecule via two new bonds.

Some authors use the term "annelation" for the fusion of an additional ring to an already existing one, and "annulation" for the formation of a ring from one or several acyclic precursors, but this distinction is not made generally.

Annulene Mancude (i.e. having formally the maximum number of noncumulative double bonds) monocyclic hydrocarbon without side chains of the general formula C_nH_n (n is an even number) or C_nH_{n+1} (n is an odd number). Note that in systematic nomenclature an annulene with seven or more carbon atoms may be named (n)annulene, where n is the number of carbon atoms, e.g. 9.annulene for cyclonona-1,3,5,7-tetraene

Anode An anode is the positive electrode in electrolysis experiments, and it is at this electrode that the ions lose electrons. Thus, oxidation reactions occurs at the anode.

Anode reactions Anode reactions are the chemical reactions that occur at the anode of an electrochemical cell, involve the negative ions in solution which migrate to the anode, where they lose an electron and are deposited as neutral atoms.

Anode reactions are oxidation reactions, as they involve the transfer of an electron from the negatively charged ion.

Anomeric effect Originally the thermodynamic preference for polar groups bonded to C-1 (the anomeric carbon of a glycopyranosyl derivative) to take up an axial position. This effect is now considered to be a special case of a general preference (the generalised anomeric effect) for synclinal (gauche) conformations about the bond C-Y in the system X-C-Y-C where X and Y are heteroatoms having nonbonding electron pairs, commonly at least one of which is nitrogen, oxygen or fluorine. For example in chloro(methoxy)methane the anomeric effect stabilises the synclinal conformation.In alkyl glycopyranosides the anomeric effect operates at two sites (i) along the endocyclic C-1 oxygen bond (endo-anomeric effect) and (ii) along the exocyclic C-1 oxygen bond (exo-anomeric effect)

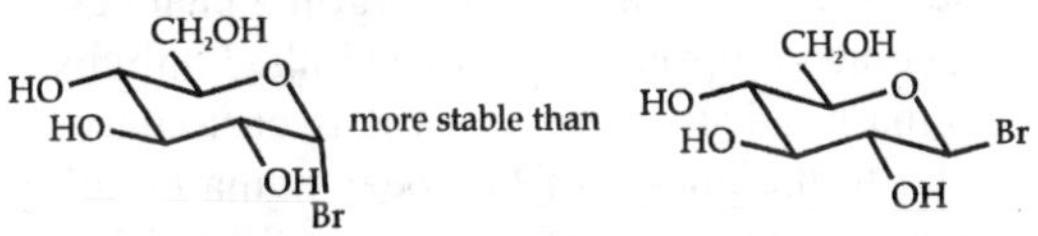

The opposite preference is claimed for some systems e.g. glycopyranosyltrialkylammonium salts, and has been referred to as the reverse anomeric effect.

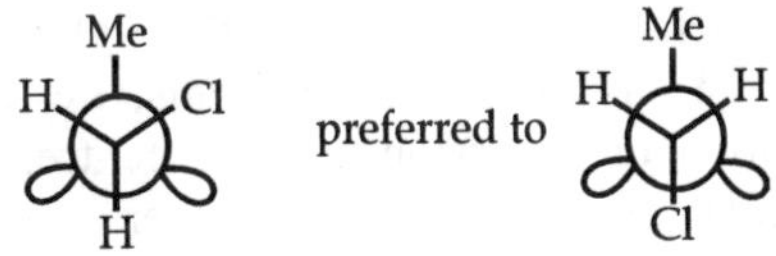

Anomers Diastereoisomers of glycosides, hemiacetals or related cyclic forms of sugars, or related molecules differing in configuration only at C-1 of an aldose, C-2 of a 2-ketose, etc.

Anoxic The lack of oxygen such as the inadequate oxygenation of the blood (anoxia). In aquatic environmental chemistry it refers to water that has become oxygen poor due to the bacterial decay of organic matter.

Antagonist A molecule that blocks the ability of a given chemical to bind to its receptor, preventing a biological response.

Antagonistic effect This is the consequence of one chemical (or a group of chemicals) interacting: the situation in which the combined effect of two or more chemicals is less than the simple sum of their independent effects. In bioassay, the term may be used to refer to the situation when a specified response is produced by exposure to either of two factors but not by exposure to both together

Antarafacial, suprafacial When a part of a molecule ("molecular fragment") undergoes two changes in bonding (bond-making or bond-breaking), either to a common centre or to two related centres, external to itself, these bonding changes may be related in one of two spatially different ways. These are designated as "antarafacial" if opposite faces of the molecular fragment are involved, and "suprafacial" if both changes occur at the same face. The terms antarafacial and suprafacial are, however, also employed in cases in which the essential part of the molecular fragment undergoing changes in bonding comprises two atoms linked only by a sigma bond. In these cases it is customary to refer to the phases of the local sigma-bonding orbital: occurrence of the two bonding changes at sites of like orbital phase is regarded as suprafacial, whereas that at two sites of opposite phase is antarafacial. There may be two distinct and alternative stereochemical outcomes of a suprafacial process involving a sigma bond between saturated carbon atoms, i.e. either retention or inversion at both centres. The antarafacial process results in inversion at one centre and retention at the second.

Antarctic ozone hole Recent data suggest that due to the release into the atmosphere of CFCs by human activities, the ozone hole can as large as North America during a given astral spring. The CFCs are broken apart by UV light—to form free radicals of chlorine—after they have diffused into the upper stratosphere from the troposphere. This long term movement from the troposphere to the stratosphere is possible for these chlorine containing chemicals because of there long atmospheric lifetimes; however, this type of movement is not important for more reactive species such as tropospheric ozone because of their reactivity and therefore short atmospheric lifetime. In the Antarctic stratosphere, the reaction that converts reservoir species of chlorine into an active form—which destroys ozone—takes place on the surface of particles in polar stratospheric clouds as the temperature drops below about200K. This is possible because of the unique isolation of the south polar vortex during the austral winter. The surfaces of these (nitric acid/water) clouds act as catalysts for reactions that release molecular chlorine which quickly photolyzes to chlorine's (radical) active state. It is this radical which destroys ozone.

Antarctic vortex The combination of a drastic temperature and corresponding pressure drop along with the rotation of the Earth on its axis produces a spinning/rotating volume of air. The rotational speed of the winds commonly reaches as high as 180 mph. The motion of these winds form an impenetrable barrier such that the trapped air inside is unmixed, as it is separated from the air outside, and remains quite cold until October. Inside the whirling volume of freezing air, the cold temperatures facilitate the condensation of gases into particles that eventually form polar stratospheric clouds.

Anthropogenic forcing Influence exerted on a habitat or chemical environment by humans. This obliquely refers to the idea of managed environment as opposed to a wild or non-disturbed environment. The variability of assumptions about radiative forcing can cause significant changes in the results from computer models that require an estimate of those values.

Anthropogenic Refers to something originating from humans and the impact of human activities on nature.

Anti In the representation of stereochemical relationships "anti" means "on opposite sides" of a reference plane, in contrast to "syn" which means "on the same side".

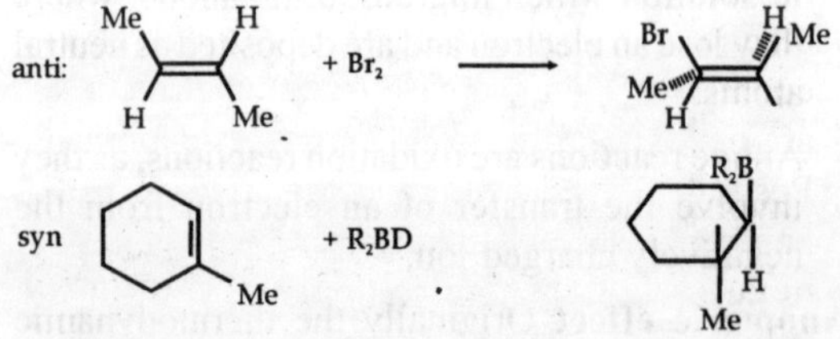

Anti addition A reaction in which th= e two groups of a reagent X-Y add on opposite faces of a carbon-carbon bond.

Anti conformation A type of staggered conformation in which the two big groups are opposite of each other in a Newman projection.

Anti-aromatic A highly unstable planar ring system with 4n pi electrons.

Antibody A protein, belonging to the class of immunoglobulins, designed to bind a specific antigen in order to remove it from the body. They are synthesised exclusively by B-lymphocytes, in millions of forms, each with a different amino acid sequence and a specific binding site for a specific antigen (antigenic determinant or epitope). Presently a world-wide interest exists in catalytic antibodies (abzymes), elicited against transition-state analogues of chemical reactions. Such an approach has permitted the design of immunoglobulins with enzymatic activity.

Anticyclonic flow The air flow produced about a high pressure center by the combination of two forces: the pressure gradient accelerating the air away from the center and the Coriolos force acting inward. Anticyclonic flows are clockwise in the northern hemisphere and counterclockwise in the southern hemisphere.

Antigen The descriptor applied to any substance that produces a specific immune response and is regognised as foreign by the immune system when it enters the tissue of an animal or human.

Anti-markownikoff rule The orientation of the addition across the multiple bond in the presence of free radicals is opposite to that specified for ionic reactions in the Markownikoff Rule.

Antimatter (Antiparticle) Species of subatomic particles that have the same mass and spin as normal particles, however, they have opposite electrical charges from their normal matter counterparts. In the case of antineutrons they are opposite of neutrons in magnetic moment. Positrons, which are the counterpart to electrons, have a positive charge and antiprotons have a negative charge. Photons are their own antimatter counterpart. When a particle of matter collides with a particle of antimatter, both particles are destroyed and their masses are converted to photons of equivalent energy.

Anti-periplanar (a.k.a. anticoplanar) The conformation in which a hydrogen and a leaving group are in the same plane and on opposite sides of a carbon-carbon single bond. The conformation required for E2 elimination.

Antipodes (Usage strongly discouraged) Obsolete synonym for enantiomers.

Anton fischer (1901-1978) Born in Pressburg (now Bratislava, Slovakia) and took his MD at the University of Vienna in 1925. He was a specialist in the study of rheumatism, but also published papers on the analysis of blood serum.

AP See torsion angle.

Apical, basal, equatorial In trigonal bipyramidal structures (e.g. a five-coordinate trigonal bipyramid with phosphorus as central atom) the term apical refers to the two positions that are collinear with the central atom or to the bonds linking these positions to the central atom. The three equivalent bonds (or positions) in a plane passing through the central atom and perpendicular to the direction of the apical bonds are described as equatorial. The term apical is also used for the bond pointing from the atom at or near the centre of the base to the apex of a pyramidal structure. The positions at or near the base of the pyramid, or the bonds linking those positions to the central atom of the base are described as basal. The apical bonds have also been called axial.

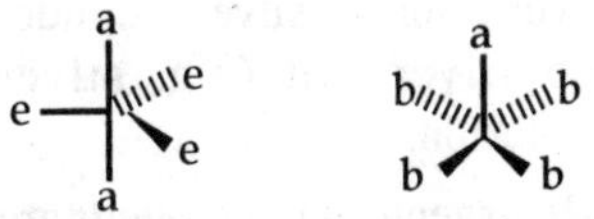

a = apical
b = basal
e = equatorial

Apoprotein A protein without its characteristic prosthetic group or metal.

Aprotic solvents Solvents that do not contain O-H or N-H bonds.

Aquation The incorporation of one or more integral molecules of water into another chemical species with or without displacement of one or more other atoms or groups. See also hydration.

Aqueous solutions Aqueous solutions are those solutions where water is the solvent. An aqueous solution found in an equation describing a chemical reaction is denoted by the state symbol, (αθ).

Ar Ar is symbol for element argon.

Archaea A group of prokaryotes, which can be subdivided into three groups (methanogenic, halophilic, thermoacidophilic), that are characterized by special constituents such as ether-bonded lipids and special coenzymes. The archaea are members of a separate kingdom that falls in between eubacterial and eukaryotic organisms.

Area Measures the size of a surface using length measurements in two dimensions.

ArenesArenes are monocyclic and polycyclic aromatic hydrocarbons.

Arenes The arenes are a homologous series of aromatic hydrocarbons, which have a ring structure of six carbon atoms, with alternating single and double bonds. This closed ring structure has a very high stability and these aromatic compounds undergo substitution reactions in order to preserve the configuration of the ring structure.

Argentic compounds Argentic compounds are ionic compounds of silver in its higher oxidation state (i.e. +2 oxidation state), Ag(++).

Argentometric titrations Argentometric titrations are those in which an aqueous solution of silver nitrate is used to quantitatively precipitate a silver halide (e.g. silver chloride, silver bromide) or other insoluble silver salt (e.g. silver thiocyanate) from solution.

Argentous compounds Argentous compounds are ionic compounds of silver in its lower oxidation state (i.e. +1 oxidation state) , Ag(+).

ArgoGel Trademark of Argonaut Technologies, San Carlos, California, USA. Beaded solid support with a crosslinked polystyrene core and grafted linear poly (ethylene glycol) (PEG) chains with terminal functional groups.

Argon An element that is a member of the 8A group (Noble Gases) of the periodic table. Argon is represented by the atomic symbol Ar, has an atomic number of 18, and an atomic weight of 39.948. It is colorless, odorless, and a very inert gas. It comprises about 1% of the Earth's atmosphere.

Arnold orville beckman (b.1900) Born in Cullom, Illinois, he took his BS degree in chemical engineering at the University of Illinois before taking his Ph.D. at California Institute of Technology in Pasadena. After teaching at Caltech, he founded National Technical Laboratories in South Pasadena in 1935, to make pH meters. NTL became Beckman Instruments in 1950. Beckman Instruments merged with SmithKline in 1982 (it was later demerged), by which time Beckman had already started a new career as a philanthropist, distributing over $300 million through his charitable foundation and thereby establishing numerous institutions including the Beckman Center for the History of Chemistry (now part of the Chemical Heritage Foundation). His name is especially associated with the Model G pH meter and the DU ultraviolet spectrophotometer.

Aromatic A planar ring system that contains uninterrupted p orbitals around the ring and a total of 4n+2 pi electrons. Aromatic compounds are unusually stable compounds.

Aromatic compounds Aromatic compounds are major group of unsaturated cyclic hydrocarbons containing one or more rings, typified by benzene, which has a 6-carbon ring containing three double bonds. All the bonds in benzene (C_6H_6) are the same length intermediate between double and single C-C bonds. The properties arise because the electrons in the p-orbitals are delocalised over the ring, giving extra stabilization energy of 150 kJ/mol over the energy of Kekulé structure. Aromatic compounds are unsaturated compounds, yet they do not easily partake in addition reactions. Historical use of the term implies a ring containing only carbon (e.g., benzene, naphthalene), but it is often generalized to include heterocyclic structures such as pyridine and thiophene.

Aromatic hydrocarbons Hydrocarbon compounds in which the carbon atoms are connected by a ring structure that is planar and joined by sigma and pie bonds between the carbon atoms. An example of an aromatic compound is benzene, C6H6.

Aromatic stability Aromatic stability is the resistance of aromatic compounds to undergo addition reactions on the ring structure which would destroy the aromatic character (i.e.

alternating single and double bonds) of the aromatic ring. Aromatic compounds readily undergo substitution reactions that preserve the aromatic structure of the ring. Aromatic stability is explained by resonance.

Aromatic, aromaticity 1. In the traditional sense, "having a chemistry typified by benzene".

2. A cyclically conjugated molecular entity with a stability (due to delocalization) significantly greater than that of a hypothetical localized structure (e.g. Kekulé structure) is said to possess aromatic character. If the structure is of higher energy (less stable) than such a hypothetical classical structure, the molecular entity is "antiaromatic".

3. The terms aromatic and antiaromatic have been extended to describe the stabilization or destabilization of transition states of pericyclic reactions. The hypothetical reference structure is here less clearly defined, and use of the term is based on application of the Hückel (4n + 2) rule and on consideration of the topology of orbital overlap in the transition state. Reactions of molecules in the ground state involving antiaromatic transition states proceed, if at all, much less easily than those involving aromatic transition states. Dewar (1971); Zimmermann (1971).

Array synthesis Form of parallel synthesis in which the reaction vessels are maintained in a specified spatial distribution, e.g. the wells of a 96-well plate or pins held in a rack.

Arrhenius concept of bases The Arrhenius concept of bases regard substance which provides hydroxyl ions as a result of dissociation and ionisation in aqueous solution as being bases.

Arrhenius concept of neutralisation The Arrhenius concept of neutralisation involves the reaction in which a hydrogen ion and a hydroxide ion combine to form water.

Arrhenius equation See energy of activation.

Arthur fischer (1878-1922) Born in Wiesbaden and took his doctorate in engineering under Classen at Aachen in 1924. He worked in the official materials testing bureau in Berlin and published papers on inorganic chemical analysis.

Arthur michael (1855-1942) Born in Buffalo and studied chemistry at Heidelberg under Robert Bunsen (1811-1899) and at Berlin under August Wilhelm Hofmann (1818-1892). He then studied under Adolphe Wurtz (1817-1884) in Paris and Dimitri Ivanoviè Mendeleev (1834-1907) in St Petersburg, but never bothered to take a degree. He was made Professor of Chemistry at Tufts College near Boston. In 1889, he married one of his most brilliant students, Helen Cecilia DeSilver Abbott (1857-1904), one of the few women organic chemists in this period. After a failed attempt to run the chemistry department at Clark University in Worcester, Massachusetts, in 1891, he spent three years working with his wife in his private laboratory on the Isle of Wight before returning to Tufts. After he retired from Tufts in 1907, Michael set up another private laboratory at Newton Center near Boston. Five years later, he was appointed a Professor of Chemistry, without lecturing duties, at Harvard University.

Aryne A hydrocarbon derived from an arene by abstraction of two hydrogen atoms from adjacent carbon atoms; thus 1,2-didehydroarene. Arynes are commonly represented with a formal triple bond. The analogous heterocyclic compounds are called heteroarynes or hetarynes. Arynes are usually transient species.

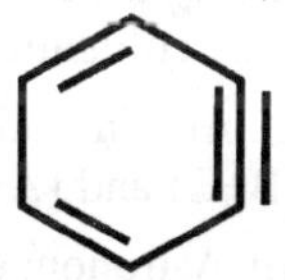

benzyne

Asbestos A naturally occurring fibrous mineral found in certain types of rock formations. The primary use of asbestos was as a fireproofing material, applied to structural steel members in multistory commercial buildings to attain proper fire protection. With its resistance to heat, asbestos was able to keep the building structure from bending and warping if exposed to fire. After widespread use of asbestos, over a 15 to

25 year time span, a pattern of illness gradually began to occur in asbestos workers. Three diseases linked to asbestos exposure are asbestosis, a fibrous scarring of the lungs, lung cancer, and mesothelioma, a cancer of the lining of the chest or abdominal cavity.

Asbestosis A lung disease (pneumoconiosis) resulting from inhaling fibers of asbestos and marked by interstitial fibrosis of the lung.

Assay A biological test, measurement or analysis to determine whether compounds have the desired effect either in a living organism, outside an organism, or in an artificial environment.

Assay equivalent An aliquot of a library which will allow the library to be screened in a single assay. Particularly applicable to libraries prepared by split/pool procedure, where it pertains to the number of particles required to sample a library. Generally consists of a specified number of library equivalents.

Assimilation cycle A natural process by which a water body uses microbes, which convert nonliving substances into protoplasm or cells, to purify itself from pollutants.

Association The assembling of separate molecular entities into any aggregate, especially of oppositely charged free ions into ion pairs or larger and not necessarily well-defined clusters of ions held together by electrostatic attraction. The term signifies the reverse of dissociation, but is not commonly used for the formation of definite adducts by colligation or coordination.

Associativity A property in math which states that: (A+B)+C=A+(B+C) and (A*B)*C=A*(B*C)

Astronomical unit Astronomical unit (AU) is a unit of length employed in astronomy for describing planetary distance. It is the mean distance of the earth from the sun, equal to $1.49597870 \times 10^{11}$ m.

Asymmetric Lacking all symmetry elements (other than the trivial one of a one-fold axis of symmetry), i.e. belonging to the symmetry point group C_1. The term has been used loosely (and incorrectly) to describe the absence of an rotation-reflection axis (alternating axis) in a molecule, i.e. as meaning chiral, and this usage persists in the traditional terms asymmetric carbon atom, asymmetric synthesis, asymmetric induction, etc.

Asymmetric carbon atom The traditional name (van't Hoff) for a carbon atom that is attached to four different entities (atoms or groups) e.g. Cabcd. See also chirality centre.

Asymmetric induction The traditional term describing the preferential formation in a chemical reaction of one enantiomer or diastereoisomer over the other as a result of the influence of a chiral feature in the substrate, reagent, catalyst or environment. The term also refers to the formation of a new chiral feature preferentially in one configuration under such influence.

Asymmetric synthesis A traditional term for stereoselective synthesis. A chemical reaction (or reaction sequence) in which one or more new elements of chirality are formed in a substrate molecule and which produces the stereoisomeric (enantiomeric or diastereoisomeric) products in unequal amounts.

Asymmetry parameter In nuclear quadrupole resonance spectroscopy this parameter, η, is used for describing non-symmetric fields. It is defined as $\eta = (q_{xx} - q_{yy})/q_{zz}$ in which q_{xx}, q_{yy} and q_{zz} are the components of the field gradient q (which is the second derivative of the time-averaged electric potential) along the x-, y- and z-axes. By convention q_{zz} refers to the largest field gradient, q_{yy} to the next largest and q_{xx} to the smallest when all three values are different.

Atmosphere The sum total of all the gases surrounding the Earth, extending several hundred kilometers above the surface in a mechanical mixture of various gases in fluid-like motion. The permanent constituents are molecular nitrogen; 78.1%, molecular oxygen; 20.9%, argon; 0.934%, and approximately 0.036% carbon dioxide. Various other components exist in trace amounts. Not to be under emphasized, these trace components are where the interesting atmospheric chemistry occurs. The atmosphere can also be artificially divided into layers. Example: the troposphere (the layer closest to the earth) and the stratosphere (the layer above the troposphere).

Atmospheres Common units for measuring pressure.

Atmospheric carbon dioxide Atmospheric carbon dioxide results from the combustion of fossil fuels, and it is an important contributor to the greenhouse effect.

Atmospheric convection Process in which heat energy is transported through a medium, usually a gas or liquid. An example, in the atmosphere occurs when warm air with a lower density experiences an upward force until it cools and its density matches the surrounding air, generally termed convection cells.

Atmospheric pollution Atmospheric Pollution is caused by the emission of substances which are not natural constituents of the air, including soot, smoke, carbon monoxide, hydrocarbons, lead compounds, carbon dioxide, chlorofluorocarbons, sulphur dioxide, and nitrogen oxides.

Soot and smoke, which are particulates released during the combustion of solid fuels, can be controlled by using smokeless solid fuels. The smoke emitted from diesel engines can be controlled by the use of properly adjusted fuel injectors.

Levels of sulphur dioxide can be reduced by using low sulphur fuels.

The quantity of lead compounds which are emitted from petrol engines has been reduced significantly since the introduction of lead-free petrol.

The quantity of carbon monoxide and of hydrocarbons which are emitted from petrol engines can be reduced by using properly adjusting and tuning the engines of cars and by fitting catalytic converters.

International agreements has lead to the reduction of the quantities of chloro-fluorocarbons which can be manufactured. Alternative new compounds are being developed to replace the CFCs.

Atom An Atom is the smallest part of an element which can take part in a chemical reaction. The atom consists of three fundamental particles, 1. The Proton which has mass of approximately 1 atomic mass unit and a positive charge,

2. The Neutron which has a mass of approximately 1 atomic mass unit and no charge, and

3. The Electron which has a mass 1/1840 of the proton and a negative charge.

Each Elements is made up of one kind of atom only.

Just over one hundred different kinds of atoms are known.

Ninety two of these elements are naturally occurring (i.e. they are found on the earth), and the remainder have been made artificially in nuclear reactors.

Atomic configuration The Atomic Configuration is the arrangement of atoms or groups in a molecule.

Atomic mass (Atomic Mass Average) The average mass of all nuclides of an element determined by the proportions in which each nuclide of the element are present within the earth and its atmosphere.

Atomic mass number The Atomic Mass Number of an atom is the total number of protons and neutrons in the nucleus of the atom.

Atomic mass of chlorine Chlorine consists of a mixture of two Isotopes.

The first isotope has a mass number of 35, and its nucleus contains 17 protons and 18 neutrons and

The second isotope has a mass number of 37, , and its nucleus contains 17 protons and 20 neutrons.

The observed atomic weight of chlorine is 35.46.

This indicates that chlorine is a mixture of these two isotopes, and that the isotope with mass number 35 is the more abundant atom present.

Indeed, the isotope of mass number 35 is approximately three times more abundant than the isotope of mass number 37.

Atomic Mass Unit (AMU) A mass unit that is exactly 1/12th the mass of a carbon 12 (C12) atom (approximately $1.67E^{-24}$g).

Atomic nucleus nucleus; nuclei; atomic nuclei.A tiny, incredibly dense positively charged mass at the heart of the atom. The nucleus is composed of protons and neutrons (and other particles). It contains almost all of the mass of the atom but occupies only a tiny fraction of the atom's volume.

Atomic number The number of protons in the nucleus of an atom. This determines an element's structure, properties and location on the periodic table of elements.

Atomic orbital A one-electron wavefunction describing an electron in the effective field provided by a nucleus and the other electrons present.

Atomic orbitals Atomic Orbitals are the specific regions in space around the nucleus of an atom in which the electrons are most likely to be found. An atomic orbital is fully described by its Quantum Numbers.

To explain why the electrons do not loose energy continuously Niels Bohr suggested that the electrons are confined to specific shells which have fixed energy levels. However, because of the Heisenberg Uncertainty Principle, the locations of these shells about a nucleus cannot be rigidly defined. Thus, the atomic orbitals are diffuse regions in space around the nucleus of an atom in which the electrons are most likely to be found.

The exact nature of each atomic orbital is described in detail its set of Quantum Numbers, each of which describes a different aspect of the shell.

Atomic particles Chemistry is concerned with the structure of atom only in terms of the three atomic particles (i.e. the proton, the neutron and the electron).

It is now known that these atomic particles are themselves composed of sub-atomic particles (i.e. the quarks and leptons). These sub-atomic particles are also called the fundamental particles or elementary particles.

However, this further detail of the structure of the nucleus of the atom has no chemical consequences and will not be considered further in this text.

Atomic radii The Atomic Radii is the calculated distance between the atomic nuclei in a crystalline solid.

Atomic radius One-half the distance between two adjacent atoms in crystals of elements. This varies according to interatomic forces.

Atomic spectra The Atomic Spectra of an element are the lines which are present at characteristic wavelengths in the light emitted from an electric arc between electrodes of that element. Each line (i.e. each specific wavelength) in the spectrum corresponds to the energy difference between two orbitals in the atom. According to the quantum theory, the electrons in an atom can only reside in specific orbitals, each of which has a fixed energy level, and one photon is emitted for each electronic transition between orbitals.

When an electron drops from an orbital of high energy to one of lower energy, a photon (i.e. a packet of light) is emitted corresponding to the energy difference between the orbitals.

Atomic theory An explanation of chemical properties and processes that assumes that tiny particles called atoms are the ultimate building blocks of matter.

Atomic theory of matter Greek philosophers were the first to propose that all matter consists of very small particles called atoms (Greek atomas, meaning indivisible). No further progress was made with the development of this atomic theory of matter until the eighteenth and nineteenth centuries, when the experimental evidence needed to support the theory began to accumulate.

In 1803AD, John Dalton proposed his atomic theory in which he held that the atom was simple, solid and impenetrable. All elements were made up of these minute particles 1. which could not be created nor destroyed,

2. which were indivisible,

3. which were identical in the same element, and

4. which combined in small whole numbers when forming compounds.

The tenets of his atomic theory are known as Dalton's Postulates.

Ninety years later, Pierre and Marie Curie were to discover and isolate radium, a new element which spontaneously disintegrated into other elements. This proved that the atoms of one element at least were not indivisible.

From about 1850AD onwards, the study of electric discharge in high vacuum by Crookes, Goldstein and others, which was made possible by Callan's work at Maynooth, led to the discovery of the particles which Stoney named Electrons. These electrons emanated from any metal used as a cathode in a discharge tube and for this reason were called rays. Thus, the electron was established as a fundamental particle of matter.

By convention the charge on the electron was considered negative and in 1909AD,R A Millikan showed by his classic oil-drop experiment that the electron possessed definite charge and mass.

Since the atom was electrically neutral, the existence of negative charge on the electron mandated the existence of an equal amount of positive charge. The problem arose as to the distribution of charge throughout the atom. J JThompson had proposed that the negative electrons were embedded in the atom at random.

In 1911AD, Sir Ernst Rutherford used a radioactive source to bombard a thin gold leaf with alpha particles, which are positively charged (i.e. helium atoms without electrons). Most of these alpha-particles passed through the thin gold leaf undeflected, some were deflected outward and a few were deflected back along their own path. He concluded that the positive charge was concentrated in a small region, called the nucleus, round which the electrons revolved. Those alpha particles which were deflected through large angles had come close to nucleus and were repelled by its concentrated positive charge. Rutherford's model of the atom contained a revolving electron which according to classical physics should lose energy continuously and spiral into the nucleus.

Neils Bohr discovered the way in which the electrons travel about the nucleus. He suggested that an electron could revolve in only certain allowed orbits. As long as an electron remained in an allowed orbit, it does not lose or gain energy. The Bohr model was satisfactory in explaining the behaviour of the electron in the hydrogen atom, but failed with multi-electron elements.

Atomic volume The atomic volume of an element is the relative atomic mass of the element divided by is density.

Atomic weight The atomic weight of an element is its relative atomic mass, which is the ratio of the average mass per atom of the naturally occurring form of an element to one twelfth (i.e. 1 / 12) of the mass of a carbon-12 atom.

Atrophy The process which is observed during the wasting of a tissue or an organ.

Atropisomers A subclass of conformers which can be isolated as separate chemical species and which arise from restricted rotation about a single bond (see rotational barrier) e.g. ortho-substituted biphenyl, 1,1,2,2-tetra-tert-butylethane

Attachment A transformation by which one molecular entity (the substrate) is converted into another by the formation of one (and only one) two-centre bond between the substrate and another molecular entity and which involves no other changes in connectivity in the substrate. For example, the formation of an acyl cation by attachment of carbon monoxide to a carbenium ion (R^+): $R^+ + CO (RCO)^+$

The product of an attachment may also be the adduct of the two reactants, but not all adducts can be represented as the products of an attachment.

$CH_2=CH-CH=CH_2 + H_2C=CH_2 \longrightarrow$ 

August wilhelm (von) hofmann (1818-1892) Born in Giessen and studied under Liebig. Professor at the Royal College of Chemistry in London (1845-1865), and then Berlin (1865-1892). Associated with the Hofmann reaction (and various other double-barreled reactions), Hofmann degradation (sometimes mistakenly given to the unrelated Hofmann reaction), Hofmann primary amine test (carbylamine reaction), Hofmann rule, Hofmann's violet, Hofmann method of vapour densities, and the Hofmann voltanmeter, which is surely more than enough for one person, even one as ambitious as AW.

Austral spring Related to the Antarctic ozone depletion, south polar austral spring denotes the season of spring in the southern hemisphere when the greatest amount of ozone is lost, generally beginning in September.

Astral Relates to the hemisphere that the observer is currently in and the season that relates to the observer, while austral refers to the south.

Autocatalytic reaction A chemical reaction in which a product (or a reaction intermediate) also functions as catalyst. In such a reaction the observed rate of reaction is often found to increase with time from its initial value. See order of reaction.

Autoignition temperature Autoignition temperature is the minimum temperature required to initiate or cause self-sustained combustion in any substance in the absence of a spark or flame. This varies with the test method.

Automerization Synonymous with degenerate rearrangement.

Autoprotolysis A proton (hydron) transfer reaction between two identical molecules (usually a solvent), one acting as a Brønsted acid and the other as a Brønsted base. For example: 2 H_2O $H_3O^+ + OH^-$

Autoprotolysis constant The product of the activities (or, more approximately, concentrations) of the species produced as the result of autoprotolysis. For solvents in which no other ionization processes are significant the term is synonymous with "ionic product". The autoprotolysis constant for water, K_W, is equal to the product of activities

$a(H_3O^+)a(OH^-) = 1.0 \times 10^{-14}$ at 25 °C.

Autotrophic organisms Organisms that are capable of using carbon dioxide as the sole carbon source for growth and product formation. Organisms that use light as a source of energy are said to be photoautotrophs, those that use the energy from chemical reactions are chemoautotrophs.

Autumnal equinox Point at which on the celestial sphere that the equator and the Ecliptic intersect. Generally the autumnal equinox occurs on or about September 23 in the northern hemisphere; this also signifies spring in the southern hemisphere.

Auxotroph An organism which requires a particular organic compound for growth.

Avagadro's number Number representing the number of molecules in one 1. mole: 6.023 * 10^{23}.

Avogadro constant Avogadro constant (N_A or L) is the number of elementary entities in one mole of a substance.

L = N / n

It has the value $(6.022\ 045 \pm 0.000\ 031) \times 10^{23}$ mol^{-1}.

Avogadro, amadeo (1776-1856) An Italian chemist who first stated the principle behind stoichiometry in 1811.

Axial bond A bond perpendicular to the equator of the ring (up or down), typically in a chair cyclohexane.

Axial chirality Term used to refer to stereoisomerism resulting from the non-planar arrangement of four groups in pairs about a chirality axis. It is exemplified by allenes abC=C=Ccd (or abC=C=Cab) and by the atropisomerism of ortho-substituted biphenyls.

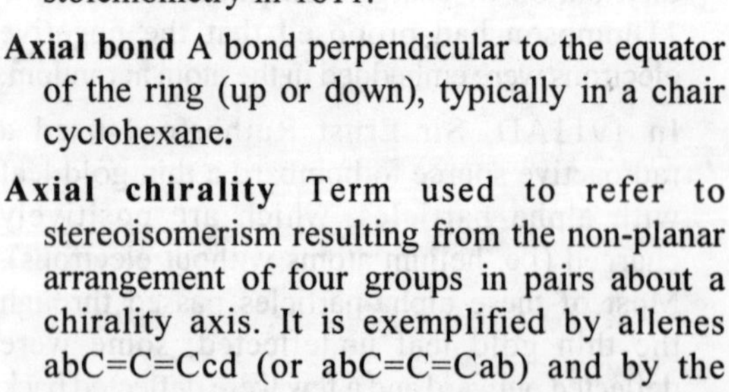

The configuration in molecular entities possessing axial chirality is specified by the stereodescriptors Ra and Sa (or by P or M).

Axial, equatorial In the chair form of cyclohexane ring bonds to ring atoms (and molecular entities attached to such bonds) are termed axial or equatorial according to whether the bonds make a relatively large or small angle, respectively, with the plane containing or passing closest to a majority of the ring atoms. Thus the axial bonds are approximately parallel to the C_3 axis and the equatorial bonds approximately parallel to two of the ring bonds. These terms are also used for the chair form of other saturated six-membered rings. The corresponding bonds occurring at the allylic positions in mono-unsaturated six-membered rings are termed pseudo-axial (or quasi-axial) and pseudo-equatorial (or quasi-equatorial). The terms axial and equatorial have similarly been used in relation to the puckered conformation of cyclobutane, crown conformer of cyclooctane, etc. and the terms pseudo-axial and pseudo-equatorial in the context of the non-planar structures of cyclopentane and cycloheptane. (See apical, basal, equatorial for an alternative use of axial and equatorial with bipyramidal structures).

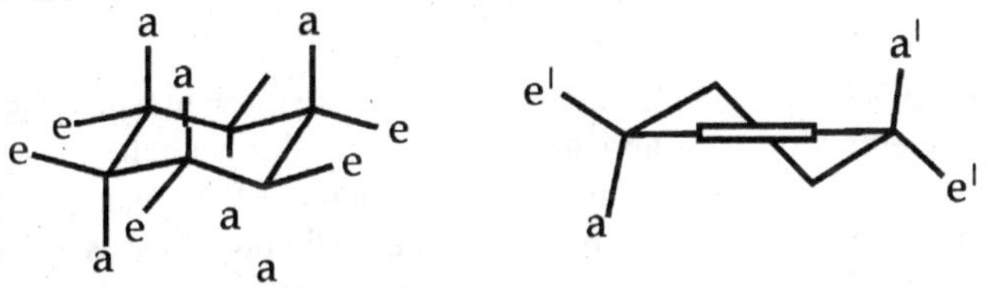

a = axial
e = equatorial
a' = pseudo-axial
e' = pseudo-equatorial

Azeotrope Azeotrope is a mixture of two liquids that boils at constant composition, i.e. the composition of the vapour is the same as that of the liquid. Azeotropes occur because of deviations in Raoult's law leading to a maximum or minimum in the boiling point - composition diagram. The composition of an azeotrope depends on the pressure.

Azo compounds Azo compounds are organic compounds containing the group -N=N- linking two other groups. They can be formed by reaction of a diazonium ion with a benzene ring.

Azurin An electron transfer protein containing a type 1 copper site, isolated from certain bacteria.

Azylene See nitrene.

B

Backscattering Process by which up to 25% of radiant energy from the sun is reflected or scattered away from the surface by clouds. Serves the greatest importance in the Atmospheric Heat Budget. Large errors in the assumed value of this variable may have important effects on computer models of the atmosphere.

Bacteriochlorin 7,8,17,18-Tetrahydroporphyrin. A reduced porphyrin with two pairs of non-fused saturated carbon atoms (C-7, C-8 and C-17, C-18) in two of the pyrrole rings.

Bacteriochlorophyll See chlorophyll.

Baeyer strain See angle strain.

Balanced chemical equations Balanced chemical equations describe the chemical changes which occur during the course of a chemical reaction, and 1. Identifies reactants and products,

2. Gives the correct composition of each reactant and product,

3. Gives relative numbers of atoms, ions and molecules involved, and

4. Gives relative weights of substances.

For example, the following equation indicates that two moles of hydrogen react with one mole of oxygen to form two moles of water.

$$2\ H_2 + O_2 \Rightarrow 2\ H_2O$$

Baldwin's rules A set of empirical rules for certain formations of 3- to 7-membered rings. The predicted pathways are those in which the length and nature of the linking chain enables the terminal atoms to achieve the proper geometries for reaction. The disfavoured cases are subject to severe distortions of bond angles and bond distances.

Ball mills Ball mills, also known as centrifugal or planetary mills, are devices used to rapidly grind materials to colloidal fineness (approximately 1 micron and below) by developing high grinding energy via centrifugal and/or planetary action.

Bar Bar (bar) is a unit of pressure equal to 10^5 Pa. Its use is temporarily maintained with the SI. The milibar (100 Pa) is commonly used in meteorology.

Barns A measurement of area equal to E^{-24} cm^2.

Barometric pressure The downward pressure, at any given point in the atmosphere, of the gases directly above that point. Average pressure globally at sea level is 1,013,000 dynes per centimeter squared or 760 torr. This is defined as one atmosphere.

Barrel Barrel (US, petrol) is an American unit of capacity usually employed in the petroleum industry and trading in fuels (barell = 158.9872949 L).

Base A base is a substance, which reacts with an acid to form salt and water only. Arrhenius in his researches into the behaviour of electrolytes in solution proposed that a base is a substance which provides hydroxide ions in aqueous solution as a result of dissociation and ionisation. Later, Lewis introduced the idea that acids and bases had a reciprocal relationship and he introduced the concept of Lewis conjugate acid/ base pairs.

A Lewis Acid (conjugate) is an electron pair acceptor and a Lewis Base (conjugate) is an electron pair donor.

$NH_3 + H_3O(+) \Rightarrow NH_4(+) + H_2O$

Lewis Lewis Lewis Lewis

Base Acid Acid Base

In particular, water and the Hydrated Proton, H3O(+), are a

Conjugate acid-base pair, as the hydrated proton is able to donate a proton to water and water is able to receive the proton.

$H(+) + H_2O \Rightarrow H_3O(+)$

Lewis

Base Acid

Base pairing The specific association between two complementary strands of nucleic acidsthat results from the formation of hydrogen bonds between the base components of the nucleotides of each strand: A=T and G≡C in DNA, A=U and G≡C (and in some cases G≡U) in RNA (the lines indicate the number of hydrogen bonds). Single-stranded nucleic acid molecules can adopt a partially double-stranded structure through intrastrand base pairing..

Basic Having the characteristics of a base.

Basic oxides Basic oxides are the oxides of metals (e.g. sodium oxide and copper oxide), and they react with water to form alkaline solutions.

$Na_2O + H_2O \Rightarrow 2\ NaOH$

Basic oxides form salts and water on reaction with acids.

$Na_2O + 2HCl \Rightarrow 2\ NaCl + H_2O$

$CuO + 2\ HNO_3 \Rightarrow Cu(NO_3)_2 + H_2O$

Strongly basic oxides form compounds with water containing a metal combined directly to a hydroxyl group, they are therefore called hydroxides. The hydroxides of sodium and potassium are also called alkalis.

$K_2O + H_2O \Rightarrow 2\ KOH$

Basicity For Brønsted bases it means the tendency of a compound to act as hydron (proton) acceptor. The basicity of a chemical species is normally expressed by the acidity of the conjugate acid (see conjugate acid-base pair). For Lewis bases it relates to the association constants of Lewis adducts and -adducts.

Bathochromic shift (effect) Shift of a spectral band to lower frequencies (longer wavelengths) owing to the influence of substitution or a change in environment. It is informally referred to as a red shift and is opposite to hypsochromic shift (blue shift). (1992).

Becquerel Becquerel (Bq) is the SI derived unit, with a special name, for radioactivity, equal to s^{-1}. It describes a radioactivity of an amount of radionuclide decaying at the rate, on average, of one spontaneous nuclear transition per second. The unit is named after the French scientist A. H. Becquerel (1852-1908) (disintegrations per unit time), equal to s^{-1}.

Beer's law Beer's law is the functional relationship between the quantity measured in an absorption method (A) and the quantity sought, the analyte concentration (c). As a consequence of interactions between the photons and absorbing particles, the power of the beam is attenuated from P_0 to P. Beer's law can be written

$A = \log (P_0/P) = abc$

where a is a proportionality constant called the absorptivity and b is the path length of the radiation through the absorbing medium. Also called the Beer-Lambert law.

Bell-evans-polanyi principle The linear relation between energy of activation (E_a) and enthalpy of reaction (H_r) sometimes observed within a series of closely related reactions.

$E_a = A + BH_r$

Benign This adjective is applied to any growth which does not invade surrounding tissue

Benthic Term denoting the lower section of a surface or bottom dwelling. The largest part of all biological life are benthic organisms in the greater part of the atmosphere.

Benzene is an aromatic hydrocarbon. It can be found in the air by several different ways of transmission. It can be produced for use with plastic or produced through the burning of fossil fuels. Benzene can also be found in the soil as well as some areas of groundwater pollution. Exposure to benzene has been linked to leukemia.

Benzyl group A benzene ring plus a methylene (CH_2) unit (C_6H5—CH_2).

Benzylic position The position of a carbon attached to a benzene ring.

Benzyne 1,2-Didehydrobenzene (the aryne derived from benzene) and its derivatives formed by substitution. The terms m- and p-benzyne are occasionally used for 1,3- and 1,4-didehydrobenzene, respectively.

Bernd fistert (1902-1978) Born in Ohlau, in lower Silesia, and entered the local university at Breslau. After working with Fritz Arndt (1885-1969) He joined I.G. Farben in Ludwigshafen (later became BASF), where he made a particular study of tautomerism. After retiring from BASF in 1957, he became a professor at Saarbrücken. His name is associated with the Arndt-Eistert synthesis (1935).

Bernhard fischer (1956-1905) Born in Hultschin near Breslau (now Wroclaw), and later director of a chemical research institute there. Not to be confused with Bernhard Fischer (1852-1915), who was an army surgeon (he died at Ypres) and marine bacteriologist.

Berry pseudorotation See pseudorotation.

Beta particle(ß-) A charged particle emitted from the nucleus of some atoms as part of those atoms' radioactive decay. Positively charged beta particles are positrons and negatively charged beta particles are electrons. Beta particles can cause burns and are harmful to living tissues, however, protection is afforded by a thin sheet of metal.

Beta particle decay A radioactive process in which the atomic number, Z, of the radioactive nuclide changes but the mass number, A, does not. In electron emission, Z increases. In positron emission or in electron capture, Z decreases.

Beta radiation Streams of beta particles are known as beta ray or beta radiation. Beta rays may cause skin burns and are harmful within the body. A thin sheet of metal can afford protection to the skin.

Beta(β) sheet Preferentially called a beta pleated sheet; a regular structure in an extended polypeptide chain, stabilised in the form of a sheet by hydrogen bonds between CO and NH groups of adjacent (parallel or antiparallel) chains.

Beta(β) strand Element of a beta sheet. One of the strands that is hydrogen bonded to a parallel or antiparallel strand to form a beta sheet.

Beta(β) turn A hairpin structure in a polypeptide chain reversing its direction by forming a hydrogen bond between the CO group of amino-acid residue n with the NH group of residue (n+3).

Bias Deviation of results or inferences from the truth, or processes leading to such deviation. Any trend in the collection, analysis, interpretation, publication, or review of data that can lead to conclusions that are systematically different from the truth. Among the ways in which deviation from the truth can occur, are the following: 1. Systematic (one-sided) variation of measurements from the true values (synonym: wsystematic error).

2. Variation of statistical summary measures (means, rates, measures of association, etc.) from their true values as a result of systematic variation of measurements, other flaws in data collection, or flaws in study design or analysis.

3. Deviation of inferences from the truth as a result of flaws in study design, data collection, or the analysis or interpretation of results.

4. A tendency of procedures (in study design, data collection, analysis, interpretation, review or publication) to yield results or conclusions

that depart from the truth.

5. Prejudice leading to the conscious or unconscious selection of study procedures that depart from the truth in a particular direction, or to one-sidedness in the interpretation of results.

(from Last, 1988).

Bicyclic A molecule with two rings that share at least two carbons.

Bifunctional catalysis Catalysis (usually for hydron transfer) by a bifunctional chemical species involving a mechanism in which both functional groups are implicated in the rate-controlling step, so that the corresponding catalytic coefficient is larger than that expected for catalysis by chemical species containing only one of these functional groups.

The term should not be used to describe the concerted action of two different catalysts ("concerted catalysis").

Bifunctional ligand A ligand that is capable of simultaneous use of two of its donor atoms to bind to one or more central atoms (see also ambidentate).

Bimolecular See molecularity.

Binary code Relationship between a set of tags and their corresponding ligands where building block identity is denoted by the presence or absence of a given tag or set of tags (i.e. the two "bits" 1 and 0).

Binary compound A compound that contains two different elements. NaCl is a binary compound; NaClO is not.

Binding constant See stability constant.

Binding site A specific region (or atom) in a molecular entity that is capable of entering into a stabilizing interaction with another molecular entity. An example of such an interaction is that of an active site in an enzyme with its substrate. Typical forms of interaction are by hydrogen bonding, coordination, and ion pair formation. Two binding sites in different molecular entities are said to be complementary if their interaction is stabilizing.

Binuclear Less-frequently used term for the IUPAC recommended:dinuclear. See nuclearity.

Bioaccumulation The process by which the amount of a substance in a living organism (or its parts) increases with time (WHO, 1979).

Bioassay A procedure for determining the concentration or biological activity of a substance (e.g.vitamin, hormone, plant growth factor, antibiotic, enzyme) by measuring its effect on an organism or tissue compared with a standard preparation.

Bioavailability (synonym:biological availability, physiological availability) The extent to which a chemical substance to which the body is exposed (by ingestion, inhalation, injection, or skin contact) reaches the systemic circulation, and the rate at which this occurs. It is recognized that the bioavailability (for gastrointestinal absorption) of, for example, both essential and non-essential metals, depends on various factors including the composition of the diet and the type of the chemical compound and its state of dispersion. For instance, the absorption of lead and cadmium is increased if the food is deficient in calcium or iron (WHO, 1979).

Biocatalyst A catalyst of biological origin, typically an enzyme.

Biochemical mechanism This is the general term for any chemical reaction or series of reactions, usually enzyme catalysed, which produces a given physiological effect in a living organism.

Biochemical oxygen demand Biochemical oxygen demand, BOD, is the amount of oxygen taken up by microorganisms to decompose organic waste matter which is present in water.

Biochemical oxygen demand is therefore used as an empirical measure of the amount of certain biologically degradable waste types of organic pollutant that are present in water.

Biochemical oxygen demand is calculated by keeping a sample of water containing a known amount of oxygen for five days at 20 degC in the dark. The oxygen content is measured again after this time.

A high biochemical oxygen demand indicates the presence of a large number of microorganisms, consuming large amounts of organic matter, which suggests a high level of pollution.

There is an empirical relationship between the two determinants that describe pollution load in a sample (i.e. between biochemical oxygen demand and chemical oxygen demand).

Biochemistry The chemistry of living things, including the structure and function of biological molecules and the mechanism and products of their reactions.

Bioconcentration A process leading to a higher concentration of a chemical in the organism relative to its environment (WHO, 1979).

Bioconjugate A molecular species produced by livingsystems of biological origin when it is composed of two parts of different origins, e.g. a conjugate of a xenobiotic with some groups, e.g. glutathione, sulfate or glucuronic acid, to make it soluble in water or compartmentalized within the cell.

Bioconversion The conversion of one substance to another by biological means. The fermentation of sugars to alcohols,catalyzed by yeasts, is an example of bioconversion.

Biogas Biogas is a mixture of methane and carbon dioxide resulting from the anaerobic decomposition of such waste materials as domestic, industrial, and agricultural sewage. Methanogenic bacteria carry out the decomposition; these obligate anaerobes produce methane, the main component of biogas, which can be collected and used as an energy source for domestic processes, such as heating, cooking, and lighting.

Biogeochemical carbon The biological and geographic study of the properties of carbon's chemical properties in relation to other reactive gases in the atmosphere.

Bioleaching Extraction of metals from ores or soil by biological processes, mostly by microorganisms.

Biological amplification or bioaccumulation Increase in concentration of DDT, PCB, and other slowly degradable, fat-soluble chemicals in organisms at successively higher levels.

Biological assessment of exposure Exposure to chemicals may be assessed by the analysis of specimens taken in the environment (air, water, food, etc.) or of specimens of biological material. Most often, urine and blood are analyzed, but other materials such as expired air, faeces, saliva, bile, hair, and biopsy or autopsy material are sometimes analyzed. In these samples, the content of the xenobiotic(s) or its metabolite(s) is determined and, on this basis, the exposure level (concentration in the air, absorbed amount of the substance) or the probability of health impairment due to exposure is derived. Biochemical changes in the components of an organism can also be used for this purpose (e.g., changes in enzyme activity or in the excretion of metabolic intermediates) if they show a relationship to the exposure (WHO, 1979).

Biological cycle The process through which a chemical substance passes in the biosphere. It may involve transport through the various media (air, water, soil), followed by environmental transformation, and carriage through various ecosystems. Chemical compounds that occur naturally have a natural biological cycle (WHO, 1979).

Biological half-life (synonym: biological half-time) The time required for the amount of a particular substance in a biological system to be reduced to one-half of its value by biological processes when the rate of removal is approximately exponential (ISO, 1972). For a one-compartment system describing an exponential biological process, biological half-life = log 2/f where f = elimination or decay constant.

Biological monitoring The periodic examination of biological specimens (in accordance with the definition of monitoring). It is usually applied to exposure monitoring but can also apply to effect monitoring (WHO, 1979).

Biological oxidation Biological oxidation involved the mediation of biological species in the Oxidation of organic matter. In general, biological oxidation proceeds by enzyme catalysis.

Biomagnification (Or ecological magnification) A sequence of processes in an ecosystem by which higher concentrations are attained in organisms of higher trophic level, i.e., of higher levels in the food chain .

Biomass burning The process of oxidizing living material. This process produces atmospheric particulates as well as the production of greenhouse and reactive tropospheric gases. These gases include CO_2, CO, NO_x, CH_4, CH_3Cl along with the addition of black carbon. All of these chemical species can be lofted relatively high in the atmosphere due to the convective heating of a fire.

Biomembrane Organized sheet-like assemblies consisting mainly of proteins and lipids (bilayers), acting as highly selective permeability barriers, containing specific molecular pumps and gates, receptors and enzymes.

Biomimetic Refers to a laboratory procedure designed to imitate a natural chemical process. Also refers to a compound that mimics a biological material in its structure or function.

Biomineralization The synthesis of inorganic crystalline or amorphous mineral-like materials by living organisms. Among the minerals synthesized biologically in various forms of life are: fluoroapatite, ($Ca_5(PO_4)_3$(F, OH)), hydroxyapatite, magnetite (Fe_3O_4) and calcium carbonate ($CaCO_3$).

Biopolymers Macromolecules (including proteins, nucleic acids and polysaccharides) formed by living organisms.

Biosensor A device that uses specific biochemical reactions mediated by isolated enzymes, immunosystems, tissues, organelles or whole cells to detect chemical compounds, usually by electrical, thermal or optical signals.

Biosphere A volume including the lower part of the troposphere (as high as living organisms can fly or be lofted) and the surface of the earth including the oceans. This region, by definition, encompasses all the living matter of the earth. Some very important atmospheric chemicals are produced in this region and pass into the atmosphere. This region exchanges chemicals and particulate matter with the atmosphere and soils and waters of the earth.

Biota Living organisms .

Biotechnology The industrial application of living organisms and/or biological techniques developed through basic research. Biotechnology products include pharmaceutical compounds and research materials.

Biotransformation A chemical transformation mediated by living organisms or enzyme preparations.

Biradical An even-electron molecular entity with two (possibly delocalized) radical centres which act nearly independently of each other, e.g.

Species in which the two radical centres interact significantly are often referred to as "biradicaloids". If the two radical centres are located on the same atom, the species are more properly referred to by their generic names: carbenes, nitrenes, etc.

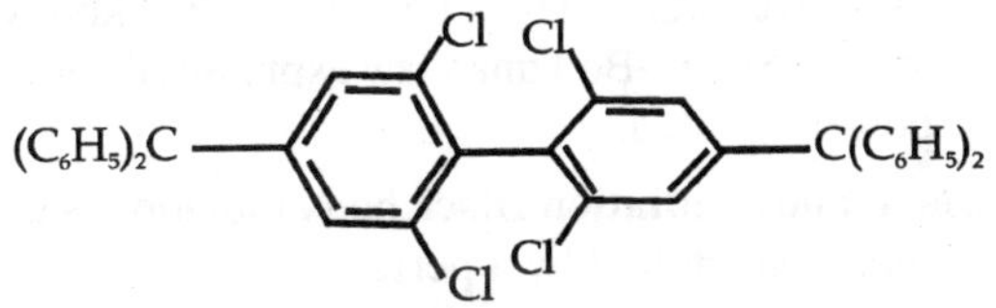

The lowest-energy triplet state of a biradical lies below or at most only a little above its lowest singlet state (usually judged relative to k_BT, the product of the Boltzmann constant k_B and the absolute temperature T). The states of those biradicals whose radical centres interact particularly weakly are most easily understood in terms of a pair of local doublets.

Theoretical descriptions of low-energy states of biradicals display the presence of two unsaturated valences (biradicals contain one fewer bond than permitted by the rules of valence): the dominant valence bond structures have two dots, the low energy molecular orbital configurations have only two electrons in two approximately nonbonding molecular orbitals, two of the natural orbitals have occupancies close to one, etc.

The term is synonymous with "diradical". See also carbene, nitrene.

Bisecting conformation, eclipsing conformation For a structure containing the grouping R_3C-C(Y)=X (with identical or different groups R) the conformation in which the torsion angle is such that X is antiperiplanar to one of the groups R, and, in a Newman projection, the double bond C=X bisects one of the R-C-R angles. In this conformation the bond C-Y eclipses one of the C-R bonds. The other conformation, in which X is synperiplanar to one of the groups R, is called an eclipsing conformation.

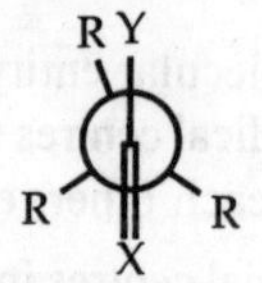

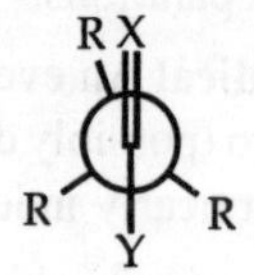

bisecting conformation eclipsing conformation

Blackbody In radiation physics, an ideal blackbody is a theoretical object that absorbs all the radiant energy falling upon it and emits it in the form of thermal radiation. Planck's radiation law gives the power radiated by a unit area of blackbody, and the Stefan-Boltzman law expresses the total power radiated.

Black body radiation Black body radiation is the radiation emitted by a perfect black body, i.e., a body which absorbs all radiation incident on it and reflects none. The wavelength dependence of the radiated energy density ρ (energy per unit volume per unit wavelength range) is given by the Planck formula

$$\rho = \frac{8\pi hc}{\lambda^5 \left(e^{hc/\lambda kT} - 1\right)}$$

where λ is the wavelength, h is Planck's constant, c is the speed of light, k is the Boltzmann constant, and T is the temperature.

Black carbon Emitted during the burning of coal, diesel fuel, natural gas and biomass and is part of the composition of soot.⇑ Black Carbon can absorb and reflect sunlight cooling the Earthís surface, but also increase solar energy absorbed in the atmosphere warming it.⇑ These effects are thought to effect global climate and rainfall cycles.⇑ Black Carbon increases the effect of global warming, visibility problems, and health problems.

Bleomycin (BLM) A glycopeptide molecule that can serveas a metal chelating ligand. The Fe(III) complex of bleomycin is an antitumor agent and its activity isassociated with DNA cleavage.

Blotting A technique used for transferring DNA, RNAor protein from gels to a suitable binding matrix, such as nitrocellulose or nylon paper, while maintaining the same physical separation.

Blue copper protein An electron transferprotein containing a type 1 copper site.Characterized by a strong absorption in the visible region, and an EPR signal with an unusually small hyperfine coupling to the copper nucleus. Both characteristics are attributed to coordination of the copper by a cysteine sulfur.

BOD (Biochemical Oxygen Demand) This is defined as the mass of dissolved oxygen required by a specific volume of solution of the substance for process of biochemical oxidation under prescribed conditions. The measurement of BOD indicates the ability of micro-organisms to metabolise an organic substance in the presence of oxygen and thus indicates the potential for depletion of oxygen by the substance. Refers to degradation.

Bohr magneton Bohr magneton (μ_B) is the atomic unit of magnetic moment, defined as

$$\mu_B = eh/4\pi m_e = 9.274\times10^{-24}\ \text{A m}^2$$

where h is Planck's constant, m_e the electron mass, and e the elementary charge. It is the moment associated with a single electron spin.

Bohr's atom Bohr made significant contributions to the atom. He understood the line spectra— the reason why only certain wavelengths are emitted when atoms jump down levels.

Boiling point The temperature at which the vapor pressure of a liquid is equal to or slightly greater than the atmospheric pressure of the environment. For water at sea level, its boiling point is 100C (212F).

Bolide impacts Asteroids or comets striking the earth; a possible cause of major climate changes and mass extinctions in the Earth's history.

Boltzmann constant The Boltzmann constant, k, named after the Austrian physicist Ludwig Boltzmann (1844-1906), is the ratio of the universal gas constant, R, to the Avogadro constant, Na.

The Boltzmann constant, k, is as the gas constant per molecule.

k = R / NaNA = 1.380622 * 10^-23 JK-1

Bomb calorimeter A bomb calorimeter is an apparatus used for measuring heats of combustion. It consists of a strong container in which the sample is sealed with excess oxygen and ignited electrically. The heat of combustion at constant volume can be calculated from the resulting rise in temperature of the calorimeter and its contents.

Bond (pi bond) See sigma, πι.

Bond (sigma bond) See sigma, πι.

Bond A bond (i.e. chemical bond) is the attractive force between the atoms in a molecule, which keeps the molecule together.

A chemical bond between atoms involves the overlap of an atomic orbital from each of the two atoms involved in the bond and keeps the molecule intact as an entity.

Bond dissociation See heterolysis, homolysis.

(In ordinary usage the term refers to homolysis only.)

Bond energy Bond Energy is the amount of energy associated with a bond in a chemical compound.

For example, in methane the bond energy of the carbon to hydrogen bond is one quarter of the Enthalpy of the synthesis process.

$$CH_{4(g)} \Rightarrow C_{(g)} + 4\ H_{(g)}$$

Bond energies can be calculated from the standard enthalpy of formation of the compound and from the enthalpies of atomisation of the elements. Energies calculated in this way are called average bond energies or bond energy terms. They depend to some extent on the molecule chosen; the C-H bond energy in methane will differ slightly from that in ethane. The bond dissociation energy is a different measurement, being the energy require to break a particular bond.

Bond number The number of electron-pair bonds between two nuclei in any given Lewis formula. For example in ethene the bond number between the carbon atoms is two, and between the carbon and hydrogen atoms is one.

Bond order A theoretical index of the degree of bonding between two atoms relative to that of a normal single bond, i.e. the bond provided by one localized electron pair. In the valence-bond theory it is a weighted average of the bond numbers between the respective atoms in the contributing structures. In molecular-orbital theory it is calculated from the weights of the atomic orbitals in each of the occupied molecular orbitals. For example, in valence-bond theory (neglecting other than Kekulé structures) the bond order between adjacent carbon atoms in benzene is 1.5; in Hückel molecular orbital theory it is 1.67. Other variants of molecular orbital theory provide other values for bond orders.

Bond pairs Bond pairs are the pairs of electrons involved in covalent bonds, or the unbonding lone pair of electrons which exist in a single sub-orbital of the central atom of a molecule.

Bond stretching vibrations The oscillatory motion of two bonded atoms relative to each other. This motion consist of the two bonded atoms stretching passed their equilibrium position, then returning to their equilibrium position, and finally contracting passed their equilibrium position. The result is an increase in frequency of the bond which allows it to absorb light at that same frequency.

Bond-dissociation energy, D (SI unit: kJ mol^{-1}, or J (per molecule))

The enthalpy (per mole) required to break a given bond of some specific molecular entity by homolysis, e.g. for CH_4 $H_3C\cdot$ + $H\cdot$ symbolized as $D(CH_3\text{-}H)$

Bonding and structure in group VI oxides Silica, SiO_2, forms a giant tetrahedral molecule, which is bonded covalently. Every silicon atom is attached to four oxygen atoms, and every oxygen atom is common to two SiO_4 tetrahedral.

Carbon, C, can form double bonds and Carbon Dioxide, CO_2, is an example of a compound that exist as discrete molecules. However, since silicon cannot form double bonds, Silicon Dioxide, SiO_2, (i.e. Silica) only forms an infinite three-dimensional tetrahedral structure.

Bone imaging The construction of bone tissue images from the radiation emitted by radionuclides that have been absorbed by the bone. Radionuclides such as ^{18}F, ^{85}Sr and ^{99m}Tc are introduced as complexes with specific ligands (very often phosphonate ligands) and are absorbed in the bones by metabolic activity.

Borderline mechanism A mechanism intermediate between two extremes, for example a nucleophilic substitution intermediate between S_N1 and S_N2, or intermediate between electron transfer and S_N2.

Boundary layer An area in the troposphere which is affected by the solar heating, radioactive cooling, and surface friction of the earth. The height of the boundary layer can range anywhere from 100m to 300 km and coincides with the height at which pollutants are mixed.

Bowsprit, flagpole In the boat form of cyclohexane and related structures there are two ring atoms lying out of the plane of the other four; exocyclic bonds to these two atoms pointing in a direction roughly parallel to that plane are called bowsprit, the other two are called flagpole.

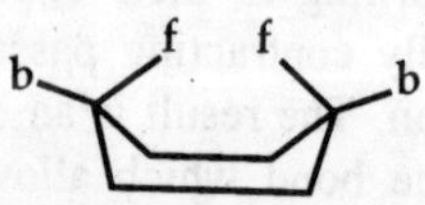

Boyle's law Boyle's law is the empirical law, exact only for an ideal gas, which states that the volume of a gas is inversely proportional to its pressure at constant temperature.

Bragg angle Bragg angle (θ) is defined by the equation

$$n\lambda = 2d\sin\theta$$

which relates the angle θ between a crystal plane and the diffracted x-ray beam, the wavelength λ of the x-rays, the crystal plane spacing d, and the diffraction order n (any integer).

Brain imaging In addition to magnetic resonance imaging that is based on the absorption by the brain of electromagnetic radiation, brain images can be acquired by scintillation counting (scintigraphy) of radiation emitted from radioactive nuclei that have crossed the blood-brain barrier. The introduction of radionuclides into brain tissue is accomplished with the use of specific $^{99m}Tc(V)$ complexes with lipophilic ligands.

Breathalyser test The breathalyser Test is based on the colour change which occurs when the dichromate ions in a glass tube through which the person blows is reduced to chromate ion by the presence of alcohol in the breath. This is a test given to suspected drunken drivers, who are required to blow through a tube containing a solution of acidified potassium dichromate on an inert support.

When the ethanol, C_2H_5OH, vapour is passed through the tube it is oxidised to carbon dioxid, and water. Correspondingly, the dichromate ions which are yellow in colour, are reduced to chromate ions which are green in colour. This results in the observed colour change from yellow to green when alcohol is present in the exhaled breath.

Bredt's rule A double bond cannot be placed with one terminus at the bridgehead of a bridged ring system unless the rings are large enough to accommodate the double bond without excessive strain. For example, while bicyclo(2.2.1)hept-1-ene is only capable of existence as a transient, its higher homologues having a double bond at the bridgehead position have been isolated.

Bridged carbocation A carbocation (real or hypothetical) in which there are two (or more) carbon atoms that could in alternative Lewis formulae be designated as carbenium centres but which is instead represented by a structure in which a group (a hydrogen atom or a hydrocarbon residue, possibly with substituents in non-involved positions) bridges these potential carbenium centres. One may distinguish "electron-sufficient bridged carbocations" and "electron-deficient bridged carbocations". Examples of the former are phenyl-bridged ions (for which the trivial name "phenonium ion" has been used), such as (A). These ions are straightforwardly classified as carbenium ions. The latter type of ion necessarily involves three-centre bonding. Structures (C) and (D) contain five-coordinate carbon atoms. The "hydrogen-bridged carbocation" (B) contains a two-co-ordinate hydrogen atom. Hypercoordination, which includes two-coordination for hydrogen and five- but also higher coordination for carbon is generally observed in bridged carbocations. See also carbonium ion, multi-centre bond, neighbouring group participation.

Bridging ligand A bridging ligand binds to two or more central atoms, usually metals, thereby linking them together to produce polynuclear coordination entities. Bridging is indicated by the Greek letter μ appearing before the ligand name and separated by a hyphen.

Bromochlorodifluoromethane Chemical formula $CBrClF_2$ Synonyms: Halon 1211, Freon 12B1, chlorodifluorobromomethane. This chemical belongs to the freon family. Halon 1211 was introduced in the 1960s as an effective gaseous fire suppression agent for application in the protection of computer control rooms, museums, telecommunication switches and other areas containing highly valuable materials. Results of studies done in the late 1980s indicated that the agent was an ozone depleting chemical. This chemical is stable in the troposphere but photodissociates to yield halogen radicals, which can catalytically destroy stratospheric ozone.

Bromomethane A volatile compound with a high mixing rate in the atmosphere. Methyl bromide escapes easily into the atmosphere where it contributes to the depletion of ozone. It works similarly to CFCs in its reaction with ozone molecules (O_3). In the stratosphere, with the Sunìs UV light as a catalyst, methyl bromide breaks down and exchanges a bromide ion for oxygen from ozone. By breaking down the ozone molecules in the stratosphere, methyl bromide thins the ozone layer and lets more UV light pass through.

Bromotrifluoromethane Also known as Halon-1301, this chemical with formula $CBrF_3$ is one of the most ozone destructive substances known to man. This chemical is commonly used in fire fighting equipment used around sensitive electronic equipment such as computer rooms, telecommunication centers, and aviation equipment. It is know to cause headache and unconsciousness in humans.

Brønsted acid (Brönsted acid) A molecular entity capable of donating a hydron (proton) to a base, (i.e. a "hydron donor") or the corresponding chemical species. For example: H_2O, H_3O^+, CH_3CO_2H, H_2SO_4, HSO_4^-, HCl, CH_3OH, NH_3.

Brønsted base (Brönsted base) A molecular entity capable of accepting a hydron (proton) from an acid (i.e. a "hydron acceptor") or the corresponding chemical species. For example: OH^-, H_2O, $CH_3CO_2^-$, HSO_4^-, SO_4^{2-}, Cl^-.

Brösted acid A material that gives up hydrogen ions in a chemical reaction.

Brownian motion Brownian motion is the continuous random movement of small particles suspended in a fluid, which arise from collisions with the fluid molecules. First observed by the British botanist R. Brown (1773-1858) when studying pollen particles. The effect is also visible in particles of smoke suspended in a still gas.

Brownian movement Brownian movement is the irregular zig-zag paths observed for the motion of pollen grains suspended in water, which strongly suggest that water molecules are moving about at random, and colliding with the grain pollens.

In 1828, Robert Brown observed the motion of pollen grains which were suspended in water. When these pollen grains were viewed through a microscope, the grains were seen to move in an erratic manner. They traveled along short straight paths, but the individual grains were rarely seen to collide with each other. The path of any one grain can best be described as a zigzag line, with a straight path between each zig and zag. This zig-zag motion is called Brownian Movement.

The irregular paths strongly suggest that the visible grains are pushed by invisible particles that are moving at random and which collided with the grains of pollen. If the water molecules are in motion, moving about at random, this would allow an explanation for the observations. This observation of Brownian Motion is one of the supporting pieces of evidence for the Atomic Theory and for the Kinetic theory of gases.

Buffer Buffer is a solution designed to maintain a constant pH when small amounts of a strong acid or base are added. Buffers usually consist of a fairly weak acid and its salt with a strong base. Suitable concentrations are chosen so that the pH of the solution remains close to the pK_a of the weak acid.

Buffer solutions Solutions that resist changes in their pH, even when small amounts of acid or base are added.

Building block One of a number of interchangeable reagents which can be used in combinatorial library synthesis, part of the structure of which becomes incorporated into the final product, i.e. its residue.

Burette A burette is a glass apparatus used in volumetric analysis. It is designed to deliver known volumes of liquids during titrations, and consists of a thin glass graduated tube and a stopcock to control delivery of the liquid.

C

C.I.P. system Short for cahn-ingold-prelog system (see CIP Priority).

Caffeine ($C_8H_{10}N_4O_2$) Methyltheobromine; guaranine; 1,3,7-trimethylxanthine; 1,3,7-trimethyl-2, 6-dioxopurine.

A substance found in tea, coffee, and cola that acts as a stimulant. It is extremely soluble in supercritical fluid carbon dioxide and somewhat soluble in water; aqueous solutions of caffeine quickly break down.

Cage An aggregate of molecules, generally in the condensed phase, that surrounds the fragments formed, for example, by thermal or photochemical dissociation. Because the cage hinders the separation of the fragments by diffusion, they may preferentially react with one another ("cage effect") but not necessarily to re-form the precursor species. For example,

R-N=N-R, heat (R· + N_2 + R·)cage R-R + N_2

Cage compound A polycyclic compound having the shape of a cage. The term is also used for inclusion compounds.

Calibration Calibration is correcting a measuring instrument by measuring values whose true values are known. Calibration minimizes systematic error.

Calmodulin A Ca^{2+} binding protein involved in metabolic regulation.

Calorie The amount of heat required to raise the temperature of 1 g of water at 14.5°C to 15.5°C. One calorie is equivalent to exactly 4.184 J.

Calorimeter An insulated vessel for measuring the amount of heat absorbed or released by a chemical or physical change.

Calorimetry Experimental determination of heat absorbed or released by a chemical or physical change.

Calpain A calcium-activated neutral protease, EC 3.4.22.17.

Calutron A device that separates isotopes (e. g. ^{235}U from ^{238}U) by ionizing the sample, accellerating the ions in a strong electric field, and then passing them through a strong magnetic field. The magnetic field bends the trajectories of the ions with high charge-to-mass ratio more, allowing ions to be separated by mass and collected.

Cancer Cancer is a disease which results from the development of a malignant tumour and its spread into the surrounding tissues.

Candela Candela (cd) is the SI base unit of luminous intensity.

The candela is the luminous intensity, in a given direction, of a source that emits monochromatic radiation of frequency 540×10^{12} Hz and that has a radiant intensity in that direction of 1/683 watt per steradian.

Capacitance (h) capillary electrophoresis (CE) (h) charge Charge is an amount of electrons. One coulomb is the charge of 6×10^{18} electrons.

Capacitor A device for storing electric charge, consisting of two metal plates separated by an insulating material.

Caplan's syndrome Pneumoconiosis associated with rheumatoid arthritis.

Captodative effect Effect on the stability of a carbon centred radical determined by the combined action of a captor (electron withdrawing) and a dative (electron releasing) substituent, both attached to the radical centre. The term is also used for certain unsaturated compounds.

Carbanion A negatively charged carbon atom.

Carbene Generic name for the species H_2C: and substitution derivatives thereof, containing an electrically neutral bivalent carbon atom with two nonbonding electrons. The nonbonding electrons may have antiparallel spins (singlet state) or parallel spins (triplet state). Use of the alternative name "methylene" as a generic term is not recommended.

Carbenium centre The three-coordinate carbon atom in a carbenium ion to which the excess positive charge of the ion (other than that located on heteroatoms) may be formally considered to be largely attributed, i.e., which has one vacant p-orbital. (N.B. It is not always possible to uniquely identify such an atom.) This formal attribution of charge often does not express the real charge distribution.

Carbenium ion A generic name for carbocations, real or hypothetical, that have at least one important contributing structure containing a tervalent carbon atom with a vacant p-orbital. (The name implies a protonated carbene or a substitution derivative thereof.)

The term was proposed (and rejected) as a replacement for the traditional usage of the name carbonium ion.

To avoid ambiguity, the name should not be used as the root for the systematic nomenclature of carbocations. The corresponding difficulty confused carbonium ion nomenclature for many years. For example, the term "ethylcarbonium ion" has at times been used to refer either to $CH_3CH_2^+$ (ethyl cation) or (correctly) to $CH_3CH_2CH_2^+$ (propyl cation).

Carbenoid A carbene like chemical species but with properties and reactivity differing from the free carbene itself, e.g. $R^1R^2C(Cl)M$ (M = metal)

Carbocation A cation containing an even number of electrons with a significant portion of the excess positive charge located on one or more carbon atoms. This is a general term embracing carbenium ions, all types of carbonium ions, vinyl cations, etc. Carbocations may be named by adding the word "cation" to the name of the corresponding radical Such names do not imply structure (e.g. whether three-coordinated or five-coordinated carbon atoms are present).

Carbohydrate carb A class of organic compounds including sugars and starches. The name comes from the fact that many (but not all) carbohydrates have empirical formula CH_2O.

Carbon 14 An isotope of carbon 12 (containing two more neutrons) that is radioactive and used in carbon dating.

Carbon cycle A complex cycle that circulates carbon through the atmosphere, oceans, and land which includes vegetation and soil and carbon is in various forms and oxidation states throughout the cycle.

Carbon dioxide CO_2, a compound consisting of one carbon and two oxygens. It is a reactant in photosynthesis and necessary for plant life. Abundant in the atmosphere due to anthropogenic and natural activities. It is a greenhouse gas.

Carbon disulfide Chemical formula CS_2; a compound used to manufacture products such as rayon and cellophane. Carbon disulfide is produced naturally by microbial activity in marshes and in volcanic ash. Since carbon disulfide does not adhere well to sediments, if it comes into contact with soil, it could percolate into groundwater, where it is very soluble. CS_2 can also add to photochemical smog development when it reacts with other organic substances in the atmosphere, such as methane or oxides of nitrogen.

Carbon monoxide CO, a toxic, odorless, colorless gas produced during fossil fuel or biomass burning. Compound consisting of one carbon and one oxygen. Except for carbon dioxide, it is one of the longest lived naturally occurring atmospheric carbon compounds (this wording is meant to exclude chlorofluorocarbons). The recent change in tropospheric CO content may portend a change in the balance between oxidants and reductants in the atmosphere.

Carbon monoxide dehydrogenases Enzymes that catalyze the oxidation of carbon monoxide to carbon dioxide. They contain iron-sulfur clusters and either nickel and zinc, or molybdopterin. Some nickel-containing enzymes are also involved in the synthesis of acetyl coenzyme A from CO2 and H2.

Carbon tetrachloride CCl_4 A compound consisting of a carbon and 4 chlorines that is active in ozone depletion when the compound is broken down and releases chlorine atoms (radicals). Chlorine reacts with the ozone creating diatomic oxygen and chlorine monoxide which cycles back to chlorine radicals.

Carbon tetrafluoride (CF_4) Known as Freon 14, carbon tetrafluoride is a stable, non-flammable, colorless gas. It absorbs light at 8 mm (in the infrared), is very stable, and does not react with water. This compound is theoretically considered a likely candidate for warming the earth during the next ice age.

Carbonate (CO_3^{2-}) 1. an inorganic ion with a charge of -2, containing carbon bound directly to three oxygens in a in a flat triangular arrangement.

2. A compound containing CO_3^{2-} ions.

Carbonate hardness Carbonate water hardness. Compare with water hardness.

Water hardness due to the presence of calcium and magnesium carbonates and bicarbonates. The "noncarbonate hardness" is due mostly to calcium and magnesium sulfates, chlorides, and nitrates.

Carbonic anhydrase A zinc-containing enzyme (carbonate hydro-lyase, carbonate dehydratase, EC 4.2.1.1) that catalyzes the reversible decomposition of carbonic acid to carbon dioxide and water.

Carboniferous period The time period between 280-345 Myr BP of Earth's geologic history. Characterized by glacial onsets and melting and massive migration and extinctions of species during this period.

Carbonium ion The term should be used with great care since several incompatible meanings are currently in use. It is not acceptable as the root for systematic nomenclature for carbocations. 1. In most of the existing literature the term is used in its traditional sense for what is here defined as carbenium ion.

2. A carbocation, real or hypothetical, that contains at least one five-coordinate carbon atom.

3. A carbocation, real or hypothetical, whose structure cannot adequately be described by two-electron two-centre bonds only. (The structure may involve carbon atoms with a coordination number greater than five.).

Carbonyl A carbon double bonded to oxygen (C=O).

Carbonyl group A divalent group consisting of a carbon atom with a double-bond to oxygen. For example, acetone (CH_3-(C=O)-CH_3) is a carbonyl group linking two methyl groups. Also refers to a compound of a metal with carbon monoxide, such as iron carbonyl, $Fe(CO)_5$.

Carbonyl sulfide COS, a gas that is very stable and unreactive in the troposphere, but, it is thought, photolyzes to form carbon monoxide, CO, and sulfur, S, in the stratosphere. Through stratospheric chemical reactions, the sulfur atoms are converted to SO_2 and H_2SO_4 which form sulfate aerosol and cloud condensation nuclei, but which eventually settle into the troposphere and reacts to form sulfuric acid, a component in acid rain. The major biospheric sources of COS are thought to be biological.

Carboplatin A "second generation" platinum drug effective in cancer chemotherapy named: cis-diammine(cyclobutane-1,1-dicarboxylato) platinum(II). Carboplatin is less toxic than the "first generation" antitumor drug, cisplatin.

Carboxylic acid A molecule containing a carboxyl (COOH) group.Also a functional group.

Carboxylic acids Carboxylic acids are organic compounds characterized by the presence of one or more RC(=O)OH groups (the carboxyl group). In systematic chemical nomenclature carboxylic acids names end in the suffix -oic (e.g. ethanoic acids, CH_3COOH). The carbon of the terminal group being counted as part of the chain. They are generally weak acids. Carboxylic acids include the large and important class of fatty acids and may be either saturated or unsaturated. There are also some natural aromatic carboxylic acids (benzoic, salicylic).

Carboy A very large bottle. Glass carboys are usually encased in a wire mesh or wooden box for protection.

Carbyne Generic name for the species HC and substitution derivatives thereof, such as EtO_2C-C containing an electrically neutral univalent carbon atom with three non-bonding electrons. Use of the alternative name "methylidyne" as a generic term is not recommended.

Carcinogen An agent, chemical, physical or biological, that can act on living tissue in such a way as to cause a malignant neoplasm (WHO, 1980).

Carcinogenesis The induction by chemical, physical, or biological agents, of neoplasms that are usually not observed, an earlier induction of neoplasms that are usually observed, and/or the induction of more neoplasms than are usually found although fundamental differences in the mechanisms may be involved.

Cardiotech A species radiolabeled with ^{99m}Tc with formula $(Tc(CNR)_6)^+$(R=tert-butyl) known for imaging the heart after a heart attack.

Carl Johann Philipp; Noé richard anschütz (1852-1937) Born in Darmstadt and initially studied engineering there before switching to chemistry. He took his doctorate at Heidelberg, then worked under Fittig at Tübingen and Kekulé at Bonn, where he remained for the rest of his career. He was interested in stereochemistry and studied the isomerism of unsaturated acids with Kekulé. He also proved the Graebe-Liebermann formula for anthracene by synthesis. Anschütz made a point of teaching chemistry to British POWs in the First World War because of his admiration for Couper's work on chemical structure. In later life he became interested in the history of chemistry. His name is associated with the Anschütz synthesis of anthracenes from substituted benzoyl chlorides.

Carl magnus von hell (1849-1926) Studied chemistry at the polytechnic in his home town Stuttgart and Munich Polytechnic. He worked at Stuttgart with Hermann von Fehling (1812-1885) and succeeded him as Professor of Chemistry in 1883. His greatest achievement was the synthesis of then largest known hydrocarbon ($C_{60}H_{122}$) in 1889. His name is associated with the Hell-Volhard-Zelinski bromination (1881-1887).

Carnot cycle Carnot cycle is the most efficient cycle of operations for a reversible heat engine. Published in 1824 by N. L. S. Carnot (1796-1832), it consists of four operations on the working substance in the engine: a) Isothermal expansion at thermodynamic temperature T_1 with heat q_1 taken in.

b) Adiabatic expansion with a fall of temperature to T_2.

c) Isothermal compression at temperature T_2 with heat q_2 given out.

d) Adiabatic compression at temperature back to T_1.

According to the Carnot principle, the efficiency of any reversible heat engine depends only on the temperature range through which it works, rather than the properties of the working substances.

Carotene Carotene is an unsaturated hydrocarbon pigment found in many plants. Carotene is the basic building block of vitamin A.

Case control study (synonyms: case comparison study, case history study, case referent study, retrospective study) A study that starts with the identification of persons with the disease (or other outcome variable) of interest, and a suitable control (comparison, reference) group of persons without the disease. The relationship of an attribute to the disease is examined by comparing the diseased and nondiseased with regard to how frequently the attribute is present or, if quantitative, the levels of the attribute, in each of the groups.

Such a study can be called "retrospective" because it starts after the onset of disease and looks back to the postulated causal factors. Cases and controls in a case control study may be accumulated "prospectively;" that is, as each new case is diagnosed it is entered in the study. Nevertheless, such a study may still be called "retrospective" because it looks back from the outcome to its causes. The terms "cases" and "controls" are sometimes used to describe subjects in a randomized controlled trial but, the term "case control study" should not be used to describe such a study

Catabolism Reactions involving the breaking down of organic substrates, typically by oxidative breakdown, to provide chemically available energy (e.g. ATP) and/or to generate metabolic intermediates used in subsequent anabolic reactions.

Catalase A hemeprotein which catalyzes the disproportionation of dihydrogen peroxide to O_2 and water ; it also catalyzes the oxidation of other compounds, such as ethanol, by dihydrogen peroxide. A non-heme protein containing a dinuclear manganese cluster with catalase activity is often called pseudocatalase.

Catalysis law See Brønsted relation.

Catalysis The action of a catalyst.

Catalyst A substance that participates in a particular chemical reaction and thereby increases its rate but without a net change in the amount of that substance in the system. At the molecular level, the catalyst is used and regenerated during each set of microscopic chemical events leading from a molecular entity of reactant to a molecular entity of product. See also autocatalytic reaction, bifunctional catalysis, catalytic coefficient, electron-transfer catalysis, general acid catalysis, general base catalysis, intramolecular catalysis, micellar catalysis, Michaelis-Menten kinetics, phase-transfer catalysis, pseudo-catalysis, rate of reaction, specific catalysis.

Catalytic antibody An antibody which catalyzes a chemical reaction analogous to an enzymatic reaction, such as an ester hydrolysis. It is obtained by using a hapten which mimics the transition state of the reaction. Also known as anabzyme.

Catalytic coefficient If the rate of reaction (v) is expressible in the form

$$v = (k_0 + k_i(C_i)^{n_i})\ (A)\ (B)...$$

where A, B, ... are reactants and Ci represents one of a set of catalysts, then the proportionality factor k_i is the catalytic coefficient of the particular catalyst Ci. Normally the partial order of reaction (n_i) with respect to a catalyst will be unity, so that k_i is an (++...+1)th order rate coefficient. The proportionality factor k_0 is the (++...)th order rate coefficient of the uncatalysed component of the total reaction.

Catalytic converter An air pollution control device using the exhaust system of cars. The converter helps complete combustion of any fuel that was not burned in the engine and reduce the presence of other harmful emission concentrations. The converter changes the unburned hydrocarbons and carbon monoxide in the exhaust into carbon dioxide and water vapor. The converter use chemical catalysis to create this change.

Cathode Cathode is a negative electrode of an electrolytic cell; to which positively charged ions (cations) migrate when a current is passed as in electroplating baths.

In a primary or secondary cell (battery or accumulator) the cathode is the electrode that spontaneously becomes negative during discharge, and form which therefore electrons emerge.

In vacuum electronic devices, electrons are emitted by the cathode and flow to the anode.

Cathode ray A negatively charged beam that emanates from the cathode of a discharge tube. Cathode rays are streams of electrons.

Cation A cation is a positively charged ion. Metals typically form cations.

Cation exchangeA cationic resin has positive ions built into its structure and therefore exchanges negative ions. In cation exchange, the side groups are ionized acidic groups, such as ($-SO_3H$, -COOH, -OH) to which cations H^+ are attached. The exchange reaction is one in which different cations in the solution displace the H^+ from the solid.

Ceiling The exposure limit for a substance that should (or in the case of OSHA PEL's "shall") not be exceeded. If instantaneous monitoring is not feasible, then a 15-minute time weighted average is used.

Ceiling value (CV) The maximum permissible airborne concentration of a potentially toxic substance and is a concentration that should never be exceeded in the breathing zone.

Cellulose A polysaccharide made of linked glucose molecules that strengthens the cell walls of most plants.

Centi. (c) Prefix used in the SI system meaning "one hundredth of". For example 1 cm means "one hundredth of a meter"; 2.3 cg could also be written "2.3×10^{-2} g" or "0.023 g".

Centigrade (*Celsius*) The scale for measuring temperature used internationally where the freezing point of water is zero and the boiling point of water at sea level is 100 degrees. To convert from centigrade into Fahrenheit, multiply the centigrade temperature by 1.8 then add 32 to the product (C*1.8+32=F). To convert from Fahrenheit to centigrade, subtract 32 and then divide by 1.8 ((F-32)/1.8).

Central atom The atom in a coordination entity that binds other atoms or group of atoms (ligands) to itself, thereby occupying a central position in the coordination entity.

Centre of chirality See chiral centre.

Ceramics Ceramics are an inorganic material of very high melting point. Ceramics are metal silicates, oxides, nitrides, etc.

Ceruloplasmin A copper protein present in blood plasma, containing type 1, type 2 and type 3 copper centers, where the type 2 and type 3 are close together, forming a trinuclear copper cluster.

Cgs system of units Cgs system of units is a system of units based upon the centimetre, gram, and second. The International System (SI) has supplanted the cgs system.

Chain reaction A reaction in which one or more reactive reaction intermediates (frequently radicals) are continuously regenerated, usually through a repetitive cycle of elementary steps (the "propagation step"). For example, in the chlorination of methane by a radical mechanism, Cl· is continuously regenerated in the chain propagation steps: $Cl\cdot + CH_4$ $HCl + H_3C\cdot$

$H_3C\cdot + Cl_2$ $CH_3Cl + Cl\cdot$

In chain polymerization reactions, reactive intermediates of the same types, generated in successive steps or cycles of steps, differ in relative molecular mass, as in

$RCH_2C\cdot HPh + H_2C{=}CHPh$ RCH_2CHPh $CH_2C\cdot HPh$

See also chain transfer, initiation, termination.

Chain transfer The abstraction, by the radical end of a growing chain-polymer, of an atom from another molecule. The growth of the polymer chain is thereby terminated but a new radical, capable of chain propagation and polymerization, is simultaneously created. For the example of alkene polymerization cited for a chain reaction, the reaction

$RCH_2C\cdot HPh + CCl_4$ $RCH_2CHClPh + Cl_3C\cdot$

represents a chain transfer, the radical $Cl_3C\cdot$ inducing further polymerization: $H_2C{=}CHPh + Cl_3C\cdot$ $Cl_3CCH_2C\cdot HPh$

$Cl_3CCH_2C\cdot HPh + H_2C{=}CHPh$ Cl_3CCH_2 $CHPhCH_2C\cdot HPh$

The phenomenon occurs also in other chain reactions such as cationic polymerization.

Chair conformation Typically, the most stable cyclohexane conformation. Looks like a chair.

Chair, boat, twist If carbon atoms 1, 2, 4 and 5 of cyclohexane occupy coplanar positions and when carbon atoms 3 and 6 are on opposite sides of the plane the conformation (of symmetry group D_{3d}) is called a chair form.

The same term is applied to similar conformations of analogous saturated six-membered ring structures containing hetero-atoms and/or bearing substituent groups, but these conformations may be distorted from the exact D_{3d} symmetry. For cyclohexane and most such analogues, the chair form is the most stable conformation. If the cyclohexane conformation has no centre of symmetry but possesses two planes of symmetry, one of them bisecting the bonds between atoms 1 and 2 and between 4 and 5 and the other plane passing through atoms 3 and 6 (which lie out of the plane and on the same side of the plane containing 1, 2, 4 and 5), that conformation (of symmetry group C_{2v}) is called a boat form and it is generally not a stable form. Again, this term is also applied to structural analogues.

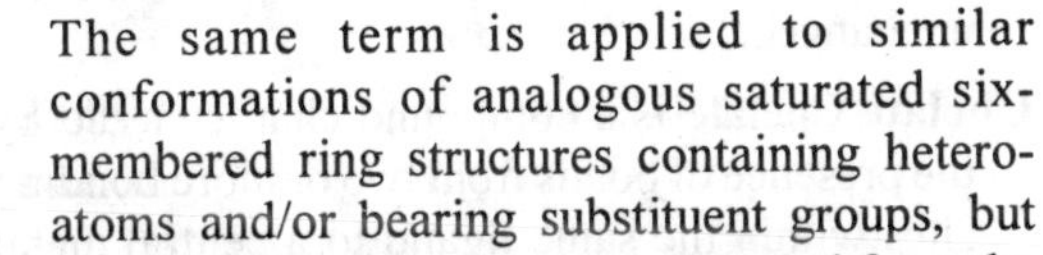

chair (D_3d) boat ($C2_v$) twist (D_2)

The conformation of D_2 symmetry passed through in the interconversion of two boat forms of cyclohexane is called the twist form (also known as skew boat, skew form and stretched form).

In a five-membered ring a conformation in which two adjacent atoms are maximally displaced, in opposite directions, relative to the plane containing the other three carbon atoms has been called a half-chair but is better called a twist conformation (see also envelope conformation).

In carbohydrate chemistry the term twist refers to a five-membered ring and the D_2 symmetry six-membered ring is referred to as skew.

Chalcogens Chalcogens are the Group 16 elements: oxygen (O), sulphur (S), selenium Se), tellurium (Te), and polonium (Po). Compounds of these elements are called chalcogenides.

Changing pressure *Example:* Decrease the volume of the container in the above example from 4.0 L to 1.0 L (Remember PV = nRT).

New partial pressures: P_{NO2} = 0.546 atm x 4 = 2.18 atm P_{N2O4} = 0.027 atm x 4 = 0.108 atm

In which direction will the reaction proceed to reestablish equilibrium? Calculate the reaction quotient, Q, to find out.

Q = (2.18 atm)^2/0.108 atm = 44 atm Q > K so the reaction will proceed in the reverse direction.

How do we find the new equilibrium partial pressures?

Start with the new initial conditions: P_{NO2} = 2.18 atm and P_{N2O4} = 0.108 atm

Q tells us the reaction will shift in the reverse direction, $N_2O_{4\,(g)} \leftarrow 2\,NO_{2\,(g)}$. The problem is set up in the following: N_2O_4 NO_2

P_o 0.108 atm 2.18 atm

ΔP +0.5x atm -x atm

P_{eq} (0.108+0.5x) atm (2.18-x) atm

K = 11 = (2.18-x)^2 / (0.108+0.5x)

Find x, then find P_{NO2} and P_{N2O4}

Changing temperature Changing the temperature of a chemical system changes K. Changing concentrations or pressures shifts the equilibrium conditions, but does not change K.

Example: $N_2O_{4\,(g)} \rightleftharpoons 2\,NO_{2\,(g)}$ ΔH° = + 57.2 kJ/mol

What happens if T increases?

The equilibrium conditions will shift to absorb heat.

Since ΔH°, this reaction shifts to the right, $N_2O_{4\,(g)} \rightarrow 2\,NO_{2\,(g)}$, until equilibrium partial pressures are reached.

Spectroscopic Lineshapes(h)

Introduction

The width and shape of spectroscopic transitions will affect the ability to extract qualitative and quantitative information from a spectrum. The lineshapes of spectroscopic transitions depend on the broadening mechanisms of the initial and final states, and include natural broadening, collisional broadening, power broadening, and Doppler broadening. Natural, collisional, and power broadening are homogeneous mechanisms and produce Lorentzian lineshapes, and Doppler broadening is a form of inhomogeneous broadening and has a Gaussian lineshape. Combinations of Lorentzian and Gaussian lineshapes can be approximated by a Voigt profile.

Chaperonin Member of the set of molecular chaperones, located in different organelles of the cell and involved either in transport of proteins through biomembranes by unfolding and refolding the proteins or in assembling newly formed polypeptides.

Chappuis band of ozone absorption The wavelengths of light absorbed by ozone in the visible region, approximately 380 to 750 nm. Compare to ozone's ultraviolet absorption region, the Hartley-Huggins band.

Charge Describes an object's ability to repel or attract other objects. Protons have positive charges while electrons have negative charges. Like charges repel each other while opposite charges, such as protons and electrons, attract one another.

Charge density See electron density.

Charge population The net electric charge on a specified atom in a molecular entity, as determined by some prescribed definition. see also electron density.

Charge transfer complex An aggregate of two or more molecules in which charge is transferred from a donor to an acceptor.

Charge transfer transition An electronic transition in which a large fraction of an electronic charge is transferred from one region of a molecular entity, called the electron donor, to another, called the electron acceptor (intramolecular charge-transfer) or from one molecular entity to another (intermolecular charge-transfer).

Charles' law The volume of a gas is directly proportional to its temperature in kelvins, if pressure and amount of gas remain constant. Doubling the kelvin temperature of a gas at constant pressure will double its volume. If V_1 and T_1 are the initial volume and temperature, the final volume and temperature ratio $V_2/T_2 = V_1/T_1$ if pressure and moles of gas are unchanged.

Chelate Chelate is a compound characterized by the presence of bonds from two or more bonding sites within the same ligand to a central metal atom.

Chelating agent Chelator.

A ligand that binds to a metal using more than one atom; a polydentate ligand.

Chelation The formation or presence of bonds (or other attractive interactions) between two or more separate binding sites within the same ligand and a single central atom. A molecular entity in which there is chelation (and the corresponding chemical species) is called a "chelate". The terms bidentate (or didentate), tridentate, tetradentate... multidentate are used to indicate the number of potential binding sites of the ligand, at least two of which must be used by the ligand in forming a "chelate". For example, the bidentate ethylenediamine forms a chelate with CuI in which both nitrogen atoms of ethylenediamine are bonded to copper. (The use of the term is often restricted to metallic central atoms.)

The phrase "separate binding sites" is intended to exclude cases such as $(PtCl_3(CH_2{=}CH_2))^-$, ferrocene, and (benzene)tricarbonylchromium in which ethene, the cyclopentadienyl group, and benzene, respectively, are considered to present single binding sites to the respective metal atom, and which are not normally thought of as chelates (see hapto).

Chelation therapy The judicious use of chelating (metal binding) agents for the removal of toxic amounts of metal ions from living organisms. The metal ions are sequestered by the chelating agents and are rendered harmless or excreted. Chelating agents such as 2,3-dimercaptopropan-1-ol, ethylenediaminetetraacetic acid, desferrioxamine and D-penicillamine have been used effectively in chelation therapy for arsenic, lead, iron and copper, respectively.

Cheletropic reaction A form of cycloaddition across the terminal atoms of a fully conjugated system with formation of two new sigma bonds to a single atom of the ("monocentric") reagent. There is formal loss of one pi bond in the substrate and an increase in coordination number of the relevant atom of the reagent.

Chelotropic reaction Alternative (and etymologically more correct) name for cheletropic reaction.

Chemical 1. of or pertaining to chemistry.

2. a substance.

Chemical bond. Bond; bonding; chemical bonding. A chemical bond is a strong attraction between two or more atoms. Bonds hold atoms in molecules and crystals together. There are many types of chemical bonds, but all involve electrons which are either shared or transferred between the bonded atoms.

Chemical change Reaction; chemical reaction. compare with physical change. A chemical change is a dissociation, recombination, or rearrangement of atoms.

Chemical changes Processes or events that have altered the fundamental structure of something.

Chemical equation Chemical equation is a way of denoting a chemical reaction using the symbol for the participating particles (atoms, molecules, ions, etc.); for example,

$aA + bB \rightleftarrows cC + dD$

The single arrow is used for an irreversible reaction; double arrows are used for reversible reactions. When reactions involve different phases it is usual to put the phase in brackets after the symbol.

s = solid l = liquid g = gas aq = aqueous

The numbers a, b, c, and d, showing the relative numbers of molecules reacting, are called the stoichiometric coefficients. The convention is that stoichiometric coefficients positive for reactants and negative for products. If the sum of the coefficients is zero the equation is balanced.

Chemical lifetime The length of time a chemical species can survive without reacting, photolyzing, dissociating, or otherwise changing into another chemical species. Highly reactive chemicals have short lifetimes.

Chemical potential (μ) The chemical potential is a partial molar Gibbs free energy, defined as $\mu_i = (G/n_i)_{T,P,nj}$. The definition means that the chemical potential is the change in Gibbs free energy when one mole of a substance is added to a very large amount of a sample. Chemical potential is a measure of chemical stability that can be used to predict and interpret phase changes and chemical reactions. Substances with higher chemical potential will react or move from one phase to another to lower the overall Gibbs free energy of the system. For example, consider a mixture of ice and water. If the ice melts, the chemical potential of the water was lower than that of the ice. If the water freezes, the chemical potential of the ice was lower.

Chemical potential For a mixture of substances, the chemical potential of constituent B (μ_B) is defined as the partial derivative of the Gibbs energy G with respect to the amount (number of moles) of B, with temperature, pressure, and amounts of all other constituents held constant.

$$\mu_B = \frac{\partial G}{\partial n}$$

Also called partial molar Gibbs energy. Components are in equilibrium if their chemical potentials are equal.

Chemical property Chemical properties. Compare with physical property.

Measurement of a chemical property involves a chemical change. For example, determining the flammability of gasoline involves burning it, producing carbon dioxide and water.

Chemical reaction A process that results in the interconversion of chemical species. Chemical reactions may be elementary reactions or stepwise reactions. (It should be noted that this definition includes experimentally observable interconversions of conformers.)

Detectable chemical reactions normally involve sets of molecular entities, as indicated by this definition, but it is often conceptually convenient to use the term also for changes involving single molecular entities (i.e. "microscopic chemical events").

Chemical relaxation If the equilibrium mixture of a chemical reaction is disturbed by a sudden change, especially of some external parameter (such as temperature, pressure, or electrical field strength), the system will readjust itself to a new position of the chemical equilibrium or return to the original position, if the perturbation is temporary. The readjustment is known as chemical relaxation.

In many cases, and in particular when the displacement from equilibrium is slight, the progress of the system towards equilibrium can be expressed as a first-order law

$$(C_t - (C_{eq})_2) = ((C_{eq})_1 - (C_{eq})_2)\exp(-t/)$$

where $(C_{eq})_1$ and $(C_{eq})_2$ are the equilibrium concentrations of one of the chemical species involved in the reaction before and after the change in the external parameter, and C_t is its concentration at time t. The time parameter t, named relaxation time, is related to the rate constants of the chemical reaction involved.

Measurements of the relaxation times by relaxation methods (involving a temperature jump (T-jump), pressure jump, electric field jump or a periodic disturbance of an external parameter, as in ultrasonic techniques) are commonly used to follow the kinetics of very fast reactions.

Chemical species An ensemble of chemically identical molecular entities that can explore the same set of molecular energy levels on the time scale of the experiment. The term is applied equally to a set of chemically identical atomic or molecular structural units in a solid array.

For example, two conformational isomers may be interconverted sufficiently slowly to be detectable by separate NMR spectra and hence to be considered to be separate chemical species on a time scale governed by the radiofrequency of the spectrometer used. On the other hand, in a slow chemical reaction the same mixture of conformers may behave as a single chemical species, i.e. there is virtually complete equilibrium population of the total set of molecular energy levels belonging to the two conformers.

Except where the context requires otherwise, the term is taken to refer to a set of molecular entities containing isotopes in their natural abundance.

The wording of the definition given in the first paragraph is intended to embrace both cases such as graphite, sodium chloride, or a surface oxide, where the basic structural units may not be capable of isolated existence, as well as those cases where they are.

Chemiluminescence The emission of absorbed energy as light as the result of a chemical reaction. This occurs in thousands of compounds both organic and inorganic. Chemical glow sticks and fireflies are examples of this.

Chemistry The study of matter and its transformations.

Chemobiokinetics The process of the uptake of chemical substances by the body, the biotransformation they undergo, the distribution of the substances and their metabolites in the tissues, and the elimination of the substances and their metabolites from the body. Both the amounts and the concentrations of the substances and their metabolites are studied. The term has essentially the same meaning as pharmacokinetics, but the latter term should be restricted to the study of pharmaceutical substances (WHO, 1979).

Chemoselective, chemoselectivity Chemoselectivity is the preferential reaction of a chemical reagent with one of two or more different functional groups. A reagent has a high chemoselectivity if reaction occurs with only a limited number of different functional groups. For example, sodium tetrahydroborate is a more chemoselective reducing agent than is lithium tetrahydroaluminate. The concept has not been defined in more quantitative terms. The term is also applied to reacting molecules or intermediates which exhibit selectivity towards chemically different reagents.

Some authors use the term chemospecificity for 100% chemoselectivity. However, this usage is discouraged.

Chemset A collection of two or more library members, building blocks or reagents; preferred notation in the Journal of Combinatorial Chemistry and convenient for describing synthetic procedures on pools of compounds.

Chiral center Asymmetric center.

An atom in a molecule that causes chirality, usually an atom that is bound to four different groups. A molecule can have chirality without having a chiral center, and a molecule may also have more than one chiral centers.

Chiral Having the property of chirality.

As applied to a molecule the term has been used differently by different workers. Some apply it exclusively to the whole molecule, whereas others apply it to parts of a molecule. For example, according to the latter view, a meso-compound is considered to be composed of two chiral parts of opposite chirality sense; this usage is to be discouraged.

In its application to an assembly of molecules, some restrict the term to an assembly in which all of the molecules have the same chirality sense, which is better called enantiopure. Others extend it to a racemic assembly, which is better just called a racemate. Use of the term to describe molecular assemblies should be avoided.

Chiral molccule Chiral molecule is a molecule which cannot be superimposed on its mirror image. A common example is an organic molecule containing a carbon atom to which four different atoms or groups are attached. Such molecules exhibit optical activity, i.e., they rotate the plane of a polarized light beam.

Chirality A term describing the geometric property of a rigid object (or spatial arrangement of points or atoms) that is non-superimposable on its mirror image; such an object has no symmetry elements of the second kind (a mirror plane, a centre of inversion, a rotation reflection axis). If the object is superimposable on its mirror image the object is described as being achiral.

Chirality axis An axis about which a set of ligands is held so that it results in a spatial arrangement which is not superposable on its mirror image. For example with an allene abC=C=Ccd the chiral axis is defined by the C=C=C bonds; and with an ortho-substituted biphenyl the atoms C-1, C-1', C-4 and C-4' lie on the chiral axis.

Chirality centre An atom holding a set of ligands in a spatial arrangement which is not superposable on its mirror image. A chirality centre is thus a generalised extension of the concept of the asymmetric carbon atom to central atoms of any element, for example N^+abcd, Pabc as well as Cabcd.

Chirality element General name for a chirality axis, chirality centre or chirality plane. Also referred to as an element of chirality.

Chirality plane A planar unit connected to an adjacent part of the structure by a bond which results in restricted torsion so that the plane cannot lie in a symmetry plane. For example with (E)-cyclooctene the chiral plane includes the double bond carbon atoms and all four atoms attached to the double bond; with a monosubstituted paracyclophane the chiral plane includes the monosubstituted benzene ring with its three hydrogen atoms and the three other atoms linked to the ring (i.e. from the substituent and the two chains linking the two benzene rings).

Chirality sense The property that distinguishes enantiomorphs. The specification of two enantiomorphic forms by reference to an oriented space e.g. of a screw, a right threaded one or a left threaded one. The expression 'opposite chirality' is short for 'opposite chirality sense'.

Chiroptic/chiroptical A term referring to the optical techniques (using refraction, absorption or emission of anisotropic radiation) for investigating chiral substances (e.g. measurements of optical rotation at a fixed wavelength, optical rotatory dispersion (ORD), circular dichroism (CD), and circular polarisation of luminescence (CPL)).

Chirotopic The description of an atom (or point, group, face, etc. in a molecular model) that resides within a chiral environment. One that resides within an achiral environment has been called achirotopic.

Chloracne A painful, disfiguring condition similar to common acne that it caused by people being exposed to high concentrations of PCB's (polychlorinated biphenyls). It is a biological response to the exposure of different types of organochlorine compounds.

Chlorine Cl_2, in the stratosphere, atomic (radical) chlorine is very destructive because it depletes the greatly needed ozone layer which protects the earth from ultraviolet radiation from the sun. in the Antarctic stratosphere, molecular chlorine along with nitric acid are formed by the reaction of hydrogen chloride and chlorine nitrate—both stratospheric chlorine reservoir species. This process occurs on polar stratospheric clouds which serve as the reaction sites. Once formed, Cl_2 vaporizes into the surrounding air as nitric acid—also formed in that process—binds with the ice matrix. Cl_2 is then photodissociated in sunlight (lamda <= 450 nm) into chlorine radicals. These chlorine radicals then catalytic destroy ozone.

Chlorine atoms Cl, the seventeenth element in the periodic table of elements. It has a atomic weight of 35.453 grams per mole. It has 17 protons in its nucleus and 7 electrons in its outer shell and is a very effective catalyst in the reaction that breaks down ozone in the stratosphere over Antarctica (see chlorine monoxide).

Chlorine dioxide ClO_2, a radical, undergoes photodecomposition in the stratosphere where the products of this reaction react with ozone. Since this is a photochemical reaction it only takes place while the sun is up. Experiments over Antarctica have shown a direct relation between polar ozone loss and the increase in halocarbon chemistry, which comes from anthropogenic sources. Scientist are currently looking at the molecular behavior of chlorine dioxide in the atmosphere in order to understand its role in depletion of ozone more thoroughly.

Chlorine monoxide ClO Aradical species which plays an important role in the breakdown of stratospheric ozone over Antarctica. Formed by the photolysis of CFCs in the stratosphere and the subsequent destruction of an ozone molecule, these radicals can act as a catalyst in the destruction of ozone while not being destroyed themselves. ClO, reacting with a oxygen atom (present from the Chapman Mechanism), releases a free chlorine radical once again. As a result, one Cl atom can destroy thousands of ozone molecules before being sequestered as HCl or another reservoir species (see chlorine nitrate).

Chlorine nitrate $ClONO_2$ This is a stratospheric reservoir species for chlorine and nitrogen, two of the catalysts in the breakdown of ozone. Frankly, it is named in a confusing manner; it is formed from the reaction of chlorine monoxide and nitrogen dioxide (not chlorine atoms with nitrate). It reacts with HCl at low temperatures on the surfaces of polar stratospheric clouds (PSCs over Antarctica and probably also in the stratosphere over the Arctic). That normally slow reaction heterogeneously produces molecular chlorine and nitric acid. The former outgases from the PSC surface and is quickly photolyzed by 450 nm or shorter wavelength light to form chlorine radicals which rapidly catalyze the breakdown of ozone (see chlorine monoxide).

Chlorobenzene C_6H_5Cl A colorless liquid that is manufactured for use as a solvent. It quickly evaporates in the air and is degraded by hydroxyl radicals that are produced photochemically. The gas acts as a source of ClOx, which helps in the breakdown of stratospheric ozone.

Chlorodifluoromethane Formula $CHClF_2$, this chemical is released into the atmosphere directly from aerosol cans. There are research efforts to find chemical replacements for these compounds because they are know to reach the stratosphere and destroy ozone.

Chloroethane This manmade VOC is highly reactive in the atmosphere. It is a gas at room temperature and when released, it readily reacts with oxidizing agents. This releases the chlorine atoms to circulate and break down ozone in the troposphere. The carbon and hydrogen atoms are released to be further oxidized to create CO_2 and HO. Chloroethane is often used in the manufacturing and production of insecticides, dyes and drugs

Chlorofluorocarbons CFCs Very stable chemical compound, used in refrigerants, solvent, and (in the past in the U.S.) aerosols, which release chlorine (important) and fluorine (less important) into the upper atmosphere. In the stratosphere, CFCs are photolyzed (by incoming solar UV) to form carbon dioxide, CO_2, hydrogen fluoride, HF, and ultimately (after multiple UV absorption events) chlorine radicals. These chlorine species are crucial in the destruction of the ozone layer over Antarctica and probably elsewhere (see chlorine).

Chloroform CHCl3 A colorless liquid that evaporates easily into the air. The compound is released into the air by direct and indirect sources and breaks down in the lower atmosphere into carbon dioxide, phosgene (carbonic dichloride), and hydrogen chloride. The degradation occurs in the troposphere by the reaction of the compound with hydroxyl radicals.

Chlorophyll Part of the photosynthetic systems in green plants. Generally speaking, it can be considered as a magnesium complex of a porphyrin in which a double bond in one of the pyrrole rings (17-18) has been reduced. A fused cyclopentanone ring is also present (positions 13-14-15).In the case of chlorophyll a, the substituted porphyrin ligand further contains four methyl groups inpositions 2, 7, 12 and 18, a vinyl group in position 3, an ethyl group in position 8 and a $-(CH_2)_2CO_2R$ group (R = phytyl, (2E)-(7R, 11R)-3,7,11,15-tetramethylhexadec-2-en-1-yl) in position 17. In chlorophyll b, the group in position 7 is a -CHO group. In bacteriochlorophyll a the porphyrin ring is further reduced (7-8), andthe group in position 3 is now a $-COCH_3$ group. In addition, in bacteriochlorophyll b, the group in position 8 is a $=CHCH_3$ group.

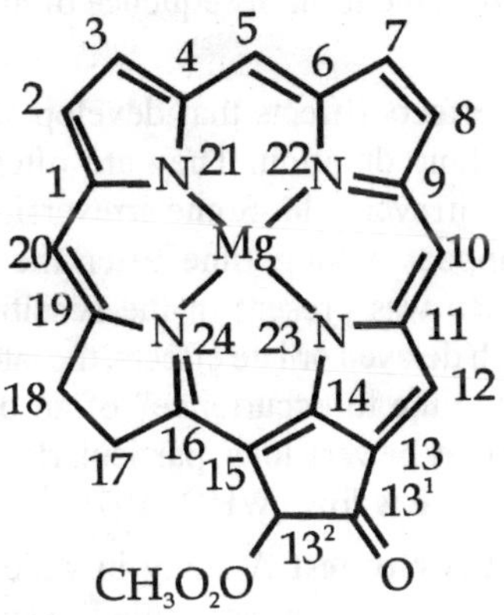

Chloroplast The organelle of eukaryotic photosynthesis.

Cholinesterase and pseudocholinesterase inhibitors These are substances which inhibit the cholinesterase-enzyme activity and thus enhance and subsequently prevent transmission of nerve impulses from one nerve cell to another or to a muscle.

ChromatographyChromatography is a method for separation of the components of a sample in which the components are distributed between two phases, one of which is stationary while the other moves. In gas chromatography, the gas moves over a liquid or solid stationary phase. In liquid chromatography, the liquid mixture moves through another liquid, a solid, or a gel. The mechanism of separation of components may be adsorption, differential solubility, ion-exchange, permeation, or other mechanisms.

Chromophore The part (atom or group of atoms) of a molecular entity in which the electronic transition responsible for a given spectral band is approximately localized. The term arose in the dyestuff industry, referring originally to the groupings in the molecule that are responsible for the dye's colour.

Chromosomal aberration Any abnormality of chromosome number or structure may be described as an aberration.

Chromosome This is a structure in the nucleus of the cell composed of deoxyribonucleic acid (DNA) andprotein; the chromosome forms the basis of heredity and carries genetic information in DNA in the form of sequence of nitrogenous bases.

Chronic effects Effects that develop slowly and have a long duration. They are often, but not always, irreversible. Some irreversible effects may appear a long time after the chemical substance was present in the sensitive tissue. For such delayed or late effects, the latent period (or the "time to occurrence" of an observable effect) may be very long, particularly if the level of exposure is low (WHO, 1979).

Chronic toxicity test A study in which animals are observed during the whole life span (or the major part of the life span) and in which exposure to the test material takes place over the whole observation time or a substantial part thereof. The term "long-term toxicity study" is sometimes used as a synonym for "chronic toxicity study" and sometimes to signify a study that falls in between subacute (short- term toxicity studies) and chronic toxicity studies (WHO, 1978a).

CIDNP (chemically induced dynamic nuclear polarization) Non-Boltzmann nuclear spin state distribution produced in thermal or photochemical reactions, usually from colligation and diffusion, or disproportionation of radical pairs, and detected by NMR spectroscopy by enhanced absorption or emission signals.

Cine-substitution A substitution reaction (generally aromatic) in which the entering group takes up a position adjacent to that occupied by the leaving group.

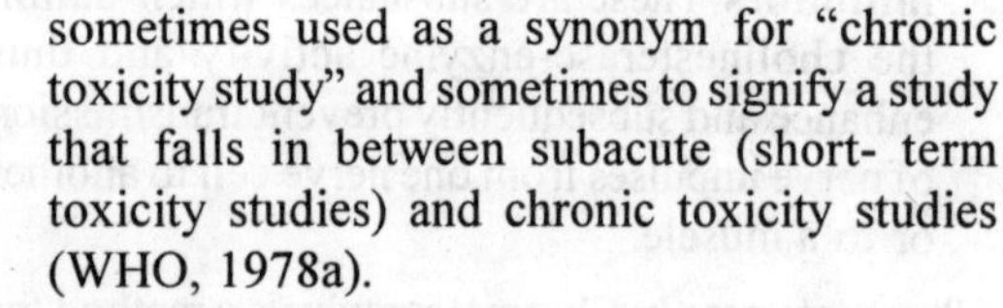

CIP priority In the CIP rules the conventional order of ligands established for the purpose of unambiguous designation of stereoisomers.

Circular dichroism (CD) A spectroscopic method which measures the difference in absorbance of left- and right-handed circularly polarized light by a material, as a function of the wavelength. Most biological molecules, including proteins and nucleic acids, are chiral and show circular dichroism in their ultraviolet absorption bands, which may be used as an indication of secondary structure. Metal centers that are bound to such molecules, even if they have no inherent chirality, usually exhibit CD in absorption bands associated with ligand-based or ligand-metal charge-transfer transitions. CD is frequently used in combination with absorption and MCD studies to assign electronic transitions.

Cirrus clouds High clouds that are formed entirely from ice crystals. They appear delicate and wispy and can reach a height of 35,000 feet (10,700 meters). Other types of cirrus clouds include cirrostratus and cirrocumulus. Cirrostratus is a thin sheet of cloud that often causes a halo to appear around the sun or moon. Cirrocumulus look like many small tufts of cotton; however, these clouds rarely form.

Cis In inorganic nomenclature a structural prefix designating two groups occupying adjacent positions (not generally recommended for precise nomenclature purposes of complicated systems).

Cis, trans Descriptors which show the relationship between two ligands attached to separate atoms that are connected by a double bond or are contained in a ring. The two ligands are said to be located cis to each other if they lie on the same side of a plane. If they are on opposite sides, their relative position is described as trans. The appropriate reference plane of a double bond is perpendicular to that of the relevant ?-bonds and passes through the double bond. For a ring (the ring being in a conformation, real or assumed, without re-entrant angles at the two substituted atoms) it is the mean plane of the ring(s). For alkenes the terms cis and trans may be ambiguous and have therefore largely been replaced by the E, Z convention for the nomenclature of organic compounds. If there are more than two entities attached to the ring the use of cis and trans requires the definition of a reference substituent .

Cisoid conformation (Usage strongly discouraged) The terms cisoid and transoid are also used to describe the stereochemistry of fused ring systems .

Cisplatin cis-diamminedichloroplatinum(II). An antitumor drug highly effective in the chemotherapy of many forms of cancer. Of major importance in the antitumor activity of this drug is its interaction with the nucleic acid bases of DNA.

Cis-trans isomers Stereoisomeric olefins or cycloalkanes (or hetero-analogues) which differ in the positions of atoms (or groups) relative to a reference plane: in the cis-isomer the atoms are on the same side, in the trans-isomer they are on opposite sides.

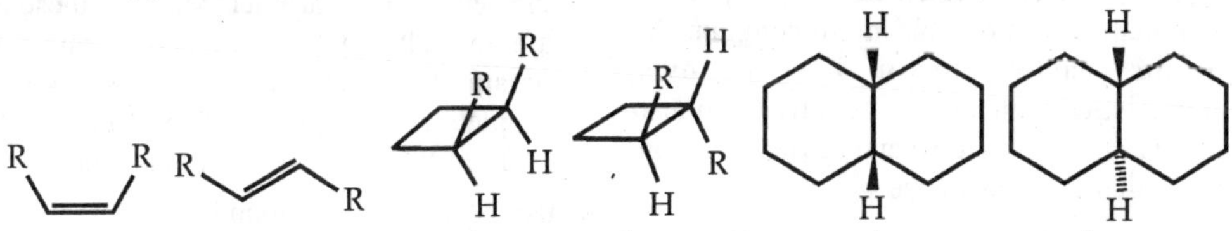

Clapeyron equation Clapeyron equation is a relation between pressure and temperature of two phases of a pure substance that are in equilibrium,

$$\frac{dp}{dT}\frac{\Delta_{trs}S}{\Delta_{trs}V}$$

where $\Delta_{trs}S$ is the difference in entropy between the phases and $\Delta_{trs}V$ the corresponding difference in volume.

Class (a) metal ion A metal ion that combines preferentially with ligands containing ligating atoms that are the lightest of their Periodic Group. See also class (b) metal ion, hard acid.

Class (b) metal ion A metal ion that combines preferentially with ligands containing ligating atoms other than the lightest of their Periodic Group.

See also class (a) metal ion, hard acid.

Clastogen A clastogen is any substance which causes chromosomal breaks.

Clathrates Also called gas hydrates, formed by or having molecules which are interlaced in a lattice-like geometrical pattern.

Clausius-clapeyron equation The Clausius-Clapeyron equation predicts the temperature dependence of vapor pressures of pure liquids or solids: ln (P/P°) = HR (1T° - 1 T) where P is the vapor pressure, P° is a vapor pressure at a known temperature T°, H is an enthalpy of vaporization if the substance is a liquid or an enthalpy of sublimation if it's a solid, R is the ideal gas law constant, and T is the temperature (in kelvins).

Cleavage Process of releasing a compound from a solid support, thereby permitting assay or analysis of the compound by solution-phase methods. Dissolution of the compound following cleavage, rather than the cleavage step itself, may be rate-limiting.

Climate Determined by the daily weather interactions over many years. Characteristics used in determining climate are temperature, precipitation, humidity, sunshine, and cloudiness, wind, and air pressure. Climatologists describe climate in terms of average temperature and precipitation amounts.

Clinal See torsion angle.

Clone 1. A population of genetically identical cells produced from a common ancestor.

2. Sometimes, "clone" is also used for a number of recombinant DNA molecules all carrying the same inserted sequence.

Close-up of the graphite furnace | View of the control box Atomic-emission detector (AED) As capillary column based gas chromatography takes its place as the major, highest resolution separation technique available for volatile, thermally stable compounds, the requirements for the sensitive and selective detection of these compounds increases. Since more and more complex mixtures can be successfully separated, subsequent chromatograms (output of a chromatographic separation) are increasingly more complex. Therefore, the need to differentiate between the sample components *using the GC detector* as a means of compounds discriminating is more and more common. In addition, each detector has its own characteristics (selectivity, sensitivity, linear range, stability, cost, etc.) that helps in a decision about which detector to use.

One of the newest additions to the gas chromatographer's arsenal is the atomic emission detector (AED). This detector, while quite expensive compared to other commercially available GC detectors, is an extremely powerful alternative. For instance, Instead of measuring simple gas phase (carbon containing) ions created in a flame as with the flame ionization detector, or the change in background current because of electronegative element capture of thermal electrons as with the electron capture detector, the AED has a much wider applicability because it is based on the detection of atomic emissions.

The strength of the AED lies in the detector's ability to simultaneously determine the atomic emissions of many of the elements in analytes that elute from a GC capillary column (called eluants or solutes in some books). As eluants come off the capillary column they are fed into a microwave powered plasma (or discharge) cavity where the compounds are destroyed and their atoms are excited by the energy of the plasma. The light that is emitted by the excited particles is separated into individual lines via a photodiode array. The associated computer then sorts out the individual emission lines and can produce chromatograms made up of peaks from eluants that contain only a specific element.

Closure A mathmatical term which says that if you operated on any two real numbers A and B with +,-,* or /, you get a real number

Cloud Condensed water vapor floating in air. They can take many different shapes due to wind patterns and moisture content. They play an important part in the world's weather because of the water they bring and because of their radiative properties vis a vis global warming.

Cloud condensation nuclei (CCN) Condensed water vapor that is so small that it can only be seen through a microscope. CCN are actually the center of the droplet. Many of these nuclei are tiny salt particles, sulfate or nitrate aerosol, or small particles present in smoke. Most droplets measure from 1/2,500 to 1/250 inch (0.01 to 0.1 millimeter) in diameter.

Cluster Group of compounds which are related by structural or behavioral properties. Organizing a set of compounds into clusters is often used in assessing the diversity of those compounds, or in developing SAR (structure-activity relationship) models.

Cluster sampling (i) A method of sampling in which the population is divided into aggregates (or clusters) of items bound together in a certain manner. A sample of these clusters is taken at random and all the items which constitute them are included in the sample (ISO, 1977); (ii) a sampling method in which each unit selected is a group of persons (all persons in a city block, a family, etc.) rather than an individual.

Coarse mode particles ñ are particles that are larger than two ?m in diameter.Ý They come form sea sprays, volcanoes, crushing or grinding of rocks, wind blown soil, and account for about 95% of the aerosol particles in ambient air.

Cobalamin Vitamin B-12 or B12.A vitamin synthesized by microorganisms and conserved in animals in the liver. Deficiency or collective uptake of vitamin B-12 leads to pernicious anemia. Cobalamin is a substituted corrin-Co(III) complex in which the cobalt atom is bound to the four nitrogen atoms of the corrin ring, an axial group R and 5,6-dimethylbenzimidazole.The latter is linked to the cobalt by the N-3 nitrogen atom and is bound to the C-1 carbon of a ribose molecule by the N-1 nitrogen atom. Various forms of the vitamin are known with different R groups such as R = CN, cyanocobalamin; R = OH, hydroxocobalamin; R = CH_3, methylcobalamin; R = adenosyl, coenzyme B-12.

COD (chemical oxygen demand) The amount of oxygen required to oxidize the substance chemically. Refers to degradation.

Codon A sequence of three consecutive nucleotides that occurs in mRNA and directs the incorporation of a specific amino acid into a protein, or represents the starting or termination signal of protein synthesis.

Coefficient of lineal thermal expansion The ratio of change in length per degree centigrade compared to the base length at zero degrees centigrade. The unit of measurement is centimeter per centimeter per degree centigrade (cm/cm/°C). This measurement means that for every centimeter of base length, the length will change X centimeters for every change of one degree centigrade.

Coenzyme A low-molecular-weight, non-protein organic compound (often a nucleotide) participating in enzymatic reactions as dissociable acceptor or donor of chemical groups or electrons.

Cofactor An organic molecule or ion (usually a metal ion) that is required by an enzyme for its activity. It may be attached either loosely (coenzyme) or tightly (prosthetic group).

Cohesion Compare with adhesion.

Attraction between like molecules.

Cohort study (synonyms: concurrent, follow up, incidence, longitudinal, prospective study) The method of epidemiologic study in which subsets of a defined population can be identified who are, have been, or in the future may be exposed or not exposed, or exposed in different degrees, to a factor or factors hypothesized to influence the probability of occurrence of a given disease or other outcome. The alternative terms for a cohort study, i.e., follow-up, longitudinal, and prospective study, describe an essential feature of the method, which is observation of the population for a sufficient number of person-years to generate reliable incidence or mortality rates in the population subsets. This generally implies study of a large population, study for a prolonged period (years), or both .

Coking ñ is the process of heating coal, usually lignite, in an oven to burn off unwanted impurities.Ý The unwanted impurities and waste gas escape from a chimney or a stack and the coal is then used as an industrial fuel.

Colligation The formation of a covalent bond by the combination or recombination of two radicals (the reverse of unimolecular homolysis). For example:

$HO\cdot + H_3C\cdot\ CH_3OH$

Colligative properties Properties of a solution that depend only on the number of particles dissolved in it, not the properties of the particles themselves. The main colligative properties addressed at this web site are boiling point elevation and freezing point depression.

Collision frequency. Collision frequencies; frequency of collision.

The average number of collisions that a molecule undergoes each second.

Collision theory Collision model.

A theory that explains reaction rates in terms of collisions between reactant molecules.

Collisional or pressure broadening Collisions broaden spectroscopic linewidths by shortening the lifetime of the excited states.

Colloid Colloids are systems in which there are two or more phases, with one (the dispersed phase) distributed in the other (the continuous phase). Moreover, at least one of the phases has small dimensions, in the range between 1 nanometer and 1 micrometer (10^{-9} - 10^{-6} m). Dimension, rather than the nature of the material, is characteristic. In this size range, the surface area of the particle is large with respect to its volume so that unusual phenomena occur, e.g., the particles do not settle out of the suspension by gravity and are small enough to pass through filter membranes. Macromolecules (proteins and other high polymers) are at the lower limit of this range; the upper limit is usually taken to be the point at which the particles can be resolved in an optical microscope.

Colloidal particles may be gaseous, liquid, or solid, and occur in various types of suspensions:

Sols - dispersions of small solid particles in a liquid.

Emulsions - colloidal systems in which the dispersed and continuous phases are both liquids.

Gels - colloids in which both dispersed and continuous phases have a three-dimensional network throughout the material.

Aerosols - colloidal dispersions of liquid or solid particles in a gas.

Foams - dispersions of gases in liquids or solids.

Colloid mills Colloid mills are machines used to grind aggregates into very fine particles or to apply very high shearing within a fluid to produce colloid suspensions or emulsions in which the particle sizes are less than 1 micrometer. One type of colloid mill is called a disc mill, in which a mixture of a solid and liquid (or two liquid) is passed between two discs a small distance apart, which rotate very rapidly relative to each other. Applications of colloid mills occur in food processing, in paint manufacture, and in the pharmaceutical industry.

Colorimetry A method for chemical analysis that relates color intensity to the concentration of analyte.

Column chromatography Column chromatography is a method for separating mixtures. A solution containing the mixture is passed through a narrow tube packed with a stationary phase. Different substances in the mixture have different affinities for the stationary phase, and so move through the tube at different rates. This allows the substances in the mixture to be detected or collected separately as they reach the end of the tube.

Column ozone The total amount of ozone that is found in a column of air. The majority of this amount of O_3 is typically found in the stratosphere.

Combination reaction A reaction in which two or more substances are chemically bonded together to produce a product. For example, 2 Na(s) + Cl_2(g) 2 NaCl(s) is a combination reaction.

Combinatorial chemistry Using a combinatorial process to prepare sets of compounds from sets of building blocks.

Combinatorial library A set of compounds prepared by combinatorial chemistry. May consist of a collection of pools or sub-libraries.

Combinatorial organic synthesis A key feature of combinatorial techniques is that compound synthesis can be designed such that a range of structures can be produced simultaneously as mixtures in the same reaction vessel or individually in parallel using semi- automated synthesis. The repetitive nature of the synthetic processes involved in most combinatorial applications lends itself to automation or semi-automation. This key feature means that the bench chemist can single- handedly prepare tens, hundreds, or thousands of compounds of known structures in the time that it would take to prepare only a few pure entities by orthodox methodology.

Combustion Combustion reaction.

A chemical reaction between a fuel and an oxidizing agent that produces heat (and usually, light). For example, the combustion of methane is represented as $CH_4(g) + 2\ O_2(g) = CO_2(g) + 2\ H_2O$.

Combustion When substances combine with oxygen and release energy.

Common-ion effect (on rates) A reduction in the rate of certain reactions of a substrate RX in solution (by a path that involves a pre-equilibrium with formation of R^+ (or R^-) ions as reaction intermediates) caused by the addition to the reaction mixture of an electrolyte solute containing the "common ion" X^- (or X^+). For example, the rate of solvolysis of diphenylmethyl chloride in acetone-water is reduced by the addition of salts of the common ion Cl^- which causes a decrease in the quasi-equilibrium oncentration of the diphenylmethyl cation in the scheme: Ph_2CHCl Ph_2CH^+ + Cl^- (free ions, not ion pairs)

$Ph_2CH^+ + OH_2\ Ph_2CHOH + H^+$(solvated)

This phenomenon is a direct consequence of the mass-law effect on ionization equilibria in electrolytic solution.

More generally, the common-ion effect is the influence of the "common ion" on the reactivity due to the shift of the dissociation equilibrium. It may also lead to an enhancement of the rate of reaction.

Commutativity A math property which states: A+B=B+A and A*B=B*A

Compartments The body is composed of a large number of organs, tissues, cells, and fluids, any one of which could be referred to as a compartment. In chemobiokinetics, a compartment often refers collectively to the organs, tissues, cells, and fluids for which the rates of uptake and subsequent distribution and elimination are sufficiently similar to preclude kinetic resolution (WHO, 1979).

Compensation 1. For internal compensation, see meso-compound.

2. See external compensation.

Compensation effect In a considerable number of cases plots of T S vs. H, for a series of reactions, e.g. for a reaction in a range of different solvents, are straight lines of approximately unit slope. Therefore, the terms H and T S in the expression partially compensate, and G = H - T S often is a much simpler function of solvent (or other) variation than H or T S separately.

Complete combustion Compare with incomplete combustion.

A combustion reaction that converts all of the fuel's carbon, hydrogen, sulfur, and nitrogen into carbon dioxide, water, sulfur dioxide, and N_2 respectively.

Complete ionic equation Total ionic equation. Compare with net ionic equation.

A balanced equation that describes a reaction occurring in solution, in which all strong electrolytes are written as dissociated ions.

Complex A molecular entity formed by loose association involving two or more component molecular entities (ionic or uncharged), or the corresponding chemical species. The bonding between the components is normally weaker than in a covalent bond.

The term has also been used with a variety of shades of meaning in different contexts: it is therefore best avoided when a more explicit alternative is applicable. In inorganic chemistry the term "coordination entity" is recommended instead of "complex".

Complex ion An ion formed by combination of simpler ions or molecules; for example, Co^{2+} combines with six molecules of water to form the complex ion $Co(H_2O)_6^{2+}$.

Complexing agent Complexant.

A ligand that binds to a metal ion to form a complex.

Complexometric titration Chelometric titration.

A titration based on a reaction between a ligand and a metal ion to form a complex. For example, free Ca^{2+} in milk powder can be determined by titrating a milk powder sample with EDTA solution, which chelates calcium ion. Endpoints in complexometric titrations are often determined using organochromic indicators.

Component 1. A substance whose concentration must be specified to describe the state of a mixture in which reactions are occurring. 2. A substance present in a mixture in which no reactions occur.

Compound Compare with element and mixture.

A compound is a material formed from elements chemically combined in definite proportions by mass. For example, water is formed from chemically bound hydrogen and oxygen. Any pure water sample contains 2 g of hydrogen for every 16 g of oxygen.

Composite reaction A chemical reaction for which the expression for the rate of disappearance of a reactant (or rate of appearance of a product) involves rate constants of more than a single elementary reaction. Examples are "opposing reactions" (where rate constants of two opposed chemical reactions are involved), "parallel reactions" (for which the rate of disappearance of any reactant is governed by the rate constants relating to several simultaneous reactions to form different respective products from a single set of reactants), and stepwise reactions.

Compound A substance composed of atoms or ions of two or more elements that are chemically combined. Elements in a compound are present in definite proportions by mass and are bonded with each other in a specific manner.

Comproportionation The reverse of disproportionation. The term "sympro-portionation" is also used. hartmanns, klenke, and metzger (1986).

Computational chemistry A branch of chemistry concerned with the prediction or simulation of chemical properties, structures, or processes using numerical techniques.

Computer-assisted drug design Using computational chemistry to discover, enhance, or study drugs and related biologically active molecules.

Concanavalin A A protein from jack beans, containing calcium and manganese, which agglutinates red blood cells and stimulates T lymphocytes to undergo mitosis.

Concentrate Compare with dilute.

To increase the amount of substance present in a unit amount of mixture. For example, allowing solvent to evaporate from a solution concentrates the solution.

Concentrated Compare with diluted.

Having a relatively large amount of substance present in a unit amount of mixture. For example, a 12 M HCl solution is more concentrated than an 0.001 M HCl solution.

Concentration Compare with dilution. 1. A measure of the amount of substance present in a unit amount of mixture. The amounts can be expressed as moles, masses, or volumes.

2. The process of increasing the amount of substance in a given amount of mixture.

Concerted process Two or more primitive changes are said to be concerted (or to constitute a concerted process) if they occur within the same elementary reaction. Such changes will normally (though perhaps not inevitably) be "energetically coupled". (In the present context the term "energetically coupled" means that the simultaneous progress of the primitive changes involves a transition state of lower energy than that for their successive occurrence.) In a concerted process the primitive changes may be synchronous or asynchronous.

Condensation (Condensed). 1. The conversion of a gas into a liquid is called condensation. Condensation usually occurs when a gas is cooled below its boiling point.

2. A reaction that involves linking of two molecules with the elimination of water (or another small molecule).

Condensation reaction A (usually stepwise) reaction in which two or more reactants (or remote reactive sites within the same molecular entity) yield a single main product with accompanying formation of water or of some other small molecule, e.g. ammonia, ethanol, acetic acid, hydrogen sulfide.

The mechanism of many condensation reactions has been shown to comprise consecutive addition and elimination reactions, as in the base-catalysed formation of (E)-but-2-enal (crotonaldehyde) from acetaldehyde, via 3-hydroxybutanal (aldol). The overall reaction in this example is known as the aldol condensation.

The term is sometimes also applied to cases where the formation of water or another simple molecule does not occur, as in "benzoin condensation".

Conditional acceptable daily intake A conditional acceptable daily intake is one that is established for a pesticide in order to limit its use to those situations where no satisfactory substitutes are avail- able. This definition will be the subject of further discussion. The allocation of conditional ADIs for intentional food additives has been superseded (Vettorazzi, 1980).

Conduction Thermal conduction the transfer of heat between two solid materials that are physically touching each other.

Electrical conduction the transfer of electrical current through a solid or liquid.

Configuration (Electronic) A distribution of the electrons of an atom or a molecular entity over a set of one-electron wavefunctions called orbitals, according to the Pauli principle. From one configuration several states with different multiplicities may result. For example, the ground electronic configuration of the oxygen molecule (O_2) is

$1_g^2, 1_u^2, 2_g^2, 2_u^2, 1_u^4, 3_g^2, 1_g^2$

resulting in the

$^3{}_g$, $^1{}_g$, and $^3{}_g^+$ multiplets .

Configuration In the context of stereochemistry, the term is restricted to the arrangements of atoms of a molecular entity in space that distinguishes stereoisomers, the isomerism between which is not due to conformation differences. See also absolute configuration; relative configuration.

Conformation The spatial arrangements of atoms affording distinction between stereoisomers which can be interconverted by rotation about formally single bonds. Some authorities extend the term to include inversion at trigonal bipyramidal centres and other "polytopal rearrangements".

Conformational analysis The assessment of the relative energies (or thermodynamic stabilities), reactivities, and physical properties of alternative conformations of a molecular entity, usually by the application of qualitative or semi-quantitative rules or by semi-empirical calculations.

Conformer One of a set of stereoisomers, each of which is characterised by a conformation corresponding to a distinct potential energy minimum.

Confounding 1. A situation in which the effects of two processes are not separated. The distortion of the apparent effect of an exposure on risk brought about by the association with other factors that can influence the outcome.

2. A relationship between the effects of two or more causal factors as observed in a set of data, such that it is not logically possible to separate the contribution that any single causal factor has made an effect.

3. A situation in which a measure of the effect of an exposure on risk is distorted because of the association of exposure with other factor(s) that influence the outcome under study.

(Last, 1988).

Confounding variable (Synonym: confounder) A variable that can cause or prevent the outcome of interest, is not an intermediate variable, and is not associated with the factor under investigation. Such a variable must be controlled in order to obtain an undistorted estimate of the effect of the study factor on risk (Last, 1988).

Congener 1. Elements belonging to the same group on the periodic table. For example, sodium and potassium are congeners.

2. Compounds produced by identical synthesis reactions and procedures.

Conjugate acid The acid that results from protonation of a base.

Conjugate acid-base pair The Brønsted acid BH^+ formed on protonation of a base B is called the conjugate acid of B, and B is the conjugate base of BH^+. (The conjugate acid always carries one unit of positive charge more than the base, but the absolute charges of the species are immaterial to the definition.) For example: the Brønsted acid HCl and its conjugate base Cl^- constitute a conjugate acid-base pair.

Conjugate base A substance which can gain a H^+ ion to form an acid.

Conjugated double bonds Double bonds separated by one carbon-carbon single bond. Alternating double bonds.

Conjugated system, conjugation In the original meaning a conjugated system is a molecular entity whose structure may be represented as a system of alternating single and multiple bonds: e.g. CH_2=CH=CH=CH_2 CH_2=CH-CN

In such systems, conjugation is the interaction of one p-orbital with another across an intervening sigma bond in such structures. (In appropriate molecular entities d-orbitals may be involved.) The term is also extended to the analogous interaction involving a p-orbital containing an unshared electron pair, e.g. :Cl-$CH=CH_2$

Conjugative mechanism See electronic effect.

Conjunctiva The term applied to the mucous membrane that covers the eyeball and undersurface of the eyelids.

Connectivity In a chemical context, the information content of a line formula, but omitting any indication of bond multiplicity.

Conrotatory See electrocyclic reaction.

Consensus sequence A sequence of DNA, RNA, protein or carbohydrate derived from a number of similar molecules, which comprises the essential features for a particular function.

Constant (sigma constant) Specifically the substituent constant for meta- and for para-substituents in benzene derivatives as defined by Hammett on the basis of the ionization constant of a substituted benzoic acid in water at 25 °C, i.e. $\lg(K_a/K_a^o)$, where K_a is the ionization constant of a m- or p-substituted benzoic acid and K_a^o that of benzoic acid itself.

The term is also used as a collective description for related electronic substituent constants based on other standard reaction series, of which, $^+$, $^-$ and ^{o} are typical; also constants which represent dissected electronic effects such as $_I$ and $_R$. For this purpose it might be better always to spell out the term in full, i.e. as "Hammett sigma constant", and restrict -constants to the scale of substituent constants which is based on benzoic acid.

A large positive -value implies high electron-withdrawing power by inductive and/or resonance effect, relative to H; a large negative -value implies high electron-releasing power relative to H. Chapman and Shorter (1972, 1978); Johnson (1973); Shorter (1973).

Constitution The description of the identity and connectivity (and corresponding bond multiplicities) of the atoms in a molecular entity (omitting any distinction from their spatial arrangement).

Constitutional isomerism Isomerism between structures differing in constitution and described by different line formulae e.g. CH_3OCH_3 and CH_3CH_2OH.

Constitutional isomers Molecules with the same molecular formula but with atoms attached in different ways.

Constructive interference Compare with destructive interference.

When the peaks and troughs of two interfering waves match, the amplitudes add to give the resultant wave a higher amplitude.

Contact ion pair See ion pair.

Contaminant In some contexts (e.g., in relation to gas cleaning equipment), used as a synonym for pollutant (ISO, 1979).

Continental drift A term applied to early theories supporting the possibility that the continents are in motion over the Earth's surface.

Continuous spectrum Compare with line spectrum and band spectrum.

A plot of the relative absorbance or intensity of emitted light vs. wavelength or frequency that shows a smooth variation, rather than a series of sharp peaks or bands.

Contrast agent Paramagnetic (or ferromagnetic) metal complex or particle causing a decrease in the relaxation times (increase in relaxivity) of nuclei detected in an image, usually of water.

Contributing structure The definition is based on the valence-bond formulation of the quantum mechanical idea of the wavefunction of a molecule as composed of a linear combination of wavefunctions, each representative of a formula containing bonds that are only single, double or triple with a particular pairing of electron spins. Each such formula represents a contributing structure, also called "resonance structure" to the total wavefunction, and the degree to which each contributes is indicated by the square of its coefficient in the linear combination. The contributing structures, also called "canonical forms", themselves thus have a purely formal significance: they are the components from which wavefunctions can be built. Structures may be covalent (or non-polar) or ionic (or polar). The representation is frequently kept qualitative so that we speak of important or major contributing structures and minor contributing structures. For example, two major non-equivalent contributing structures for the conjugate base of acetone are $CH_2=C(CH_3)-O^-$ $H_2C^--C(CH_3)=O$.

Control limit A regulatory value applied to the airborne concentration in the workplace of a potentially toxic substance which is judged to be "reasonably practicable" for the whole spectrum of work activities and which must not normally be exceeded.

Controls matched Controls (persons) who are selected so that they are similar to the study group, or cases, in specific characteristics. Some commonly used matching variables are age, sex, race and socio-economic status (Last, 1988).

Convection The movement of heat by a moving liquid or gas. Convection results from the differences in the densities of a material at different temperatures. As a liquid or gas raises in temperature, it becomes less dense and thus lighter thereby rising above its cooler and denser counterparts, which in turn sink.

Conversion factor A conversion factor is a fraction that relates one unit to another. Multiplying a measurement by a conversion factor changes the units of the measurement. For example, since 1 in = 2.54 cm, to convert 10 inches to centimeters,

(10 in)2.54 cm

1 in= 25.4 cm

Cooperativity The phenomenon that binding of an effector molecule to a biological system either enhances or diminishes the binding of a successive molecule, of the same or different kind, to the same system. The system may be an enzyme, or a protein that specifically binds another molecule such as oxygen or DNA. The effector molecule may be an enzyme substrate or an allosteric effector. The enzyme or protein exists in different conformations, with different catalytic rates or binding affinities, and binding of the effector molecule changes the proportion of these conformations. Enhanced binding is named positive cooperativity; diminished binding is named negative cooperativity. A well-known example of positive cooperativity is in hemoglobin. In biocatalysis it was originally proposed that only multi-subunit enzymes could respond in this way. However single-subunit enzymes may give such a response (so-called mnemonic enzymes).

Coordinate covalence (coordinate link) See coordination.

Coordination A coordination entity is composed of a central atom, usually that of a metal, to which is attached a surrounding array of other atoms or group of atoms, each of which is called a ligand. A coordination entity may be a neutral molecule, a cation or an anion. The ligands may be viewed as neutral or ionic entities that are bonded to an appropriately charged central atom. It is standard practice to think of the ligand atoms that are directly attached to the central atom as defining a coordination polyhedron (tetrahedron, square plane, octahedron, etc.) about the central atom. The coordination number is defined as being equal to the number of sigma-bonds between ligands and the central atom; this definition is not necessarily appropriate in all areas of (coordination) chemistry. In a coordination formula, the central atom is listed first. The formally anionic ligands appear next and they are listed in alphabetic order according to the first symbols of their formula. The neutral ligands follow, also in alphabetical order, according to the same principle. The formula for the entire coordination entity, whether charged or not, is enclosed in square brackets. In a coordination name, the ligands are listed in alphabetical order, without regard to charge, before the name of the central atom. Numerical prefixes indicating the number of ligands are not considered in determining that order. All anionic coordination entities take the ending -ate, whereas no distinguishing termination is used for cationic or neutral coordination entities.

Coordination number The coordination number of a specified atom in a chemical species is the number of other atoms directly linked to that specified atom . For example, the coordination number of carbon in methane is four, and it is five in protonated methane, CH_5^+. (The term is used in a different sense in the crystallographic description of ionic crystals.)

Coordinatively saturated A transition metal complex that has formally 18 outer shell electrons at the central metal atom.

Coordinatively unsaturated A transition metal complex that possesses fewer ligands than exist in the coordinatively saturated complex. These complexes usually have fewer than 18 outer shell electrons at the central metal atom.

CopolymerCopolymers are also known as heteropolymers. They made from two (or more) different monomers, which usually undergo a condensation reaction with the elimination of a simple molecule, such as ammonia or water. A typical example is the condensation of 1,6-diaminohexane (hexamethylenediamine) with hexanedioic acid (adipic acid) to form nylon 6,6.

The properties of a polymeric plastic can most easily be modified if it is a copolymer of two or more different monomers, e.g. acrylonitrile-butadiene-styrene copolymer (ABS). Varying the proportions of the component monomers can preselect its properties.

Core electron Compare with valence electron.

Electrons occupying completely filled shells under the valence shell.

Coriolis force A fictional source associated with the earth's rotation. It results in the deflection of all objects not at the equator to the right in the direction of motion in the northern hemisphere and to the left in the southern hemisphere.

Corphin Corphin is the F-430cofactor found in methyl-coenzyme M reductase, a nickel containing enzyme that catalyzes one step in the conversion of CO2 to methane in methanogenic bacteria. The Ni ion in F-430 is coordinated by the tetrahydrocorphin ligand. This ligand combines the structural elements of both porphyrins and corrins.

Correlation analysis The use of empirical correlations relating one body of experimental data to another, with the objective of finding quantitative estimates of the factors underlying the phenomena involved. Correlation analysis in organic chemistry often uses linear free-energy relations for rates or equilibria of reactions, but the term also embraces similar analysis of physical (most commonly spectroscopic) properties and of biological activity.

Corrin A ring-contracted porphyrin derivative that is missing a carbon from one of the mesopositions (C-20). It constitutes the skeleton $C_{19}H_{22}N_4$ upon which various B-12 vitamins, cofactors and derivatives are based.

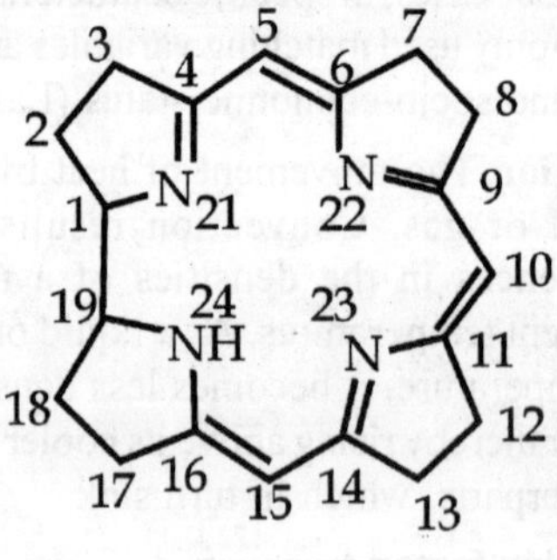

Corrosion Corrode. Corrosion is a reaction that involves action of an oxidizing agent on a metal. The oxidizing agent is often oxygen dissolved in water.

Corrosive of tissue The descriptor applied to any substance which destroys tissues on direct contact.

Cosmic rays Cosmic rays are high energy (10^{15} - 10^{17} eV) nuclear particles, electrons, and photons, originating mostly outside the solar system, which continually bombard the Earth's atmosphere.

CoulombCoulomb (C) is the SI unit of electric charge. It is equal to the charge transferred by a current of one ampere in one second (C = A s). The unit is named after French physicist C. A. Coulomb (1736-1806).

Coulombic interactions Electrostatic interactions. Attractions between opposite charges or repulsions between like charges that grow stronger as the charges become closer to each other.

Coulomb's law Coulomb's law is the statement that the force F between two electrical charges q_1 and q_2 separated by a distance r is

$$F = \frac{1}{4\pi\varepsilon_0} \cdot \frac{q_1 q_2}{r^2}$$

where ε_0 is the permittivity of a vacuum, equal to

$\varepsilon_0 = 8.8542 \times 10^{-12}$ F/m.

Coulometric titration Due to concentration polarization it is very difficult to completely oxidize or reduce a chemical species at an electrode. Coulometry is therefore usually done with an intermediate reagent that quantitatively reacts with the analyte. The intermediate reagent is electrochemically generated from an excess of a precursor so that concentration polarization does not occur. An example is the electrochemical oxidation of I^- (the precursor) to I_2 (the intermediate reagent). I_2 can then be used to chemically oxidize organic species such as ascorbic acid.

The point at which all of the analyte has been converted to the new oxidation state is called the endpoint and is determined by some type of indicator that is also present in the solution. For the coulometric titration of ascorbic acid, starch is used as the indicator. At the endpoint, I_2 remains in solution and binds with the starch to form a dark purple complex.

The analyte concentration is calculated from the reaction stoichiometry and the amount of charge that was required to produce enough reagent to react with all of the analyte.

de Broglie Equation(h)

Moving particles; such as electrons, protons, and neutrons; have wave properties as described by the de Broglie equation:

$\lambda = h / p$

where λ is wavelength, h is Planck's constant ($6.62618 x 10^{-34}$ J·s), and p is the momentum of the particle. Beams of particles can therefore show wave effects such as diffraction.

Ion Detectors(h)

Channeltron A channeltron is a horn-shaped continuous dynode structure that is coated on the inside with a electron emissive material. An ion striking the channeltron creates secondary electrons that have an avalanche effect to create more secondary electrons and finally a current pulse.

Count mean diameter The mean of the diameters of all particles in the population.

Count median diameter The diameter in the population above which there are as many particles with larger diameter as there are particles with smaller diameters.

Coupling constant (spin-spin coupling constant), J (SI unit: Hz (NMR))

A quantitative measure for nuclear spin-spin, nuclear-electron (hyperfine coupling) and electron-electron (fine coupling in EPR) coupling in magnetic resonance spectroscopy. The "indirect" or scalar NMR coupling constants are in a first approximation independent of the external magnetic field and are expressed in Hz.

Coupling protons Protons that interact with each other and split the NMR peak into a certain number of lines following the n+1 rule.

Covalent bond Covalent; covalently bound.A covalent bond is a very strong attraction between two or more atoms that are sharing their electrons. In structural formulas, covalent bonds are represented by a line drawn between the symbols of the bonded atoms.

Covalent compound Molecular compound. Compare with ionic bond and ionic compound.

A compound made of molecules- not ions. The atoms in the compound are bound together by shared electrons. Also called a molecular compound.

Covalent radius A chemical bond between atoms formed by the sharing of valence electrons.

Cox-yates equation A modification of the Bunnett-Olsen equation of the form

$lg ((SH^+))/(S)) - lg(H^+) = m^*X + pK_{SH^+}$

where X is the activity function $lg(SH^+)/SH^+)$ for an arbitrary reference base. The function X is called the excess acidity because it gives a measure of the difference between the acidity of a solution and that of an ideal solution of the same concentration. In practice X = - (Ho + $lg(H^+)$) and $m^* = 1$ - . COX .

Criteria Validated sets of data used as a basis for judgement (WHO, 1979).

Critical concentration for a cell The concentration at which undesirable (or adverse) functional changes, reversible or irreversible, occur in the cell.

Critical effect The first adverse effect that appears when the critical concentration in the critical organ is reached in an individual (or the adverse effect that occurs as a result of the lowest dose in the critical organ) (WHO, 1979).

Critical group That part of the target population most in need of protection (WHO, 1979).

Critical mass Critical mass is the minimum mass of a fissionable material (^{235}U or ^{239}Pu) that will initiate an uncontrolled chain reaction as in an atomic bomb. The critical mass of pure ^{239}Pu is about 4.5 kg, and of ^{235}U about 15 kg.

Critical micelle concentration (CMC) There is a relatively small range of concentrations separating the limit below which virtually no micelles are detected and the limit above which virtually all additional surfactant molecules form micelles. Many properties of surfactant solutions, if plotted against the concentration, appear to change at a different rate above and below this range. By extrapolating the loci of such a property above and below this range until they intersect, a value may be obtained known as the critical micellization concentration (critical micelle concentration), symbol cM, abbreviation cmc (or c.m.c.). As values obtained using different properties are not quite identical, the method by which the cmc is determined should be clearly stated.

Critical molar volume (V_c) The molar volume at the critical point.

Critical organ (critical tissue)(in toxicology) The particular organ that first attains the critical concentration (of metal) under specified circumstances of exposure and for a given population (Task Group on Metal Toxicology, 1976).

Critical organ concentration (critical tissue concentration) The mean concentration in the organ (tissue) at the time the most sensitive type of cell reaches the critical concentration (Task Group on Metal Toxicity, 1976).

Critical organ (In radiation biology) The organ whose damage (by radiation) results in the greatest injury to the individual (or his descendants). The injury may result from inherent radiosensitivity or indispensability of the organ, or from high dose, or from a combination of all three (ICRP, 1965).

Critical period A period during the development of a human, animal, or vegetable body, which is of particular importance in the life cycle if the normal full development of some anatomical, physiological, metabolic, or psychological structure or function is to be attained (WHO, 1972).

Critical point In general, critical point is the point on the phase diagram of a two-phase system at which the two coexisting phases have identical properties and therefore represent a single phase. At the liquid-gas critical point of a pure substance, the distinction between liquid and gas vanishes, and the vapour pressure curve ends. The coordinates of this point are called the critical temperature and critical pressure. Above the critical temperature, it is not possible to liquefy the substance.

Critical pressure Critical pressure is the pressure of a fluid in its critical point; i.e. when it is at its critical temperature and critical volume.

Critical temperature (T_c) The temperature at the critical point. A gas above the critical temperature will never condense into a liquid, no matter how much pressure is applied. Most substances have a critical temperature that is about 1.5 to 1.7 times the standard boiling point, in kelvin.

Critical volume Critical volume is the volume of a fixed mass of a fluid at critical temperature and pressure.

Cross section The effective size of a nucleus in capturing a thermal neutron (slow neutron). Larger cross sections have a greater probability of neutron capture.

Cross-conjugation In a system XC_6H_4GY this is conjugation involving the substituent X, the benzene ring, and the side-chain connective-plus-reaction site GY, i.e. either X is a +R group and GY is a -R group, or X is a -R group and GY is a +R group. In Hammett correlations this situation can lead to the need to apply exalted substituent constants $^+$ or $^-$, respectively, as in electrophilic or nucleophilic aromatic substitution, respectively. The term "through resonance" is synonymous. Cross conjugation has also been used to describe the interactions occurring in 2-phenylallyl and and similar systems).

Cross-linking Cross-linking is attachment of two chains of polymer molecules by bridges, composed of either an element, a group, or a compound, that join certain carbon atoms of the chains by primary chemical bonds. Cross-linking occurs in nature in substances made up of polypeptide chains that are joined by the disulfide bonds of the cysteine residue, as in keratins or insulin. Cross-linking can be effected artificially, either adding a chemical substance (cross-linking agent), or by subjecting the polymer to high-energy radiation. Examples are: vulcanisation of rubber with sulphur, cross-linking of polystyrene with divinylbenzene, or cross-linking of polyethylene by means of high-energy radiation.

Cross-linking has the effect of changing a plastic from thermoplastic to thermosetting. Thus, it also increases strength, heat and electrical resistance, and especially resistance to solvents and other chemicals.

Cross-sectional study (Synonyms: disease frequency survey, prevalence study) A study that examines the relationship between diseases (or other health-related characteristics) and other variables of interest as they exist in a defined population at one particular time. The presence or absence of disease and the presence or absence of the other variables (or, if they are quantitative, their level) are determined in each member of the study population or in a representative sample at one particular time. The relationship between a variable and the disease can be examined 1. in terms of the prevalence of disease in different population subgroups defined according to the presence or absence (or level) of the variables 2. in terms of the presence or absence (or level) of the variables in the diseased versus the nondiseased. Note that disease prevalence rather than incidence is normally recorded in a cross-sectional study. The temporal sequence of cause and effect cannot necessarily be determined in a cross-sectional study.

Crown A molecular entity comprising a monocyclic ligand assembly that contains three or more binding sites held together by covalent bonds and capable of binding a guest in a central (or nearly central) position. The adducts formed are sometimes known as "coronates". The best known members of this group are macrocyclic polyethers, such as "18-crown-6", containing several repeating units $-CR_2-CR_2O-$ (where R is most commonly H), and known as crown ethers. Pedersen (1967); CRAM et al. (1986).

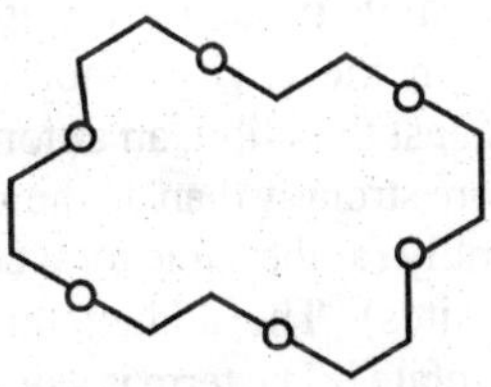

" 18-crown-6"

Crown conformation A conformation of a saturated cyclic molecular entity, containing an even number (>= 8) of atoms in the ring, in which these atoms lie alternately in each of two parallel planes and are symmetrically equivalent (D_{4d} for cyclooctane, D_{5d} for cyclodecane etc). It is analogous to the chair conformation of cyclohexane.

Crude death rate See death rate.

Crust Crust is the outer layer of the solid earth, above the Mohorovicic discontinuity. Its thickness averages about 35 km on the continents and about 7 km below the ocean floor.

Cryogen Cryogenic gas.

A gas that has been liquified by lowering temperature, usually to a temperature under about -100°C.

Cryoscopic constant Cryoscopic constant (E_f) is the constant that expresses the amount by which the freezing point T_f of a solvent is lowered by a non-dissociating solute, through the relation

$$\Delta T_f = E_f m$$

where m is the molality of the solute.

Cryosphere The portion of the earth which consists of the ice masses and snow deposits which include continental ice sheets, mountainous glaciers, sea ice, surface snow cover and lake/river ice. Alterations in the extent of snow cover are due to seasonal fluctuations and are interrelated with atmospheric circulation. Sea level and hydrologic cycle variations can affect the volume of water tied up in the glaciers and ice sheets.

Cryptand A molecular entity comprising a cyclic or polycyclic assembly of binding sites that contains three or more binding sites held together by covalent bonds, and which defines a molecular cavity in such a way as to bind (and thus "hide" in the cavity) another molecular entity, the guest (a cation, an anion or a neutral species), more strongly than do the separate parts of the assembly (at the same total concentration of binding sites). The adduct thus formed is called a "cryptate". The term is usually restricted to bicyclic or oligocyclic molecular entities.

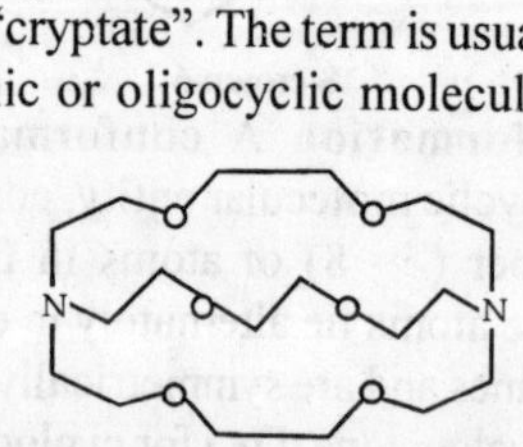

Crystal A sample of a crystalline solid that has a regular shape bound by plane surfaces (facets) that intersect at characteristic angles. The shape results from the arrangement of the substances atoms, ions, or molecules. Most crystals contain defects that can strongly affect their optical and electrical properties.

Crystal field Crystal field theory is the theory which interpretes the properties of coordination entities on the basis that the interaction of the ligands and the central atom is a strictly ionic or ion-dipole interaction resulting from electrostatic attractions between the central atom and the ligands. The ligands are regarded as point negative (or partially negative) charges surrounding a central atom; covalent bonding is completely neglected. The splitting or separation of energy levels of the five degenerate d orbitals in a transition metal, when the metal is surrounded by ligands arranged in a particular geometry with respect to the metal center, is called the crystal field splitting.

Crystal field theory Crystal field.

The color, spectra, and magnetic properties of metal-ligand complexes can be explained by modeling the effect of ligands on metal's d orbital energies.

Crystal structure (crystal system) The atomic arrangement of the atoms of an element when it is in its solid state.

Crystalline solid crystalline. Compare with amorphous.

A solid that has a repeating, regular three-dimensional arrangement of atoms, molecules, or ions.

Crystallite Compare with crystal.

A perfect crystalline part of a larger imperfect crystal. Real crystals are usually built of a large number of crystallites.

Crystallization Fractional crystallization; crystallisation.

The process of forming pure crystals by freezing a liquid, evaporating a solution, or precipitating a solid from solution. Impurities remain in the liquid, so crystallization is often to purify solid substances.

C-terminal amino-acid residue See amino-acid residue.

Cumulative effect (functional accumulation) Occurs when repeated doses of a toxic substance or harmful radiation summate to give an enhanced effect (WHO, 1979).

Cumulative incidence ratio The ratio of the cumulative incidence rate in the exposed to the cumulative incidence rate in the unexposed (Last, 1983).

Cumulative incidence, cumulative incidence rate The number or proportion of a group of people who experience the onset of a health related event during a specified time interval; this interval is generally the same for all members of the group, but, as in lifetime incidence, it may vary from person to person without reference to age .

Cumulus clouds Clouds forming in the troposphere which are vertically formed with flat bases and fluffy, rounded tops. They have often been described as cauliflower-like in structure. They occur at heights of 500-6000 meters in elevation from the earth and most often occur scattered or in dense heaped packs. They are formed due to buoyant upward convection during warm, anti-cyclonic summer weather.

Cupric (Cu2+) Cupric ion.Deprecated. 1. the copper(II) ion, Cu2+. 2. A compound that contains copper in the +2 oxidation state.

Cuprous (Cu+) Cuprous ion.Deprecated. 1. the copper(I) ion, Cu+. 2. A compound that contains copper in the +1 oxidation state.

Curie point Temperature above which a ferromagnetic material loses its ferromagnetism.

Curie relation See magnetic susceptibility.

Curie temperature For a ferromagnetic material, Curie temperature or Curie point (T_C) is the critical temperature above which the material becomes paramagnetic. For iron the Curie point is 760 °C and for nickel 356 °C. It is named after French physicist Pierre Curie (1859-1906).

Current Symbol: I, unit: ampere or amp (A)

Current is the rate of flow of charge, i. e., the number of coulombs flowing past a point per second. One amp is equal to one coulomb per second.

Curtin-hammett principle In a chemical reaction that yields one product (X) from one conformational isomer (A') and a different product (Y) from another conformational isomer (A") (and provided these two isomers are rapidly interconvertible relative to the rate of product formation, whereas the products do not undergo interconversion) the product composition is not in direct proportion to the relative concentrations of the conformational isomers in the substrate; it is controlled only by the difference in standard free energies (dG) of the respective transition states.

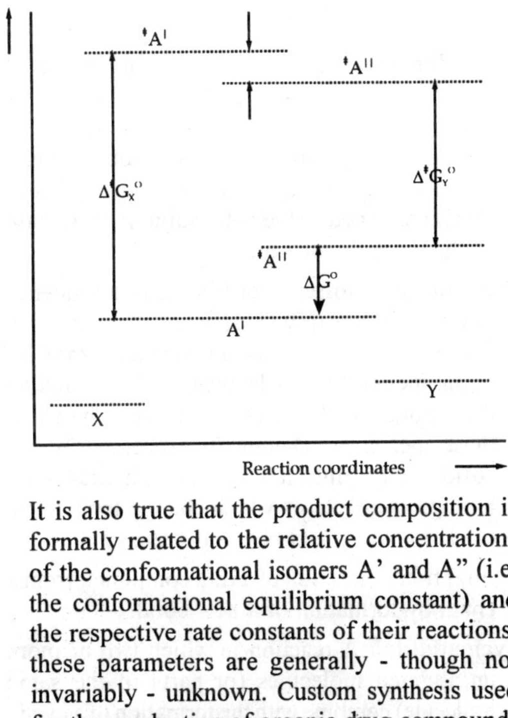

It is also true that the product composition is formally related to the relative concentrations of the conformational isomers A' and A" (i.e. the conformational equilibrium constant) and the respective rate constants of their reactions; these parameters are generally - though not invariably - unknown. Custom synthesis used for the production of organic drug compounds to the specification of the client for the their specific development and research needs.

Cyanide (CN^-) 1. An ion with a -1 charge containing one atom of carbon bound to one atom of nitrogen.

2. A compound that contains CN^- ions.

Cyanide process A method for separating a metal from an ore. Crushed ore is treated with cyanide ion to produce a soluble metal cyanide complex. The complex is washed out of the ore and reduced to metallic form using an active metal (usually zinc).

Cyanosis The pathological condition where there is an excessive concentration of reduced haemoglobin in the blood. This results in blue appearance of the skin, especially on the face and extremities, indicating the lack of sufficient oxygen in arterial blood.

Cybotactic region That part of a solution in the vicinity of a solute molecule in which the ordering of the solvent molecules is modified by the presence of the solute molecule. The term solvent "cosphere" of the solute has also been used.

Cyclative cleavage Cleavage resulting from intramolecular reaction at the linker which results in a cyclized product. The cleavage may also act as a purification if resin-bound side-products are incapable of cyclizing, and thus remain attached to the solid support on release of the desired material.

Cycle In the atmosphere or biosphere a sequence of events in repetitive motion in which the final output feeds back into the initial input. Examples of this include biogeochemical cycles, including the nitrogen, carbon, and sulfur cycles in which these species are chemically processed in gas, solid, and solution phase by physical and biological processes which change their form, oxidation state, and physical state.

Cyclization Formation of a ring compound from a chain by formation of a new bond.

Cycloaddition A reaction in which two or more unsaturated molecules (or parts of the same molecule) combine with the formation of a cyclic adduct in which there is a net reduction of the bond multiplicity.

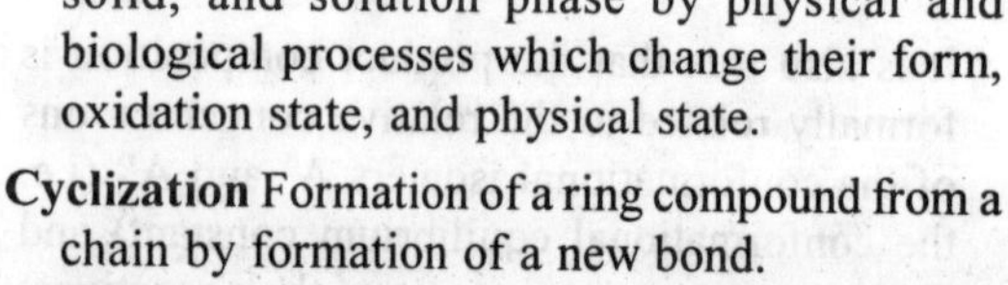

The following two systems of notations have been used for the more detailed specification of cycloadditions, of which the second, more recent system (described under 2.) is preferred: 1. An (i+j+...) cycloaddition is a reaction in which two or more molecules (or parts of the same molecule), respectively, provide units of i, j,... linearly connected atoms: these units become joined at their respective termini by new sigma bonds so as to form a cycle containing (i+j+...) atoms. In this notation, (a) a Diels-Alder reaction is a (4+2) cycloaddition, (b) the initial reaction of ozone with an alkene is a (3+2) cycloaddition, and (c) the reaction shown below is a (2+2+2) cycloaddition. (N.B.: parentheses (...) are used in the description based on numbers of atoms.) 2. The symbolism (i+j+...) for a cycloaddition identifies the numbers i, j,... of electrons in the interacting units that participate in the transformation of reactants to products. In this notation the reaction (a) and (b) of the preceding paragraph would both be described as (2+4) cycloadditions, and (c) as a (2+2+2) cycloaddition. The symbol a or s (a = antarafacial, s = suprafacial) is often added (usually as a subscript after the number to designate the stereochemistry of addition to each fragment. A subscript specifying the orbitals, viz., , (sigma, pi) with their usual significance) or n (for an orbital associated with a single atom only), may be added as a subscript before the number. Thus the normal Diels-Alder reaction is a (4s+2s) or (4s + 2s) cycloaddition, whilst the reaction would be a (14a+2s) or (14a + 2s) cycloaddition. Cycloadditions may be pericyclic reactions or (non-concerted) stepwise reactions. The term "dipolar cycloaddition" is used for cycloadditions of 1,3-dipolar compounds.

CycloalkanesCycloalkanes are cyclic saturated hydrocarbons containing a ring of carbon atoms joined by single bonds. They have the general formula C_nH_{2n}, for example cyclohexane, C_6H_{12}. In general, they behave like the alkanes but are rather less reactive.

Cycloelimination The reverse of cycloaddition. The term is preferred to the synonyms "cycloreversion", "retro-addition", and "retrocycloaddition".

Cycloreversion See cycloelimination.

Cysteine Compare with cystine.

A naturally occuring amino acid with an SH group on its side chain.

Cystine Dicysteine. Compare with cysteine.

A naturally occuring amino acid with a disulfide bridge group on its side chain, formed by condensation of two cysteine residues.

Cytochrome A heme protein that transfers electrons,and exhibits intense absorption bands (the α and β bands, α the band having the longer wavelength) between 510 and 615 nm in the reduced form. Cytochromes are designated types a, b, c or d, depending on the position of the α band, which depends on the type of heme. The iron undergoes oxidation-reduction between oxidation states Fe(II)and Fe(III). Most cytochromes are hemochromes, in which the fifth and sixth coordination sites in the iron are occupied by strong field ligands, regardless of the oxidation state of iron. Cytochromes may be distinguished by the wavelength of the α band, such as cytochrome c-550. Certain specific cytochromes with particular functions, are designated with suffixes, such as cytochrome a_1,b_2, etc., but this practice is discouraged.

Cytochrome P-450 General term for a group of heme-containing monooxygenases (EC 1.4.14.1). Named from the prominent absorption band of the Fe(II)-carbonyl complex. The heme comprises protophorphyrin IX, and the proximal ligand to iron is a cysteine sulfur. Cytochromes P-450 of microsomes in tissues such as liver are responsible for metabolism of many xenobiotics, including drugs. Others, such as the mitochondrial enzymes from adrenal glands, are involved in biosynthetic pathways such as those of steroids. The reaction with dioxygen appears to involve higher oxidation states of iron, such as Fe(IV)=O.

Cytochrome-c oxidase An enzyme, ferrocytochrome - c:dioxygen oxidoreductase, EC 1.9.3.1, cytochrome aa_3. The major respiratory protein of animal and plant mitochondria, it catalyzes the oxidation of Fe(II)-cytochrome c, and the reduction of dioxygen to water. Contains two hemes and three copper atoms, arranged in three centers. Heme a_3 and copper-B form a center that reacts with dioxygen;the second heme is cytochrome a; the third site, copper-A,is a dinuclear center.

Cytoplasm The part of protoplasm in a cell outside of and surrounding the nucleus.

Cytotoxic The adjective applied to anything that is harmful to the cell structure and function and ultimately causing cell death.

D

D, l, dl (Usage strongly discouraged) Obsolete alternatives for respectively the prefixes (+)- and (-)- (used to designate the sign of optical rotation of enantiomers under specified condition (and hence to distinguish the enantiomers)) and (±)- for a racemate.

Dalton's law Dalton's law of partial pressure.

The total pressure exerted by a mixture of gases is the sum of the pressures that each gas would exert if it were alone. For example, if dry oxygen gas at 713 torr is saturated with water vapor at 25 torr, the pressure of the wet gas is 738 torr.

Daly detector A Daly detector consists of a metal knob that emits secondary electrons when struck by an ion. The secondary electrons are accelerated onto a scintillator that produces light that is then detected by a photomultiplier tube.

Dative bond See coordination.

Daughter isotope In a nuclear equation the compound remaining after the parent isotope (the original isotope) has undergone decay. A compound undergoing decay, such as alpha decay, will break into an alpha particle and a daughter isotope.

DDT Para-dichlorodiphenyltrichloroethane was commonly used during the second World War. It was used to kill mosquitoes that carried malaria and yellow fever and it was also used to kill body lice that can transmit typhus. Many problems have been associated with the use of DDT. Some of these problems are the bioaccumulation in the food chain, decline of smaller animals, and softening of birdsI eggs. DDT is still being used extinsively in developing countries.

Debye (D) Debye unit. Compare with dipole moment.

A common non-SI unit of dipole moment, named for Dutch physical chemist Peter Debye. A charge separation equal to one electron charge placed one Ångstrom unit apart has a dipole moment of 4.8 D. In SI units, $1\ D = 3.338 \times 10^{-30}$ coulomb meters.

Decay Change of an element into a different element, usually with some other particle(s) and energy emitted.

Decimal The number of digits to the right of the decimal point in a number

Decode Use of a surrogate analyte to define the reaction path to which the solid support was exposed, and hence imply the structure of a member of a combinatorial library or the reaction sequence for its preparation.

Decomposition Decompose; decomposable; decomposition reaction. Compare with synthesis.

A reaction in which a compound is broken down into simpler compounds or elements.

Compounds sometimes decompose if heated strongly or if subjected to a strong electric current (electrolysis).

Decomposition, chemical The breakdown of a single entity (normal molecule, reaction intermediate, etc.) into two or more fragments.

De-electronation See oxidation 1..

Deflagration The extremely rapid burning of a material. This is much much faster than normal combustion, but slower than detonation.

Deforestation Type of "forestry practice" which involves the permanent removal of forests and their undergrowth so that the land can serve another purpose. This practice has had a profound effect on global environmental problems (air pollution, global warming), soil erosion, desertification, sedimentation of water courses, alteration of climate and hydrological cycles, alteration of the atmospheric oxygen and carbon dioxide balance and has caused many species to become extinct, reducing worldwide biodiversity. A common deforestation practice is to simply burn the forest while clearing the land for a seasonal crop. This also creates a large-scale release of carbon dioxide further compounding the global warming effect. Since most of the available nutrients are contained in the forest biomass the newly cleared land does not support a crop for more than a few seasons before adjacent land must be deforested.

Degenerate Degenerate orbital.

A set of orbitals are said to be degenerate if they all have the same energy. This degeneracy can sometimes be "lifted" by external electric or magnetic fields.

Degenerate chemical reaction See identity reaction.

Degenerate rearrangement A molecular rearrangement in which the principal product is indistinguishable (in the absence of isotopic labelling) from the principal reactant. The term includes both "degenerate intramolecular rearrangements" and reactions that involve intermolecular transfer of atoms or groups ("degenerate intermolecular rearrangements"): both are degenerate isomerizations. The occurrence of degenerate rearrangements may be detectable by isotopic labelling or by dynamic NMR techniques.

Dehydrogenase An oxidoreductase which catalyzes the removal of hydrogen atoms from a substrate.

Dehydrohalogenation Loss of a hydrohalic acid (like HBr, HCl, and so on) to form a double bond.

Deliquescent Deliquesce; deliquescence.

Deliquescent compounds absorb so much moisture from the air that they dissolve. Examples are calcium chloride and sodium hydroxide.

Delocalization A quantum mechanical concept most usually applied in organic chemistry to describe the pi bonding in a conjugated system. This bonding is not localized between two atoms: instead, each link has a "fractional double bond character" or bond order.

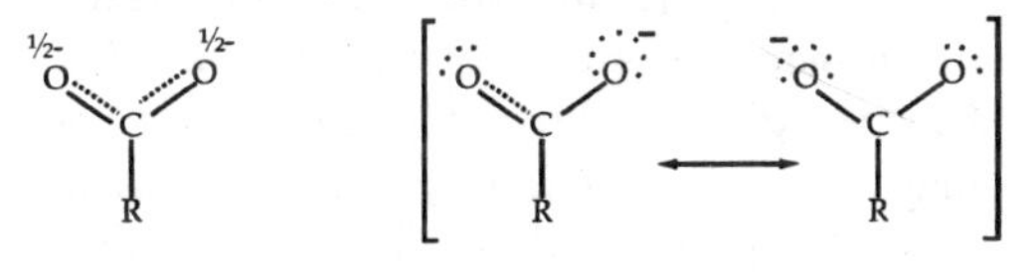

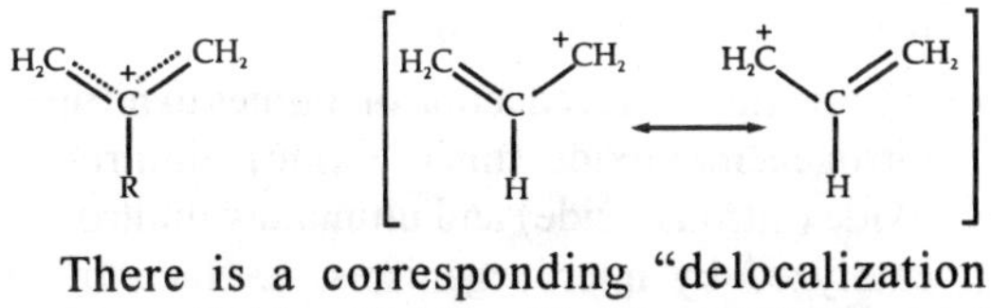

There is a corresponding "delocalization energy", identifiable with the stabilization of the system compared with a hypothetical alternative in which formal (localized) single and double bonds are present. Some degree of delocalization is always present and can be estimated by quantum mechanical calculations. The effects are particularly evident in aromatic systems and in symmetrical molecular entities in which a lone pair of electrons or a vacant p-orbital is conjugated with a double bond (e.g.

carboxylate ions, nitro compounds, enamines, the allyl cation). Delocalization in such species may be represented by partial bonds or as resonance between contributing structures.

These examples also illustrate the concomitant delocalization of charge in ionic conjugated systems. Analogously, delocalization of the spin of an unpaired electron occurs in conjugated radicals.

Delta value (a.k.a. d value) The chemical shift. The location of an NMR peak relative to the standard tetramethylsilane (TMS), given in units of parts per million (ppm).

Delta?, ?(Lambda) Designations of stereoisomers of tris(didentate) metal complexes and other octahedral complexes.

Denature Denatured; denaturation.

A loss of chemical function, usually due to some heat or chemically-induced structural change. For example, heating a protein causes it to lose its three dimensional form and it no longer functions correctly.

Dendrimer A polymer having a regular branched structure. If suitably functionalized (such as the benzyl alcohol-substituted it may be used as a soluble support, in which case the desired, dendrimer-supported, material may be isolated by size-exclusion chromatography. Dendrimers may also be attached to a polymer and used as a solid support, with significantly increased loading over the initial resin.

Denitrification The reduction of nitrates to nitrites, nitrogen monoxide (nitric oxide), dinitrogen oxide (nitrous oxide) and ultimately dinitrogen catalyzed by microorganisms, e.g.facultative aerobic soil bacteria under anaerobic conditions.

De-nol Trade name for the potassium salt or mixed ammonium potassium salt of a bismuth citrate complex, used in the treatment of ulcers.

Dense A compact substance or a substance with a high density.

DensityIn the most common usage, density (ρ) is mass density or mass per unit volume.

$$\rho = m / V$$

More generally, the amount of some quantity (mass, charge, energy, etc.) divided by a length, area, or volume.

Relative density is the ratio of the density of a substance to the density of some reference substance. For liquids or solids, it is the ratio of the density (usually at 20 °C) to the density of water at 4 °C. This quantity was formerly called specific gravity.

Density functional Density functional theory; density functional model.

A model that describes the electronic structure of an atom or molecule by approximating the total energy as a function of electron density.

Denticity The number of donor groups from a given ligandattached to the same central atom.

Deoxyribonucleic acid (DNA) A high-molecular-mass linear polymer, composed of nucleotides containing 2-deoxyribose and linked between positions 3' and 5' by phosphodiester groups; DNA contains the geneticinformation of organisms. The double-stranded form consists of a double helix of two complementary chains that run in opposite directions and are held together by hydrogen bonds between pairs of the complementary nucleotides. The way the helices are constructed may differ and is usually designated as A, B, Z, etc. Occasionally, alternative structures are found, such as those with Hoogsteen base pairing.

Dependent variable Compare with independent variable.

A dependent variable changes in response to changes in independent variables. For example, in an experiment where the vapor pressure of a liquid is measured at several different temperatures, temperature is the independent variable and vapor pressure is the dependent variable.

Deposition The process by which a certain amount of a substance arrives at a particular site (e.g., the deposition of particles on the ciliated epithelium of the bronchial airways) (WHO, 1979).

Deracemisation See asymmetric transformation.

Derived unit Derived units are units constructed from the SI system's base units. For example, the SI unit for density is kg/m^3, derived from the base units kg and m.

Dermatitis Inflammation of the skin.

Desalination Removal of dissolved salts from seawater.

Descriptor Numerical representation of a molecular property, including bulk properties (e.g. log P, molecular weight), two-dimensional (2-D) features (atom connectivity's) or three-dimensional (3-D) features (molecular shape). A complete set of descriptors comprises a fingerprint.

Desferrioxamine (dfo) Chelating agent used world-wide in the treatment of iron overload conditions, such as hemochromatosis and thalassemia.

Destructive interference Compare with constructive interference.

When the peaks of one wave match the troughs of another, the waves interfere destructively. The amplitudes of the interfering waves cancel to give the resultant wave a lower amplitude.

Desymmetrisation The modification of an object which results in the loss of one or more symmetry elements, such as those which preclude chirality (mirror plane, centre of inversion, rotation-reflection axis), as in the conversion of a prochiral molecular entity into a chiral one.

Desymmetrisation step The removal of the smallest possible number of symmetry elements from a molecule with (ref. 1) or without (ref. 2) the restriction to pointwise substitution.

Detachment The reverse of an attachment. See also electron attachment.

Detailed balancing, When equilibrium is reached in a reaction system (containing an arbitrary number of components and reaction paths), as many atoms, in their respective molecular entities, will pass forward, as well as backwards, along each individual path in a given finite time interval. Accordingly, the reaction path in the reverse direction must in every detail be the reverse of the reaction path in the forward direction (provided always that the system is at equilibrium).

Principle of detailed balancing is a consequence for macroscopic systems of the principle of microscopic reversibility.

Detection and analysis Iodine also produces a deep blue colour when it forms a complex with starch solution. Thus, a starch solution is used as an indicator for free elemental iodine in solution during titration's.

Iron

Iron, Fe, is a silvery malleable and ductile transition metal element, found in Group VIII of the periodic table.

Atomic Number : 26

Relative Atomic Mass : 56

Melting Point : 1535 degC

Boiling Point : 2750 degC

Relative Density : 7.8

Iron, Fe, normally consists of four isotopes. Six other isotopes are known.

Detection and analysis Lead is identified by the characteristic precipitate of lead chloride which precipitates out on the addition of hydrochloric acid to a solution of its salts. The precipitate is soluble in hot solution and recrystallises as fine white needles.

Mercury

Mercury, Hg, is a heavy silvery liquid transition metallic element, found in Group IIb of the periodic table.

Atomic Number : 80

Relative Atomic Mass : 200.59

Melting Point : -38.87 degC

Boiling Point : 356 degC

Relative Density : 13.5

Mercury metal and its compounds are highly toxic and must be handled with caution.

Detection and analysis Nitrogen is a colourless, odourless, nonflammable gas neutral gas, which does not support combustion. Thus, the confirmation of the detection of nitrogen gas is based on a negative inference based on its reactive inertness.

Oxygen

Oxygen, O2, is a colourless odourless gaseous main group element which belong to Group VIb of the periodic table.

Atomic Number : 8

Atomic Mass : 15.9994

Melting Point : -214 degC

Boiling Point : -183 degC

Density : 1.429

Atmospheric oxygen is of vital importance for all aerobic organisms: For industrial purposes, oxygen is obtained by fractional distillation of liquid air. It is used in metallurgical processes, in high-temperature flames and in breathing apparatus.

Detection and analysis Zinc is detected by heating the material with sodium carbonate on a charcoal block when the zinc oxide produced is coloured yellow when hot and white when cold.

Detonation The sudden and violent release of mechanical, chemical or nuclear energy from a confined space which creates a shock wave that travels at supersonic speeds. Often used interchangeably with explosion.

Detoxify Reduce the toxicity of a substance either

1. by making it less harmful or

2. by treating patients suffering from poisoning in such away as to reduce the probability and/or severity of harmful effects.

Deuterium (D, 2H) An isotope of hydrogen that contains one neutron and one proton in its nucleus.

Dewpoint The temperature at which air is cooled enough to reach its saturation point (100% humidity) for a given volume of water vapor in the air. If the air cooling occurs because of adiabatic expansion, clouds are formed; if it is due to contact of the air with a colder object, condensation (known as dew) is formed on the object; and if it is due to mixing of a warmer, moist air parcel with a cooler, dry parcel, fogs and condensation trails are created. Unlike humidity, the dewpoint is not based upon the temperature or pressure of the air, but solely upon the water vapor content. It can be measured using a dew point hygrometer, from the known temperature and relative humidity and by measuring the difference in wet and dry bulb temperatures. It is of importance in predicting fogs, frost, dew and overnight minimum temperatures.

Dextrorotatory Dextrorotary. Compare with levorotatory.

Having the property of rotating plane-polarized light clockwise.

Dialysis Dialysis is the separation of components in a mixture by passing them across a semipermeable membrane.

Diamagnetic Substances having a negative magnetic susceptibility are diamagnetic. They are repelled out of a magnetic field.

DiamagnetismIn diamagnetism the magnetization is in the opposite direction to that of applied field, i.e. susceptibility is negative. It results from changes induced in the orbits of electrons in the atoms of a substance by the applied field, the direction of the change opposing the applied flux.

Diamond A crystalline form of carbon, made of a network of covalent, tetrahedrally bound carbon atoms.

Diastereoisomer excess/diastereoisomeric excess This is defined by analogy with enantiomer excess, as $D_1 - D_2$ (and the percent diastereoisomer excess as $100(D_1 - D_2)$), where the mole fractions of two diastereoisomers in a mixture or the fractional yields of two diastereoisomers formed in a reaction are D_1 and D_2 ($D_1 + D_2 = 1$). The term is not applicable if more than two diastereoisomers are present. Frequently this term is abbreviated to d.e.

Diastereoisomerisation The interconversion of diasteoisomers.

Diastereoisomerism Stereoisomerism other than enantiomerism. Diastereoisomers (or diastereomers) are stereoisomers not related as mirror images. Diastereoisomers are characterized by differences in physical properties, and by some differences in chemical behaviour toward achiral as well as chiral reagents. Iupac stereochemical terminology (1993).

Diastereoisomers Diastereoisomers are stereoisomers not related as mirror images.

Diastereomer Diastereomeric. Compare with enantiomer.

Stereoisomers which are not mirror images of each other. Diastereomers are chemically similar but distinguishable; they have different melting points and boiling points and they react at different rates.

Diastereomeric ratio This is defined by analogy with enantiomeric ratio as the ratio of the percentage of one diastereoisomer in a mixture to that of the other.

Diastereomers Stereoisomers that are not mirror images of each other.

Diastereomorphism The relationship between objects (or models) analogous to that between diastereoisomeric molecular entities.

Diastereoselectivity See stereoselectivity.

Diastereotopic Constitutionally equivalent atoms or groups of a molecule which are not symmetry related. Replacement of one of two diastereotopic atoms or groups results in the formation of one of a pair of diastereoisomers. In the example below the two hydrogen atoms of the methylene group C-3 are diastereotopic.

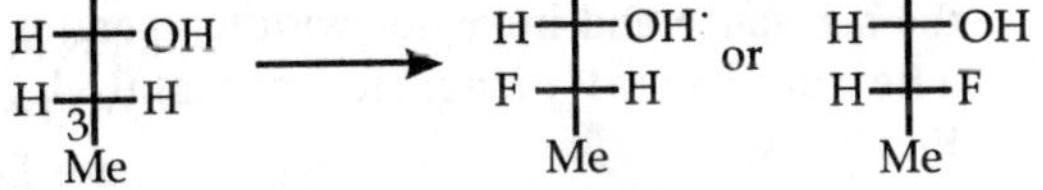

Diatomic Elements that are present in the gaseous state as molecules composed of two atoms. For example: O_2, N_2, Cl_2 and H_2 are diatomic.

Diatomic molecule Compare with binary compound and polyatomic molecule.

A molecule that contains only two atoms. All of the noninert gases occur as diatomic molecules; e. g. hydrogen, oxygen, nitrogen, fluorine, and chlorine are H_2, O_2, N_2, F_2, and Cl_2, respectively.

Diazo compounds Diazo compounds are compounds having the divalent diazo group, $=N^+=N^-$, attached to a carbon atom. The term includes azo compounds, diazonium compounds, and also such compounds as diazomethane, $CH_2=N_2$.

Diazonium salt A diazonium salt is a compound with general form $Ar\text{-}NN^+X^-$, where Ar represents a substituted benzene ring and X^- is a halide ion such as chloride. Diazonium salts are unstable and explosive in dry form. They are used to manufacture many different organic compounds, including azo dyes.

Diazotization Diazotization is a reaction that converts an $-NH_2$ group connected to a phenyl ring to a diazonium salt. For example,

Diazotization reactions are extremely useful in organic synthesis. The nitrous acid provides NO^+ which replaces a hydrogen on the $-NH_3^+$ group to produce $-NH_2NO^+$ and water; a second water is eliminated to produce the $-N_2^+$ group.

Dibromotetrafluoroethane A chemical also know as Halon-2402 which has the chemical formula $C_2F_4Br_2$. It is commonly used as a fire extinguishing agent. The reason it has so much ozone depleting potential is because it contains bromine, which has many times the ozone depleting potential of chlorine.

Dichlorodifluoromethane CCl_2F_2 This compound, also known as Freon 12, is the primary CFC that has been linked to the greenhouse effect. Even though the compound is found in very small concentrations, its properties make it a significant factor in stratospheric chemistry. CCl_2F_2 absorbs UV light in the upper stratosphere where it arrives decades after being released in the troposphere. CFCs, which were originally created by Dupont corporation, were used as refrigerants for everything from air conditioning to home refrigerators because they are non-toxic and extremely unreactive (in the troposphere). It is this unreactivity that makes them such a problem for the atmosphere, because they are able to last long enough to make it to the stratosphere, where they cause so many problems. CCl_2F_2 production, along with all CFCs, has been banned since the mid 90s, but some scientists believe that the damage has already been done.

Dichloromethane (CH_2Cl_2) Dichloromethane (CH_2Cl_2) is an organic solvent often use to extract organic substances from samples. It is toxic but much less so than chloroform or carbon tetrachloride, which were previously used for this purpose.

Dichloromethane (CH_2Cl_2) Dichloromethane is a clear, colorless organic solvent that has a sweet but mild odor and does not occur naturally in the environment. Its principle use is as a solvent in paint removers and as an aerosol propellant in paints and automotive spray products. In addition, it is used as a blowing and cleaning agent of metal, as a solvent in the production of polycarbonate resins, and in film processing, as well as many other industrial applications. In the food industry, it is used as an extraction solvent for caffeine, spices, and hops. All of these applications result in the release of dichloromethane into the environment, with the primary sources being the industries that manufacture the chemical or use it in production, such as the plastic and synthetics manufacturers, as well as the electronics industry. Upon release of the liquid chemical into the environment, dichloromethane quickly evaporates to a gas, where it is then degraded by photochemically produced products in the air. Because of its reactivity in the troposphere, it is not believed to react with ozone in the upper atmosphere. Despite its solubility in water, it has low acute toxicity to aquatic animals. In animals (as well as humans), dichloromethane is metabolized into carbon monoxide, which has a higher affinity for the heme groups in hemoglobin than does oxygen and results in deprivation of oxygen. Interestingly, dichloromethane has been found to be an antisickling agent in sickle cell hemoglobin.

Diels-alder reaction A reaction that brings together a diene and a dienophile to form bicyclic molecules and rings.

Diene A molecule that contains two alternating double bonds. A reactant in the Diels-Alder reaction.

DienesDienes are unsaturated organic compounds that contain two fixed double bonds between carbon atoms. Dienes in which the two double-bond units are linked by one single bond are termed conjugated.

Dienophile A reactant in the Diels-Alder reaction that contains a double bond. Dienophiles are often substituted with electron-withdrawing groups.

Differential thermal analysis (DTA) A technique that is often used to analyze materials that react or decompose at higher temperatures. The difference in temperature between the sample and an inert reference material is monitored as both are heated in a furnace. Phase transitions and chemical reactions taking place in the sample on heating cause the temperature difference to become larger, at temperatures that are characteristic of the sample.

Diffraction Diffract. The ability of a wave to bend around the edges of obstacles or holes. The effect is most noticeable when the obstacle or hole is comparable to the size of the wavelength.

DiffusionDiffusion is the spontaneous mixing of one substance with another when in contact or separated by a permeable membrane. Diffusion is a result of the random motions of their component atoms, molecules, ions, or other particles. Diffusion occurs most readily in gases, less so in liquids, and least in solids. The rate of diffusion is proportional to the concentration of the substance and increases with temperature. The theoretical principles are stated in Fick's laws.

Diffusion rate Rate of diffusion. Compare with effusion.

The number of randomly moving molecules that pass through a unit area per second. Diffusion rates are fastest when a large concentration difference exists on either side of the unit area. Diffusion rates increase with temperature, and decrease with increasing pressure, molecular weight, and molecular size.

Diffusion-controlled rate See encounter-controlled rate, microscopic diffusion control.

Dihedral angle The angle between two intersecting planes on a third plane normal to the intersection of the two planes..

Dihydrofolate An oxidation product of tetrahydrofolate that appears during DNA synthesis and other reactions. It must be reduced to tetrahydrofolate to be of further use.

Dilatometer A device for measuring volume changes.

Dilute Diluted; diluting. Compare with concentrate.

Having a relatively low concentration.

Dilution Adding solvent to a solution to lower its concentration.

Dimerization The transformation of a molecular entity A to give a molecular entity A_2. For example: $CH_3\cdot$ + $H_3C\cdot$ CH_3CH_3

2 CH_3COCH_3 $(CH_3)_2C(OH)CH_2COCH_3$

2 RCOOH $(RCOOH)_2$

Dimethyl sulfide Released by bacteria on the continents and in the oceans. Oxidized in the marine atmosphere to partially form cloud condensation nuclei and this may effect the formation of clouds over the oceans.

Dimroth-reichardt e_t parameter A measure of the ionizing power (loosely polarity) of a solvent, based on the maximum wavenumber of the longest wavelength electronic absorption band in a given solvent. E_T, called $E_T(30)$ by its originators, is given by

E_T = 2.859 x 10^{-3}

= 2.859 x 10^{4-1}

where E_T is in kcal/mol, is in cm^{-1} and is in nm.

The so-called normalized E_T^N scale is defined as

$T^N = (E_T(solvent) - E_T(SiMe_4))/(E_T(water) - E_T(SiMe_4)) = (E_T(solvent) - 30.7)/32.4$

Dimroth, Reichardt, Siepmann and Bohlmann (1963); Reichardt (1988).

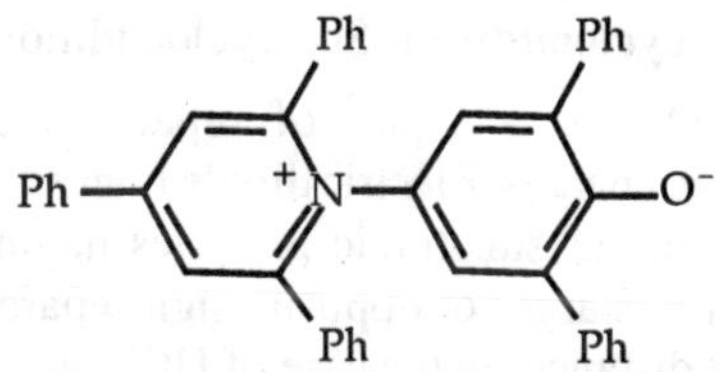

Dinitrogen pentoxide A compound that contributes to ozone depletion during springtime in the Antarctic. This occurs because ozone converts some nitrogen dioxide radical to nitrogen trioxide radical, which combines with other dinitrogen radicals to produce dinitrogen pentoxide. The reaction is reversible except in the presents of stratospheric liquid droplets where nitric acid occurs. This reaction ties up the nitrogen dioxide radical allowing more chlorine atoms to be present in the active form.

Dioxins Family of 75 different toxic chlorinated hydrocarbons formed as by-products in chemical reactions involving chlorine and hydrocarbons, usually at high temperatures.

Dioxygenase An enzyme that catalyzes the insertion of two oxygen atoms into a substrate, both oxygens being derived from O_2.

Dipolar aprotic solvent A solvent with a comparatively high relative permittivity (or dielectric constant), greater than ca. 15, and a sizable permanent dipole moment, that cannot donate suitably labile hydrogen atoms to form strong hydrogen bonds, e.g. dimethyl sulfoxide. The term (and its alternative "polar aprotic solvent") is a misnomer and is therefore discouraged. Such solvents are usually not aprotic, but protophilic (and at most weakly protogenic). In describing a solvent it is better to be explicit about its essential properties, e.g. dipolar and non-protogenic.

Dipolar bond A bond formed (actually or conceptually) by coordination of two neutral moieties, the combination of which results in charge-separated structures, e.g.,

$R_3N: + O\ R_3N^+ - O^-$

The term is preferred to the obsolescent synonyms "coordinate link", "co-ordinate covalence", "dative bond", "semipolar bond".

Dipolar cycloaddition See cycloaddition.

Dipole Dipole is a pair of separated opposite electric charges. Electric dipole is an assemblage of atoms or subatomic particles having equal electric charges of opposite sign separated by a finite distance. In the case of HCl, the electrons are attracted towards the more electronegative chlorine atom.

Dipole moment Electric dipole moment (μ) is the product of the positive charge and the distance between the charges. Dipole moments are often stated in debyes; The SI unit is the coulomb metre. In a diatomic molecule, such as HCl, the dipole moment is a measure of the polar nature of the bond; i.e. the extent to which the average electron charges is displaced towards one atom (in the case of HCl, the electrons are attracted towards the more electronegative chlorine atom). In a polyatomic molecule, the dipole moment is the vector sum of the dipole moments of the individual bonds. In a symmetrical molecule, such as tetrafluoromethane (CF_4) there is no overall dipole moment, although the individual C-F bonds are polar.

Dipole dipole forces Intermolecular forces that exist between polar molecules. Active only when the molecules are close together. The strengths of intermolecular attractions increase when polarity increases.

Dipole dipole interaction Intermolecular or intramolecular interaction between molecules or groups having a permanent electric dipole moment. The strength of the interaction depends on the distance and relative orientation of the dipoles. The term applies also to intramolecular interactions between bonds having permanent dipole moments. See also van der Waals forces.

Dipole induced dipole forces See van der Waals forces.

Diradical See biradical.

Direct effect See field effect.

Directed library Library which uses a limited number of building blocks chosen onthe basis of pre-existing information or hypothesis which defines the type of functionalities deemed important to obtain a particular activity.

Directed sorting Technique for organizing a mixture of solid-supported samples by identifying each particle (for instance, on the basis of its shape, marking or by reading a radiofrequency code) and transferring it to an appropriate position in an array.

Disaccharides Disaccharides are compounds in which two monosaccharides are joined by a glycosidic bond. For example, sucrose comprises one glucose molecule and one fructose molecule bonded together.

Discharge (Or effluent or emission) standard or release limit The maximum acceptable release of a pollutant from a given source to a specified medium under specified circumstances (WHO, 1979).

Discovery 1. Helium was discovered from the spectroscopic examination of a gas which was occluded in certain minerals including cleveite (i.e. a variety of pitchblende), by Hildebrand in 1882AD. The line spectrum of this gas was identical to that from a line in the spectrum of the sun, which was observed by P J C Janssen in 1868AD. E Frankland and J N Lockyer has suggested that this line may be caused by an unknown element which they named Helium (from the Greek, Helios, Sun). Some time later, the presence of traces of Helium in the atmosphere was established.

2. Hydrogen was discovered by Henry Cavendish in 1776AD. 3.Iodine was discovered by B Courtois in 1811AD. He treated the liquor obtained from the extraction of kelp (i.e. seaweed), which is now known to contain iodide salts, with sulphuric acid to produce a vapour with a violet colour.

2 NaI + H2SO4 ⇒ I2 + Na2SO4

In 1812AD, Guy Lussac demonstrated that iodine was an element and its chemical relationship to chlorine. About the same time, Davy confirmed many of these properties.4.Iron was known since ancient times. 5.Lead was known to the Egyptians 3000 years ago, and was used by the Romans for water pipes. 5.Mercury was reported by Theophrastus in 300BC. 6.Nitrogen was discovered by D Rutherford in 1772 AD, when he reported in a thesis that it did not support life. His experimental isolation of nitrogen was confirmed by Cavendish in 1785 AD.

Lavoisier called the gas "Azote". 7.Sulphur has been known since the beginning of history, and is described in the Bible.

Sulphur has been used by the Greeks and Romans as a fumigant and disinfectant. 8.The discovery of Oxygen was credited to Priestley in 1774 AD. However in a paper looking into Alchemy, by Richard Brzezinski, an expert in the history of science and Zbigniew Szydlo, a chemistry lecturer, published in the authoritative magazine History Today credit the discovery of Oxygen to a Polish alchemist called Michael Sendivogius who found that heated saltpeter produced "the elixir of life" and who, in 1604, described his experiments in a book regarded as so authoritative that it found its way into every major scientific library in Europe. They say that Priestley, would surely have had access to it. Cornelis Drebbel a Dutch inventor employed by the King of England James 1 in 1621 used Sendivogius work which was about 150 years before Priestley was credited with the discovery of Oxygen. Drebbel built a submarine which was manned by 12 oarsmen, made of wood and waterproofed by a coat of greases leather. It successfully traveled along the River Thames from Westminster to Greenwich, at a dept of 15 ft. The trip, and the method used to keep the oarsmen alive, was subsequently verified by Robert Boyle. Occurrence

Oxygen occurs in the free state as a gas, to the extent of 21 per cent by volume or 23 per cent by weight in the atmosphere.

Combined Oxygen also occurs in water,

In vegetable and animal tissues,

in nearly all rocks and in many minerals.

Oxygen occurs to a larger extent in the earth's crust than any other element. 9. Uranium in the form of its oxide was discovered by M H Klaproth in 1789AD.

Uranium, U, was isolated as the metal by E M Peligot in 1842AD. 10.Zinc was known from ancient times, when Brass (i.e. an alloy of Copper and Zinc) was used. Lohneyes was the first to apply the term "Zinc" correctly to the metal that we know today in 1697AD. Occurrence

Zinc Spar, ZnCO3,

Calamine, ZnCO3,

Zinc Blande, ZnS,

Zincite, ZnO,

Willemite, 2ZnO.SiO2,

Franklinite, (ZnFe)O.Fe2O3, and

Zinc Spinel, ZnO.Al2O3.

Dismutase An enzyme that catalyzes a disproportionation reaction.

Dismutation See disproportionation.

D-isomer. compare with L-. Prefix used to designate a dextrorotatory enantiomer.

Dispersion forces (also called London dispersion forces) Dispersion is an intermolecular attraction force that exists between all molecules. These forces are the result of the movement of electrons which cause slight polar moments. Dispersion forces are generally very weak but as the molecular weight increases so does their strength.

Displacement Displacement reaction; replacement reaction; replacement.

A reaction in which a fragment of one reactant is replaced by another reactant (or by a fragent of another reactant). Displacement reactions have the same number of products as reactants, and are described by equations of the form A + BC AB + C (single displacement) or AB + CD AC + BD (double displacement).

Disproportionation Any chemical reaction of the type A + A A' + A", where A, A' and A" are different chemical species. For example: 2 ArH^+ ArH + ArH^{2+}

The reverse of disproportionation is called comproportionation.

A special case of disproportionation (or "dismutation") is "radical disproportionation", exemplified by

CH2CH3 + ·CH2CH3 CH2=CH2 + CH3CH3

Reactions of the more general type

RC·HCH3 + R'C·HCH3 RCH=CH2 + R'CH2CH3

are also loosely described as radical disproportionations.

(A somewhat more restricted usage of the term prevails in inorganic chemistry.)

Disrotatory See electrocyclic reaction.

Dissimilatory Related to the conversion of food or other nutrients into products plus energy-containing compounds.

Dissociation Dissociation is the process by which a chemical combination breaks up into simpler constituents as a result of either added energy (dissociated by heat), or the effect of a solvent on a dissolved polar compound (electrolytic dissociation). It may occur in the gaseous, solid, or liquid state, or in solution.

An example of dissociation is the reversible reaction of hydrogen iodide at high temperatures

$2HI(g) \rightleftarrows H_2(g) + I_2(g)$

The term dissociation is also applied to ionisation reactions of acids and bases in water. For example $HCN + H_2O \rightleftarrows H_2O^+ + CN^-$

which is often regarded as a straightforward dissociation into ions

$HCN \rightleftarrows H^+ + CN^-$

Dissociation constant Dissociation constant is a constant whose numerical value depends on the equilibrium between the undissociated and dissociated forms of a molecule. A higher value indicates greater dissociation.

The term dissociation is also applied to ionisation reactions of acids and bases in water. For example

$HCN + H_2O \rightleftarrows H_2O^+ + CN^-$

which is often regarded as a straightforward dissociation into ions

$HCN \rightleftarrows H^+ + CN^-$

The equilibrium constant of such a dissociation is called the acid dissociation constant or acidity constant, given by

$$K_a = \frac{[H^+].[CN^-]}{[HCN]}$$

The concentration of water (H_2O) can be taken as constant.

Similarly, for a base, the equilibrium

$NH_3 \rightleftarrows NH_4^+ + OH^-$

is also a dissociation; with the base dissociation constant or basicity constant, given by

$$K_b = \frac{[NH_4^+].[OH^-]}{[NH_3]}$$

K_a (K_b) is a measure of the strength of the acid (base).

Dissolution The transfer of gas to the surface of a water droplet from the bulk atmosphere, and the gas mixing within the droplet.

Dissolved oxygen DO.

The amount of oxygen dissolved in a solvent (usually water). Dissolved oxygen levels are used as a general indicator of water quality.

Dissymmetry Obsolescent synonym for chirality.

Distillate The vapor collected and condensed from a distillation.

DistillationDistillation is the process of boiling a liquid and condensing and collecting the vapour. The liquid collected is the distillate. The usual purpose of distillation is purification or separation of the components of a mixture. This is possible because the composition of the vapour is usually different from that of liquid mixture from which it is obtained. Gasoline, kerosene, fuel oil, and lubricating oil are produced from petroleum by distillation.

Distonic radical cation A radical cation in which charge and radical sites are separated. Yates, Bouma and Radom (1986)/a>.

Distribution This is a general term for the dispersal of a applied substance and its derivatives throughout an organism or environmental system.

Distributivity A math property which states: A*(B+C)=(A*B)+(A*C)

Diurnal ozone concentration The ozone concentration in the troposphere is directly related to the pollutants that are in the air and the time of day. For example, during the early hours of the day many pollutants are released into the air. As the day progresses, the sun becomes more overhead and the concentration of the sunlight increases. This solar radiation causes chemical reactions to occur and the primary pollutants decrease while the concentration of ozone increases. Eventually the ozone breaks back down and the process is ready to proceed the next day

Divalent Binds to two other things (which may be other atoms, molecules, ions, or electrons).

Divalent anion An ion with a charge of -2.

Divalent cation An ion with a charge of +2.

Diversity The "unrelatednes" of a set of, for example, building blocks or members of a combinatorial library, as measured by their properties, such as atom connectivity, physical properties, computational measurements or bioactivity.

Diversity reagent One of a set of reagents which introduces diversity into the library products, as opposed to one which results in an identical conversion for each member of the library. Similar to building block but may be useful to distinguish from other (i.e. "non-diversity") reagents.

DNA Deoxyribonucleic acid. Compare with nucleic acid and nucleotide.

A nucleic acid with 2-deoxy-D-ribose as the sugar in its nucleotides. DNA contains encoded genetic information, specifically templates for the synthesis of all of an organism's proteins and enzymes.

Dobson units Measurement unit for determining the total amount of ozone present in a vertical column of air above the surface of the earth. An air layer at atmospheric pressure of 1013 hPa and temperature of 298 K which measures 1 mm in thickness and is equivalent to 100 dobson units.

Doldrums Calm, light winds, or squalls lying along the length of the equator. Winds which are part of the general circulation of the atmosphere, which are driven by the input of solar energy and modified by the rotation of the Earth. It is here that 17th and 18th century sailing ships had the most trouble making way and lack of fresh water or food sometimes spelled disaster .

Domain An independently folded unit within a protein, often joined by a flexible segment of the polypeptide chain.

Domoic acid Domoic acid is a toxic amino acid produced by certain species of algae. Domoic acid binds to a receptor that helps nerve cells control the flow of ions across their cell membranes. The receptor no longer works correctly, and the uncontrolled flux of ions damages and eventually kills the nerve cell.

Donor atom symbol A polydentate ligandpossesses more than one donor site, some or all of which may be involved in coordination. To indicate the points of ligation, a system is needed. The general and systematic system for doing this is called the kappa convention: single ligating atom attachments of a polyatomic ligand to a coordination centre are indicated by the italic element symbol preceded by a Greek kappa, κ. In earlier practice, the different donors of the ligand were denoted by adding to the end of the name of the ligand the italicized symbol(s) for the atom or atoms through which attachment to the metal occurs.

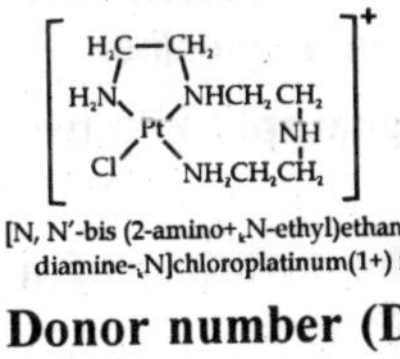

[N, N'-bis (2-amino+,N-ethyl)ethane-1,2-diamine-,N]chloroplatinum(1+) ion

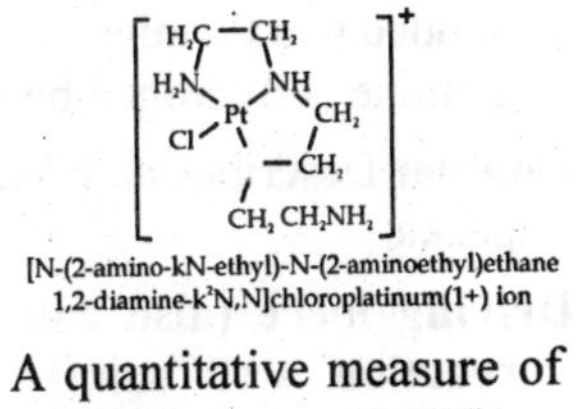

[N-(2-amino-kN-ethyl)-N-(2-aminoethyl)ethane 1,2-diamine-k^{2}N,N]chloroplatinum(1+) ion

Donor number (DN) A quantitative measure of Lewis basicity devised by Gutmann (1976).

Doppler broadening Doppler broadening is due to the distribution of atomic velocities (speed and direction), which each have a Doppler shift with respect to an observer.

Dose The amount of a chemical administered to an organism.

Dose exposure-response relationship The relationship between administered dose or exposure and the biological change in organisms. It may be expressed as the severity of an effect in one organism (or part of an organism) or as the proportion of a population exposed to a chemical that shows a specific reaction (WHO, 1979).

Dose rate (in radiation protection) The quantity of energy absorbed per unit of time (WHO, 1979).

Double bond When an atom is bonded to another atom by two sets of electron pairs.

Double displacement Double displacement reaction; double replacement; double replacement reaction; double exchange; exchange; metathesis.

A double displacement or metathesis is a reaction in which two reactants trade fragments: AB + CD = AC + BD Most commonly, the fragments are ions, e. g.

$AgNO_3(aq) + NaCl(aq) = AgCl(s) + NaNO_3(aq)$

Double helix Two strands of DNAcoiled about a central axis, usually a right-handed helix. The two sugar phosphate backbones wind around the outside of the bases (A = adenine, G = guanine, T = thymine, C = cytosine) and are exposed to the solvent. The strands are antiparallel, thus the phosphodiester bonds run in opposite directions. As a result the structure has major and minor grooves at the surface. Each adenine in one strand of DNA is hydrogen bonded to a thymine in the second strand; each guanine is hydrogen bonded to a cytosine.

Doublet Describes an NMR signal split into two peaks.

Driving force (also called the affinity) of a reaction, A (SI unit: kj mol^{-1})

The decrease in Gibbs energy on going from the reactants to the products of a chemical reaction (-G).

Drug Any substance presented for treating, curing or preventing disease in human beings or in animals. A drug may also be used for making a medical diagnosis or for restoring, correcting, or modifying physiological functions.

Drug targeting A strategy aiming at the delivery of a compound to a particular tissue of the body.

Dry adiabatic lapse rate Rate at which unsaturated air cools as it travels vertically, provided that all temperature change is adiabatic (without heat exchange), and no condensation occurs. In dry air it can be approximated as 1 degree Celsius per 100 meters of rise. It can be used as a basis of comparison for actual temperature profiles of air and can help predict smoke stack gas dispersion characteristics.

Dry cell Leclanché cell.

A electrolytic cell that uses a moist paste rather than a liquid as an electrolyte. Flashlight batteries are dry cells with a zinc cup for an anode, a carbon rod for a cathode, and a paste made of powdered carbon, NH_4Cl, $ZnCl_2$, and MnO_2 for an electrolyte.

Dry deposition The transfer of trace species (gases or particles) from the atmosphere to surfaces as a consequence of molecular diffusion, brownian diffusion, or gravitation settling, in the absence of active precipitation. The term refers to the transfer process and not to the surface themselves, which may be moist.

Dual substituent-parameter equation In a general sense, this is any equation which expresses substituent effects in terms of two parameters. However, in practice the term is used more specifically for an equation for summarizing the effects of meta- or para-substituents (i = m or p) X on chemical reactivity, spectroscopic properties, etc. of a probe site Y in benzene or other aromatic system.

$P^i = I^iI + R^iR$

P is the magnitude of the property Y for substituent X, expressed relative to the property for X=H; I and R are inductive or polar and resonance substituent constants, respectively, there being various scales for R; I and R are the corresponding regression coefficients. Ehrenson, Brownlee, and Taft (1973); TAFT and Topsom (1987).

Ductile Ductility. Compare with malleable.

Capable of being drawn into wire. Metals are typically ductile materials.

Duplicate portion sampling method (duplicate diet study) This method is frequently used for the same purposes as the total diet study technique. It implies that test persons consume their ordinary diet, but that, for each meal, they prepare a duplicate portion of all food as prepared, served, and consumed (WHO, 1979).

Dynamic equilibrium Equilibrium. Compare with position of equilibrium.

Dynamic equilibrium is established when two opposing processes are occuring at precisely the same rate, so that there is no apparent change in the system over long periods of time.

Dynamic library Collection of compounds in dynamic equilibrium. If the composition of the library is altered, for instance by the presence of a receptor which selectively binds certain library members, then shifting of the equilibrium will lead to an increase in the amount of those components which bind to the target with relatively high affinity.

Dyne (dyn) The unit of force in the obsolete cgs system of units. A dyne is the force required to accelerate a 1 g mass by 1 cm/s per second.

Dyotropic rearrangement An uncatalyzed process in which two sigma bonds simultaneously migrate intramolecularly.

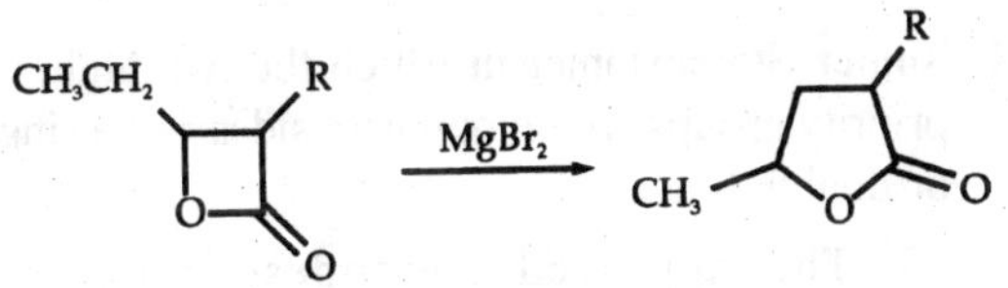

E

E isomer Stereoisomer in which the two highest priority groups are on opposite sides= of a ring or double bond.

E, Z The approved stereodescriptors of stereoisomeric alkenes $R^1R^2C=CR^3R^4$ (R^1 ??R^2, R^3 ??R^4; neither R^1 nor R^2 need be different from R^3 or R^4), cumulenes $R^1R^2C(=C=C)_n=CR^3R^4$ and related systems e.g. $R^1R^2C=NOH$, $HON=C\{(CH_2)_n\}_2$ C=NOH. The group of highest CIP priority attached to one of the terminal doubly bonded atoms of the alkene, oxime, etc. or cumulene (i.e. R^1 or R^2) is compared with the group of highest precedence attached to the other (i.e. R^3 or R^4). The stereoisomer is designated as Z (zusammen = together) if the groups lie on the same side of a reference plane passing through the double bond and perpendicular to the plane containing the bonds linking the groups to the double-bonded atoms; the other stereoisomer is designated as E (entgegen = opposite). The descriptors may be applied to structures with a fractional bond order between one and two; and to double bonds involving elements other than carbon. They are not used to describe ring substitution relationships.

E1 elimination reaction A reaction that eliminates a hydrohalic acid (like HCl, HBr, etc) to form an alkene. A first order reaction that goes through a carbocation mechanism.

E2 elimination reaction A reaction that eliminates a hydrohalic acid (like HCl, HBr, etc) to form an alkene. A second order reaction that occurs in single step, in which the double bond is formed as the hydrohalic acid is eliminated.

Earth Planet Earth orbits the sun between the planets Venus and Mars, and has three distinct layers. 1. The gaseous atmosphere,

2. The liquid hydrosphere, and

3. the solid lithosphere, which has three distinct regions

4. the crustEarthCrust>Subsid,

5. the mantleEarthMantle>Subsid and

6. the coreEarthCore>Subsid.

Earth's atmosphere Earth's atmosphere is the layer of gasses that surrounds the earth.

Earth's core Earth's core, part of which is believe to be liquid, lies below the mantle.

Earth's mantle Earth's mantle extends some 2900 km below the crust.

Ebulliometry Ebulliometric.

Determination of average molecular weight of a dissolved substance from the boiling point elevation of the solution.

Ebullioscopic constant Ebullioscopic constant (E_b) is the constant that expresses the amount by which the boiling point T_b of a solvent is raised by a nondissociating solute, through the relation

$\Delta T_b = E_b\ b$

where b is the molality of the solute.

EC nomenclature for enzymes A classification of enzymes according to the Enzyme Commission of the International Union of Biochemistry and Molecular Biology. Enzymes are allocated four numbers, the first of which defines the type of reaction catalyzed, the next two define the substrates, and the fourth is a catalogue number. Categories of enzymes are EC 1, Oxidoreductases; EC 2, Transferases; EC 3, Hydrolases; EC 4, Lyases; EC 5, Isomerases; EC 6, Ligases (Synthetases).

Eclipsed conformation Conformation about a carbon-carbon single bond in which all of the bonds off of two adjacent carbons are aligned with each other (0° apart when viewed in a Newman projection).

Eclipsed, eclipsing Two atoms or groups attached to adjacent atoms are said to be eclipsed if the torsion angle between the three bonds is zero (or approximately so).

Eclipsing conformation See bisecting conformation.

Eclipsing strain Intramolecular strain present due to non-bonding interactions between two eclipsed atoms or groups. It is, for example, one of the interactions responsible for the existence of a rotational barrier and hence restricted rotation about the C-C bond in ethane. It has also been called Pitzer strain, torsional strain, and bond opposition strain.

Ecliptic The plane of orbit of the earth around the sun. An animation showing this process is available here.

Ecology Branch of biology which focuses on the relationships among various species and among the species and their environments. The basic units for study are the species (all the organisms which are capable of interbreeding), population (all of the members of a species occupying a certain geographical area) and community (number of populations interacting within a certain area). The emphasis in environmental science, however, is on the processes of energy flows and nutrient cycles within these communities and ecosystems and involves such ideas as species abundance, biodiversity, complexity, trophic (feeding status) levels, ecological niches, and structure of communities.

Ecotoxicology The effects of chemical agents on the environment, including, in addition to effects on man, adverse events that take place in the general ecosystem. It is not necessarily related primarily to human health (WHO, 1979).

Eddy diffusion The exchange of gaseous components of the atmosphere in a turbulent flow by rapid mixing of fluid eddies.

Edgar henry summerfield bailey (1848-1933) Born in Middlefield, Connecticut, he studied chemistry at Yale. He then taught intermittently at Lehigh University in Bethlehem, Pennsylvania, before taking his Ph.D. at Illinois Wesleyan. He spent the rest of his career at the University of Kansas at Lawrence. He was interested in the chemistry of foods and the relationship between chemical composition and taste and odour. He published several books on analysis and food chemistry.

Edmond henri (or henry) fischer (b.1920) Born in Shanghai, he was educated at Geneva, Montpellier and Basel. At the University of Washington in Seattle he studied phosphorylase with Edwin Krebs (no relation of Hans Krebs). For work on the mechanism of in vivo phosphorylation, they were awarded the Nobel Prize in 1992.

EDTA Ethylenediaminetetracetic acid; versine.

A polydentate ligand that tightly complexes certain metal ions. EDTA is used as a blood preservativc by complexing free calcium ion (which promotes blood clotting). EDTA's ability to bind to lead ions makes it useful as an antidote for lead poisoning.

Educt Term used mainly in the German literature for starting material (reactant). It should be avoided in English, because there it means "something that comes out" and not "something that goes in". The German use of the term is in fact also incorrect.

Edward monroe bailey (1879-1948) Born in New London, Connecticut and studied chemistry at Yale. He spend his entire career at the Connecticut Agricultural Experimental Station at New Haven. Published research on the chemistry of foods, drugs and agricultural products.

Effect A biological change in an organism, organ, or tissue. A positive deviation of an -nucleophile (a nucleophile bearing an unshared pair of electrons on an atom adjacent to the nucleophilic site) from a Brønsted-type plot of lg k_{nuc} vs. pKa constructed for a series of related normal nucleophiles. More generally, it is the influence of the atom bearing a lone pair of electrons on the reactivity at the adjacent site. HOZ and BUNCEL (1985).

The use of the term has been extended to include the effect of any substituent on an adjacent reactive centre, for example in the case of the "-silicon effect".

Effective molarity (or effective concentration) The ratio of the first-order rate constant of an intramolecular reaction involving two functional groups within the same molecular entity to the second-order rate constant of an analogous intermolecular elementary reaction. This ratio has the dimension of concentration. The term can also apply to an equilibrium constant.

Effective nuclear charge (Z_{eff}) Compare with atomic number.

The nuclear charge experienced by an electron when other electrons are shielding the nucleus.

Effervescence Effervescence is the formation of gas bubbles in a liquid by chemical reaction.

An example of effervescence is the release of carbon dioxide which bubbles as a gas from the liquid when limestone chips, which are composed of calcium carbonate are added to dilute hydrochloric acid.

Efflorescence The reverse process of deliquescence; the drying of a salt solution when the vapor pressure of the saturated solution of a substance is greater than that of the ambient air.

Efflorescent Efflorescence; efflorescing. Efflorescent substances lose water of crystallization to the air. The loss of water changes the crystal structure, often producing a powdery crust.

Efflorescent Efflorescence; efflorescing. Efflorescent substances lose water of crystallization to the air. The loss of water changes the crystal structure, often producing a powdery crust.

Efforescence Efforescence is the process where a crystalline hydrate loses water, forming a powdery deposit on the crystals.

Effusion Effuse. Gas molecules in a container escape from tiny pinholes into a vacuum with the same average velocity they have inside the container. They also move in straight-line trajectories through the pinhole.

Effusion Effusion is the flow of a gas through a small aperture. The relative rates at which gases effuse, under the same conditions, is approximately inversely proportional to the square roots of their densities.

EF-Hand A common structure to bind Ca^{2+} in calmodulin and other Ca^{2+} binding proteins consisting of a helix (E), a loop and another helix (F).

Egor egoroviè vagner (1849-1903) Born in Kazan. He initially trained at the University of Kazan to become a lawyer like his father, but soon switched to chemistry, probably under the influence of another Kazan native Aleksandr Michailoviè Sayteff or Zajcev (1841-1910). He then went to the University of St Petersburg, and Institute of Forestry and Agriculture in Novo-Aleksandria (near Lublin) before taking up a Chair in Organic Chemistry at the University of Warsaw. He soon transferred his chair to the new polytechnic in Warsaw. Initially Vagner prepared secondary alcohols using zinc alkyls, but soon switched to terpene chemistry. In the course of this work, he suggested in 1899, that the formation of bornyl chloride from á-pinene involved an internal rearrangment. In 1914, this reaction was generalised by Hans Meerwein (1879-1965) as the Wagner-Meerwein rearrangement. Vagner used the German spelling of his name in papers published in German journals.

Ehanol (CH_3CH_2OH) Ethyl alcohol; grain alcohol.

A colorless, flammable liquid produced by fermentation of sugars. Ethanol is the alcohol found in alcoholic beverages.

Ehyl (-Et, -CH_2CH_3) Ethyl group.

A molecular fragment produced by removing a hydrogen atom from ethane (CH_3-CH_3). For example, ethyl chloride is CH_3-CH_2-Cl.

ethyl acetate ($CH_3COOCH_2CH_3$)

A flammable liquid with a fruity odor, used in flavorings and as a solvent.

Eighteen-electron rule An electron-counting rule to which an overwhelming majority of stable diamagnetic transition metal complexes adhere. The number of nonbonded electrons at the metal plus the number of electrons in the metal-ligand bonds should be 18. The 18 electron rule in transition metal chemistry is a full analogue of the "Lewis octet rule".

El nino A condition caused by the decrease in atmospheric pressure over the Eastern Pacific Ocean, weakening the prevailing westerly winds and resulting in warm waters and less nutrient replacement from cold, deeper waters into the Eastern Pacific along the coast of South America. El Nino creates strange weather patterns worldwide sometimes causing flooding in California and droughts in South America simultaneously. The effect gets it name from the Spanish for the Christ Child—the little one—because it is experienced in December by the local fisherman off the western coast of South America as a poor fishing season.

El nino-southern oscillation (ENSO) The relationship of the El Nino conditions and the Southern Oscillation pattern (The Walker Circulation). A high pressure over the Pacific Ocean and low pressure over the East Indies causes heavy rainfall across the Pacific, rise in the sea level and increased surface temperatures.

Electric charge charge.A property used to explain attractions and repulsions between certain objects. Two types of charge are possible: negative and positive. Objects with different charge attract; objects with the same charge repel each other.

Electric current Current; electrical current.

A flow of electric charges. The SI unit of electric current is the ampere.

Electric dipole Dipole.

An object whose centers of positive and negative charge do not coincide. For example, a hydrogen chloride (HCl) molecule is an electric dipole because bonding electrons are on average closer to the chlorine atom than the hydrogen, producing a partial positive charge on the H end and a partial negative charge on the Cl end.

Electric dipole moment (μ) Dipole moment.

A measure of the degree of polarity of a polar molecule. Dipole moment is a vector with magnitude equal to charge separation times the distance between the centers of positive and negative charges. Chemists point the vector from the positive to the negative pole; physicists point it the opposite way. Dipole moments are often expressed in units called Debyes.

Electric field A field of forces that act on any electric charge placed within it. The stronger the field, the stronger the force that acts on the charge. For example, the positive charge on an atomic nucleus creates an electric field that traps electrons.

Electrical conductivity Conductivity; electric conductivity; electrical conductance; conductance. Compare with resistance.

A measure of how easily an electric current can pass through a material. The conductivity is the reciprocal of the resistance. The SI unit of conductance is the siemens.

Electrical resistance Resistance. Compare with conductivity.

The ability of a material to oppose the flow of an electric current, converting electrical energy into heat. The SI unit of resistance is the ohm.

Electrical resistivity Electrical resistivity (ρ) is electric field strength divided by current density when there is no electromotive force in the conductor. Resistivity is an intrinsic property of a material. For a conductor of uniform cross section with area A and length L, and whose resistance is R, the resistivity is given by

$$\rho = R\frac{A}{l}$$

The SI unit is Ω□m.

Electroanalytical chemistry There are two categories of electrochemical processes that are applied to quantitative measurements: potentiometry - measurement of a potential (voltage)

Electrolytic methods - measurement involving a current

In both of these categories, an electrochemical measurement can be related to the concentration of an analyte.

Electrolytic Methods(h) Unlike potentiometry, where the free energy contained within the system generates the analytical signal, electrolytic methods are an area of *electroanalytical chemistry* in which an external source of energy is supplied to drive an electrochemical reaction which would not normally occur. The externally applied driving force is either an applied potential or current. When potential is applied, the resultant *current* is the analytical signal; and when current is applied, the resultant potential is the analytical signal. Techniques which utilize applied potential are typically referred to as *voltammetric methods* while those with applied current are referred to as *galvanostatic methods*.

Electrophoresis(h) Electrophoresis is a separations technique that is based on the the mobility of ions in an electric field. Positively charged ions migrate towards a negative electrode and negatively-charged ions migrate toward a positive electrode.For safety reasons one electrode is usually at ground and the other is biased positively or negatively. Ions have different migration rates depending on their total charge, size, and shape, and can therefore be separated.

Electrochemical cell An Electrochemical Cell consists of a pair of electrodes immersed in solutions of electrolytes and connected via an external metallic conductor such an arrangement is called an electrochemical cell.

If the cell is used as a source of electrical energy, (i.e. if it converts the free energy of a physical or chemical change into electrical free energy), it is called a Galvanic Cells.

If the cell is used to achieve a change in the physical or chemical composition of the constituents of the cell by the application of electrical energy from some external source, it is called an Electrolytic Cell.

Electrochemical equivalent The Electrochemical Equivalent of a substance is the mass of the substance, in grams, which is liberated by the passage of 1 coulomb of electricity.

Electrochemical series The electrochemical series consists of a list of metals have been arranged in order of their standard electrode potentials. The standard electrode potentials which has been measured for most elements that form ions in solution. Metals which are higher in the electrochemical series displace metals which are lower in the sequence from solutions of their salts.

A negative value for electrode potential means that the element loses electrons more readily than hydrogen (i.e. it is better reducing agent). A positive value for electrode potential means that the element is a poor reducing agent than hydrogen.

A strong oxidising agent has a large positive potential and a strong reducing agent a large negative potential.

The electrochemical series is extremely useful for predicting the chemical behaviour of an element. An electropositive element will displace a less electropositive element from a solution of its ions (e.g. zinc will displace copper from a copper solution).

However, aluminium which is more electropositive than copper does not displace copper from copper salt solutions, due to the presence of a very stable film of aluminium oxide on the surface of the aluminium which has a lower electrode potential. This oxide film can be destroyed by the presence of added chlorine ions. Thus, in the presence of added chloride ions in the solution, copper is deposited and the aluminium dissolves.

Electrochemistry Electrochemistry is the study of the effects of electric current on the properties and reactions of chemical substances in solution, including electrolysis and electric cells.

Electrocyclic reaction A molecular rearrangement that involves the formation of a sigma bond between the termini of a fully conjugated linear pi-electron system (or a linear fragment of a pi-electron system) and a decrease by one in the number of pi bonds, or the reverse of that process.

The stereochemistry of such a process is termed "conrotatory" or antarafacial if the substituents at the interacting termini of the conjugated system both rotate in the same sense.

or "disrotatory" (or suprafacial) if one terminus rotates in a clockwise and the other in a counter-clockwise sense.

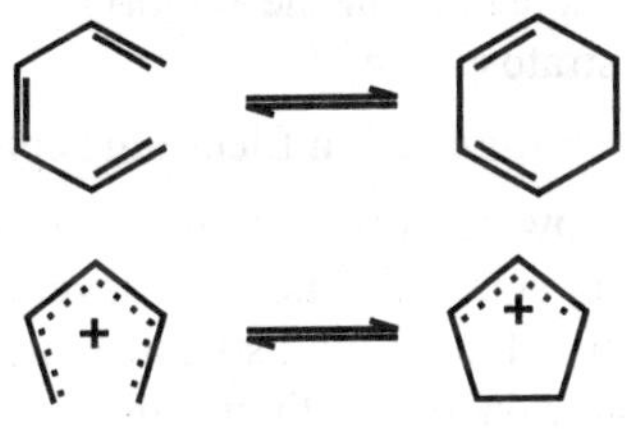

Electrode An Electrode is a conductor that emits or collects electrons in a cell. The anode is the positive electrode, where ions lose electrons and thus, oxidation occurs at the anode.

The cathode is the negative electrode, where ions gain electrons, and thus, reduction occurs at the cathode.

Electrofuge A leaving group that does not carry away the bonding electron pair. For cxample, in the nitration of benzene by NO_2^+, H^+ is the electrofuge. The adjective of electrofuge is electrofugal.

Electrolysis Electrolysis is the production of a chemical change by electrical energy. In electrolysis, positive ions migrate to the cathode and negative ions migrate to the anode. At the anode, the negative ions in solution may lose electrons to form neutral species. Alternatively, atoms of the electrode can lose electrons and go into solution as positive ions. In either case, anodic reactions are oxidation reactions. At the cathode, positive ions in solution can gain electrons to form neutral species. Thus; cathodic reactions are reduction reactions.

Electrolyte A substance that dissociates fully or partially into ions when dissolved in a solvent, producing a solution that conducts electricity.

Electrolytic cell An Electrolytic Cell is one in which a physical or chemical change is caused by electrical energy applied from some external source.

Hybridisation

When bonds are formed between atoms in the formation of molecules, there is a change in the nature of the Atomic Orbitals in each atom. Effectively, new molecular orbitals are formed, and the electrons in these orbitals are those of the molecule as a whole. The alteration of the structure of the atomic orbitals is called hydridisation, as it involves combining a number of orbitals to create an equal number of new orbitals, where each of the new hybrid orbitals have properties which are an average of those of the orbitals from which they were created.

A number of different types of hybrid orbitals are known for carbon in the organic compounds.

Hybridisation sp

Hybridisation sp2

Hybridisation sp3

The different hybrid orbitals, which have different orientations in space, account for the geometry of the organic compounds of carbon in which they appear.

Electromagnetic radiation Electromagnetic wave.

A wave that involves perpendicular oscillations in the electric and magnetic fields, moving at a speed of 2.99792458×10^8 m/s in a vacuum away from the source. gamma rays, x-rays, ultraviolet light, visible light, infrared radiation, and radio waves are all electromagnetic waves.

Electromagnetic spectrum Complete range of wavelengths which light can have. These include infared, ultraviolet, and all other types of electromagnetic radiation, as well as visible light.

Electromeric effect A molecular polarizability effect occurring by an intramolecular electron displacement (sometimes called the "conjugative mechanism" and, previously, the "tautomeric mechanism") characterized by the substitution of one electron pair for another within the same atomic octet of electrons. The term has been deemed obsolescent or even obsolete (see mesomeric effect, resonance effect). It has long been custom to use phrases such as "enhanced substituent resonance effect" which imply the operation of the electromeric effect, without using the term, and various modern theoretical treatments parametrize the response of substituents to "electronic demand", which amounts to considering the electromeric effect together with the inductomeric effect.

Electromotive force Electromotive force (e.m.f. or E_{MF}) is the difference in electric potential that exists between two dissimilar electrodes immersed in the same electrolyte or otherwise connected by ionic conductors.

Electron(e) A particle of matter that has a negative electric charge of 4.8 E^{-10} esu and a mass of 9.1E $^{-28}$g or 1/1837 the mass of a proton. They can be found as a constituent part of an atom orbiting around the nucleus or in the free state. Electrons are arranged around the nucleus of an atom in from one to seven orbitals with the number of electrons in each orbital is strictly limited by the laws of physics.

Electron acceptor 1. A substance to which an electron may be transferred; for example 1,4-dinitrobenzene or the dication 1,1'-dimethyl-4,4'-bipyridyldiium .

+ 2. A Lewis acid. This usage is discouraged.

Electron affinity Electron affinity (E_A) is the energy change occurring when an atom or molecule gains an electron to form a negative ion. For an atom or molecule X, it is the energy released for the electron-attachment reaction

$$X(g) + e^- \rightleftarrows X^- (g)$$

Often this is measured in electronvolts. Alternatively, the molar enthalpy change, ΔH, can be used.

Electron attachment The transfer of an electron to a molecular entity, resulting in a molecular entity of (algebraically) increased negative charge.

Electron capture detector One of the most sensitive gas chromatographic detector for halogen-containing compounds like chlorofluorocarbons. James Lovelock's early work with this detector led to the discovery that anthropogenic chlorofluorocarbons used most heavily in the Northern hemisphere were very long-lived and therefore well-mixed throughout the troposphere. Later work by Rowland and Molina suggested that CFCs also difussed into the stratosphere where their chlorine atoms could by freed by photolysis-CFC bond breaking by UV light. The result of this is the destruction of ozone molecules in the important ozone layer of the stratosphere.

Electron configuration Electronic configuration.

A list showing how many electrons are in each orbital or subshell. There are several notations. The subshell notation lists subshells in order of increasing energy, with the number of electrons in each subshell indicated as a superscript. For example, $1s^2\ 2s^2\ 2p^3$ means "2 electrons in the 1s subshell, 2 electrons in the 2s subshell, and 3 electrons in the 2p subshell.

Electronegativity Compare with ionization energy and electron affinity.

Electronegativity is a measure of the attraction an atom has for bonding electrons. Bonds between atoms with different electronegativities are polar, with the bonding electrons spending more time on average around the atom with higher electronegativity.

Electron density If P(x,y,z) dx dy dz is the probability of finding an electron in the volume element dx dy dz at the point of a molecular entity with coordinates x,y,z, then P(x,y,z) is the electron density at this point. For many purposes (e.g., X-ray scattering, forces on atoms) the system behaves exactly as if the electrons were spread out into a continuously distributed charge. The term has frequently been wrongly applied to negative charge population.

Electron detachment The reverse of an electron attachment.

Electron donor 1. A molecular entity that can transfer an electron to another molecular entity, or to the corresponding chemical species.

+ 2. A Lewis base. This use is discouraged.

Electron geometry Structure of a compound based on the arrangement of its electrons.

Electron magnetic resonance (EMR) spectroscopy See electron paramagnetic resonance spectroscopy.

Electron multiplier tube (EMT) Electron multiplier tubes are similar in design to photomultiplier tubes. They consist of a series of biased dynodes that eject secondary electrons when they are struck by an ion. They therefore multiply the ion current and can be used in analog or digital mode.

Electron paramagnetic resonance (EPR) spectroscopy The form of spectroscopy concerned with microwave-induced transitions between magnetic energy levels of electrons having a net spin and orbital angular momentum. The spectrum is normally obtained by magnetic field scanning. Also known as electron spin resonance (ESR) spectroscopy or electron magnetic resonance (EMR) spectroscopy. The frequency (ν) of the oscillating magnetic field to induce transitions between themagnetic energy levels of electrons is measured in gigahertz (GHz) or megahertz (MHz). The following band designations are used: L (1.1 GHz), S (3.0 GHz), X (9.5 GHz), K (22.0 GHz) and Q (35.0 GHz). The static magnetic field at which the EPR spectrometer operates is measured by the magnetic flux density (B) and its recommended unit is the tesla (T). In the absence of nuclear hyperfine interactions, B and ν are related by : $h\nu = g \mu_B B$ whereh is the Planck constant, μ_B is the Bohr magneton, and the dimensionless scalar g is called the g-factor. When the paramagnetic species exhibits an anisotropy, the spatial dependency of the g-factor is represented by a 3x3 matrix. The interaction energy between the electron spin and a magnetic nucleus is characterized by the hyperfine coupling constant A. When the paramagnetic species has anisotropy, the hyperfine coupling is expressed by a 3x3 matrix called a hyperfine coupling matrix. Hyperfine interaction usually results in splitting of lines in an EPR spectrum. The nuclear species giving rise to the hyperfine interaction should be explicitly stated, e.g."the hyperfine splitting due to ^{65}Cu". When additional hyperfine splittings due to other nuclearspecies are resolved ("superhyperfine"), the nomenclature should include the designation of the nucleus, and the isotopic number.

Electron spin Electron spin (s) is the quantum number, equal to 1/2, that specifies the intrinsic angular momentum of the electron.

Electron spin-echo (ESE) spectroscopy A pulsed technique in electron paramagnetic resonance, in some ways analogous to pulsed techniques in NMR. ESE may be used for measurements of electron spin relaxation times as they are influenced by neighbouring paramagnets or molecular motion. It may also be used to measure anisotropicnuclear hyperfine couplings. The effect is known as electron spin-echo envelope modulation (ESEEM). The intensity of the electron spin-echo resulting from the application of two or more microwave pulses is measured as a function of the temporal spacing between the pulses. The echo intensity is modulated as a result of interactions with the nuclear spins. The frequency-domain spectrum corresponds to hyperfine transition frequencies.

Electron transfer The transfer of an electron from one molecular entity to another, or between two localized sites in the same molecular entity. See also inner sphere (electron transfer), outer sphere (electron transfer), Marcus equation.

Electron transfer protein A protein, often containing a metal ion, that oxidizes and reduces other molecules by means of electron transfer.

Electron volt Electron volt (eV) is a non-SI unit of energy used in atomic and nuclear physics, equal to approximately 1.602177×10^{-19} J. The electron volt is defined as the kinetic energy acquired by an electron upon acceleration through a potential difference of 1 V.

Electron work function (Photoelectric work function) The smallest amount of photonic energy necessary to remove an electron from the boundary of an element.

Electron-deficient bond A single bond between adjacent atoms that is formed by less than two electrons, as in B_2H_6. The B-H-B bonds are called a "two-electron three-centre bonds".

Electron-donor-acceptor complex A term sometimes employed instead of charge-transfer complex or Lewis adduct. See also adduct, coordination.

ElectronegativityElectronegativity is a parameter originally introduced by L. Pauling which describes, on a relative basis, the power of an atom to attract electrons. For example, in hydrogen chloride, the chlorine atom is more electronegative than the hydrogen and the molecule is polar, with negative charge on the chlorine atom.

There are various ways of assigning values for the electronegativity of an element. Pauling electronegativities are based on bond dissociation energies using a scale in which fluorine, the most electronegative element, has value 4 and francium, the lowest electronegative element, has value 0.7.

Electronic effect of substituents: symbols and signs The inductive effect has universally been represented by the symbol I. This is now commonly taken to include both through-bonds and through-space transmission, but I is also used specifically for through-bonds transmission; through-space transmission is then symbolized as F (for field effect). The symbols for the influence of substituents exerted through electron delocalization have variously been M (mesomeric), E (electromeric), T (tautomeric), C (conjugative), K (konjugativ), and R (resonance). Since the present fashion is to use the term resonance effect, R is the most commonly used symbol, although M is still seen quite often.

Both the possible sign conventions are in use. The Ingold sign convention associates electronegativity (relative to hydrogen atom) with a negative sign, electropositivity with a positive sign. Thus the nitro group is described as electron- withdrawing by virtue of its - I and - M effects; chloro is described as a - I, +M substituent, etc. For correlation analysis and linear free-energy relationships this convention has been found inconvenient, for it is in contradiction to the sign convention for polar substituent constants (-constants). Authors concerned with these fields often avoid this contradiction by adopting the opposite sign convention originally associated with Robinson, for electronic effects. This practice is almost always associated with the use of R for the electron delocalization effect: thus the nitro group is a +I, +R substituent; chloro a +I, - R substituent, etc.

Electron-nuclear double resonance (ENDOR) A magnetic resonance spectroscopic technique for the determination of hyperfine interactions between electrons and nuclear spins. There are two principal techniques. In continuous-wave ENDOR the intensity of an electron paramagnetic resonance signal, partially saturated with microwave power, is measured as radiofrequency is applied. In pulsed ENDOR the radiofrequency is applied as pulses and the EPR signal is detected as a spin-echo. In each case an enhancement of the EPR signal is observed when the radiofrequency is in resonance with the coupled nuclei.

Electrophile, electrophilic An electrophile (or electrophilic reagent) is a reagent that forms a bond to its reaction partner (the nucleophile) by accepting both bonding electrons from that reaction partner.

An "electrophilic substitution reaction" is a heterolytic reaction in which the reagent supplying the entering group acts as an electrophile. For example

$C_6H_6 + NO_2^+$ (electrophile) $C_6H_5NO_2 + H^+$ (electrofuge)

Electrophilic reagents are Lewis acids. "Electrophilic catalysis" is catalysis by Lewis acids.

The term "electrophilic" is also used to designate the apparent polar character of certain radicals, as inferred from their higher relative reactivities with reaction sites of higher electron density.

Electrophilicity 1. The property of being electrophilic (see electrophile).

2. The relative reactivity of an electrophilic reagent. (It is also sometimes referred to as "electrophilic power".) Qualitatively, the concept is related to Lewis acidity. However, whereas Lewis acidity is measured by relative equilibrium constants, electrophilicity is measured by relative rate constants for reactions of different electrophilic reagents towards a common substrate (usually involving attack at a carbon atom). See also nucleophilicity.

Electrophoresis Electrophoresis is a technique for the analysis and separation of colloids, based on the movement of charged colloidal particles in an electric field. The migration is toward electrodes of charge opposite to that of the particles. The rate of migration of the particles depends on the field, the charge on the particles, and on other factors, such as the size and shape of the particles.

Electrophoresis is important in the study of proteins. The acidity of the solution can be used to control the direction in which a protein moves upon electrophoresis.

Electrorefining Electrorefining is a method for purifying a metal using electrolysis. An elcctric current is passed between a sample of the impure metal and a cathode when both are immersed in a solution that contains cations of the metal. Metal is stripped off the impure sample and deposited in pure form on the cathode.

Electrostatic forces Forces between charged objects.

Electrostatic precipitator A device that reduces pollution from smokestacks. It gives the particles in smoke a electrical charge. The charged particles are then collected on filters that have the opposite charge and thereby removed from the emission stream.

Element One of the 116 presently known substances that cannot be decomposed by chemical reaction into a simpler substance. Elements comprise all matter at and above the atomic level. All elements heavier than lead (Pb) are radioactive and unstable. In addition, there are no stable elements with odd numbers of both protons and neutrons heavier than nitrogen (N). Elements with even numbers of protons and an even number of neutrons make up about 90% of the earth's crust.

Element effect The ratio of the rate constants of two reactions that differ only in the identity of the element of the atom in the leaving group, e.g., k_{Br}/k_{Cl}. As for isotope effects, a ratio of unity is regarded as a "null effect".

Element of chirality See chirality element.

Element symbol An international abbreviation for element names, usually consisting of the first one or two distinctive letters in element name. Some symbols are abbreviations for ancient names.

Elementary reaction A reaction for which no reaction intermediates have been detected or need to be postulated in order to describe the chemical reaction on a molecular scale. An elementary reaction is assumed to occur in a single step and to pass through a single transition state.

Elimination The reverse of an addition reaction or transformation.

In an elimination two groups (called eliminands) are lost most often from two different centres (1/2/elimination or 1/3/elimination, etc.) with concomitant formation of an unsaturation in the molecule (double bond, triple bond) or formation of a new ring.

If the groups are lost from a single centre (-elimination, 1/1/elimination) the resulting product is a carbene or a "carbene analogue".

Elliptic An object having the shape of an ellipse: a squashed or elongated circle. The sun follows an elliptic path across the sky.

Embryo This term is applied to the earliest stages of development of a plant or an animal. The embryo is generally contained in another structure: the seed, egg or uterus.

Embryotoxicity The potential of a substance to induce adverse effects in progeny in the first period of pregnancy between conception and the fetal stage .

Emil hermann fischer (1852-1919) Perhaps the greatest organic chemist and a pioneering biochemist, was born in Euskirchen near Bonn and studied chemistry at Bonn after a commercial education. He took his doctorate at Stassburg (Strasbourg) under Baeyer. After teaching in Munich, Erlangen and Würzburg, he succeeded August Wilhelm von Hofmann in Berlin and won the Nobel Prize in 1902. His name is associated with several reactions and syntheses, notably the Fischer phenylhydrazine synthesis, Fischer indole synthesis, Fischer peptide synthesis and the Kiliani-Fischer synthesis (discovered independently not jointly). He also gave his name to the Fischer projection of carbohydrates.

Emil's cousin otto phillipp fischer (1852-1932) Was also born in Euskirchen. He studied chemistry in Berlin, Bonn and Strassburg. With Liebemann and Emil Fischer, he worked on the structure of dyes, particularly the triphenylmethane dyes. He succeeded Emil as professor of organic chemistry in Erlangen in 1885. His name is associated with the Fischer-Hepp rearrangement. Curiously, his name is sometimes given as Phillipp Otto Fischer, can anyone confirm which is the correct order?

Emission A substance discharged into the air, especially by an internal combustion engine. The majority of the substance consists of un-burned hydrocarbons that react in the atmosphere to produce ground-level ozone. Urban air gases consist of nitrogen oxides, reactive organic gases, carbon monoxide, and sulfur oxides.

Emission or exposure control The technical and administrative procedures applied for the reduction or elimination of emissions from the source or of exposure to the target (WHO, 1988).

Emission spectrum When heat is applied, electrons move from a lower state to an excited state of energy. When an atom drops back down to its lower state, photons are emitted to produce light. This light in contact with a prism forms distinct spectural lines by the separation of wavelengths. As usually observed in the visible wavelengths, infrared and UV wavelengths are also used for the determination of an element. The emission spectrum is different for every element.

Emission standard This regulatory value is a quantitative limit on the emission or discharge of a potentially toxic substance from a source. See limit values.

Emollient A substance added to a formulation that gives it softening ability. For example, oils that can soften skin are added as emollients in some skin creams.

Empirical formula. Simplest formula Empirical formulas show which elements are present in a compound, with their mole ratios indicated as subscripts. For example, the empirical formula of glucose is CH2O, which means that for every mole of carbon in the compound, there are 2 moles of hydrogen and one mole of oxygen.

Empirical temperature A property that is the same for any two systems that are in thermodynamic equilibrium with each other.

Emulsion Emulsion is colloidal system in which the dispersed and continuous phases are both liquids (e.g. oil in water or water in oil). Such systems require an emulsifying agent to stabilize the dispersed particles.

Enantioconvergence See stereoconvergence.

Enantioenriched See enantiomerically enriched

Enantiomer Enantiomeric. Compare with diasteromer.

Two molecules that are nonsuperimposable mirror images of each other. One enantiomer rotates plane-polarized light to the left; the other rotates it to the right.

Enantiomeric ratio The ratio of the percentage of one enantiomer in a mixture to that of the other e.g. 70(+) : 30(-).

Enantiomerically enriched/Enantioenriched A sample of a chiral substance whose enantiomeric ratio is greater than 50:50 but less than 100:0

Enantiomerically pure/enantiopure A sample all of whose molecules have (within limits of detection) the same chirality sense. Use of homochiral as a synonym is strongly discouraged.

Enantiomerisation The interconversion of enantiomers.

Enantiomerism The isomerism of enantiomers.

Enantiomers Enantiomers are a chiral molecule and its non-superposable mirror image. The two forms rotate the plane of polarized light by equal amounts in opposite directions. Also called optical isomers.

Enantiomorph One of a pair of chiral objects or models that are non-superposable mirror images of each other. The adjective enantiomorphic is also applied to mirror-image related groups within a molecular entity.

Enantiotopic Constitutionally identical atoms or groups in molecules which are related by symmetry elements of the second kind only (mirror plane, inversion centre or rotation-reflection axis). For example the two groups c in a grouping Cabcc are enantiotopic. Replacement of one of a pair of enantiotopic groups forms one of a pair of enantiomers. Analogously, if complexation or addition to one of the two faces defined by a double bond or other molecular plane gives rise to a chiral species, the two faces are called enantiotopic.

Encoding Strategy for pool/split synthesis whereby a surrogate analyte is associated with each member of a combinatorial library. This is often achieved by the use of tags attached to the particle of solid support on which the library members are assembled. This allows the determination of the reaction history of an individual particle.

Encounter complex A complex of molecular entities produced at an encounter-controlled rate, and which occurs as an intermediate in a reaction mechanism. When the complex is formed from two molecular entities it is called an "encounter pair". A distinction between encounter pairs and (larger) encounter complexes may be relevant in some cases, e.g. for mechanisms involving pre-association.

Encounter-controlled rate A rate of reaction corresponding to the rate of encounter of the reacting molecular entities. This is also known as "diffusion-controlled rate" since rates of encounter are themselves controlled by diffusion rates (which in turn depend on the viscosity of the medium and the dimensions of the reactant molecular entities).

For a bimolecular reaction between solutes in water at 25 ^{o}C an encounter-controlled rate is calculated to have a second-order rate constant of about 10^{10} dm^3 mol^{-1} s^{-1}.

Endo, exo, syn, anti Descriptors of the relative orientation of groups attached to non-bridgehead atoms in a bicyclo(x.y.z)alkane (x ? y > z > 0). If the group is orientated towards the highest numbered bridge (z bridge, e.g. C-7 in example below) it is given the description exo; if it is orientated away from the highest numbered bridge it is given the description endo. If the group is attached to the highest numbered bridge and is orientated towards the lowest numbered bridge (x bridge, e.g. C-2 in example below) it is given the description syn; if the group is orientated away from the lowest numbered bridge it is given the description anti.

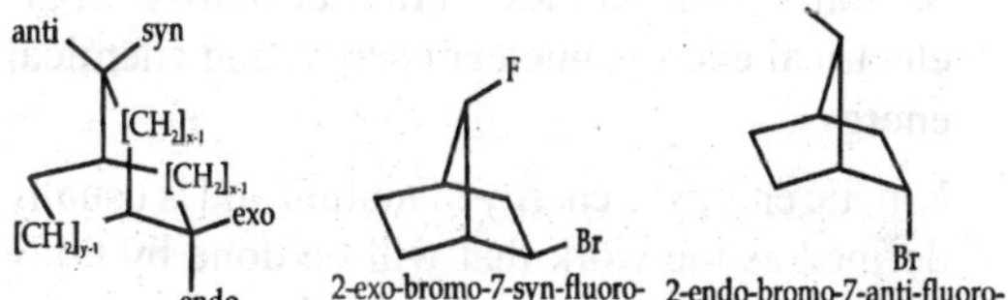

2-exo-bromo-7-syn-fluoro-bicyclo[2.2.1]heptane 2-endo-bromo-7-anti-fluoro-bicyclo[2.2.1]heptane

Endogenous Originating internally. In the description of metal ion coordination in metalloproteins, endogenous refers to internal, or protein-derived, ligands.

ENDOR Acronym for electron-nuclear double resonance.

Endothelium-derived relaxing factor (EDRF) The factor originally describedas EDRF is NO·, produced by a specific P-450-type of enzyme from arginine upon response of a cell to a biological signal (molecule). Different types of cells respond differently to the presence of NO·.

Endothermic Endothermic reaction; endothermic process. Compare with exothermic.

A process that absorbs heat. The enthalpy change for an endothermic process has a positive sign.

Endpoint End point. Compare with equivalence point.

The experimental estimate of the equivalence point in a titration.

Ene reaction The addition of a compound with a double bond having an allylic hydrogen (the "ene") to a compound with a multiple bond (the "enophile") with transfer of the allylic hydrogen and a concomitant reorganization of the bonding, as illustrated below for propene (the "ene") and ethene (the "enophile"). The reverse is a "retro-ene" reaction.

Energy Energy (E, U) is the characteristic of a system that enables it to do work. Like work itself, it is measured in joules (J).

The internal energy of a body is the sum of the potential energy and the kinetic energy of its component atoms and molecules.

Potential energy is the energy stored in a body or system as a consequence of its position, shape, or state (this includes gravitational energy, electrical energy, nuclear energy, and chemical energy).

Kinetic energy is energy of motion and is usually defined as the work that will be done by body possessing the energy when it is brought to rest. For a body of mass m having a speed v, the kinetic energy is $mv^2/2$. Kinetic energy is most clearly exhibited in gases, in which molecules have much greater freedom of motion than in liquids and solids.

Energy levels (Electron shells, shells) The possible locations around an atom where electrons having specific energy values (quantum number) may be found. The term shell has been replaced with the term energy levels because the term shell insinuated that electrons circled the atom in fixed orbits like planets circle the sun. This model however is incorrect and outdated. Although the term shell is outdated, it is still often used grade school and high school science classes, as its concepts are less intimidating than are the use of terms like quantum number.

Energy of activation (Arrhenius energy of activation; activation energy), E_a (SI unit: kJ mol^{-1})

An operationally defined quantity expressing the dependence of a rate constant on temperature according to

$E_a = RT^2(\ln k /T)_p$

as derived from the "Arrhenius equation", $k = A \exp(-E_a/RT)$, where A (SI unit: as for the corresponding rate constant) is termed the "pre-exponential factor".

Enforced concerted mechanism Variation of reaction parameters in a series of reactions proceeding in non-concerted steps may lead to a situation, where the putative intermediate will possess a lifetime shorter than a bond vibration, so that the steps become concerted. The transition state structure will lie on the coordinate of the More O'Ferrall-Jencks diagram leading to that of the putative intermediate.

Enkephalin Enkephalins are molecules produced naturally by the central nervous system to numb pain. Enkephalins lock into receptors on the surface of a nerve cell and open ion channels. Ions flow into the cell and the distribution of charge on either side of the cell membrane becomes such that the nerve cell cannot fire.

Ent A prefix used to indicate the enantiomer for natural products and related molecules where the trivial name only refers to one enantiomer, e.g. ent-kaurene is the enantiomer of kaurene.

Entatic state A state of an atom or group which due to its binding in a protein, has its geometric or electronic condition adapted for function. Derived from entasis (Greek) meaning tension.

Entering group An atom or group that forms a bond to what is considered to be the main part of the substrate during a reaction. For example: the attacking nucleophile in a bimolecular nucleophilic substitution reaction.

Enterobactin A siderophore found in enteric bacteria such as Escherichia coli; sometimes called enterochelin.

Enterohepatic circulation Intestinal reabsorption of material that has been excreted through the bile and transferred back to the liver, making it available for biliary excretion again (WHO, 1979).

Enthalpy (H) enthalpy change. Compare with heat.

Enthalpy (H) is defined so that changes in enthalpy (H) are equal to the heat absorbed or released by a process running at constant pressure. While changes in enthalpy can be measured using calorimetry, absolute values of enthalpy usually cannot be determined. Enthalpy is formally defined as H = U + PV, where U is the internal energy, P is the pressure, and V is the volume.

Enthalpy of atomization (H_{at}) Atomization enthalpy; heat of atomization.

The change in enthalpy that occurs when one mole of a compound is converted into gaseous atoms. All bonds in the compound are broken in atomization and none are formed, so enthalpies of atomization are always positive.

Enthalpy of combustion (H_c) Heat of combustion.

The change in enthalpy when one mole of compound is completely combusted. All carbon in the compound is converted to CO_2(g), all hydrogen to H_2O, all sulfur to SO_2(g), and all nitrogen to N_2(g).

Enthalpy of fusion (H_{fus}) Heat of fusion; molar heat of fusion; molar enthalpy of fusion.

The change in enthalpy when one mole of solid melts to form one mole of liquid. Enthalpies of fusion are always positive because melting involves overcoming some of the intermolecular attractions in the solid.

Enthalpy of hydration (H_{hyd}) Hydration enthalpy; heat of hydration.

The change in enthalpy for the process A(g)A(aq)

where the concentration of A in the aqueous solution approaches zero. Enthalpies of hydration for ions are always negative because strong ion-water attractions are formed when the gas-phase ion is surrounded by water.

Enthalpy of neutralization Heat of neutralization.

The heat released by an acid-base neutralization reaction running at constant pressure.

Enthalpy of reaction (H_{rxn}) Heat of reaction.

The heat absorbed or released by a chemical reaction running at constant pressure.

Enthalpy of solution (H_{soln}) Heat of solution. Compare with integral enthalpy of solution.

The heat absorbed or released when a solute is dissolved in a solvent. The heat of solution depends on the nature of the solute and on its concentration in the final solution.

Enthalpy of sublimation (H_{sub}) Heat of sublimation.

The change in enthalpy when one mole of solid vaporizes to form one mole of gas. Enthalpies of sublimation are always positive because vaporization involves overcoming most of the intermolecular attractions in the sublimation.

Enthalpy of vaporization (H_{vap}) Heat of vaporization.

The change in enthalpy when one mole of liquid evaporates to form one mole of gas. Enthalpies of vaporization are always positive because vaporization involves overcoming most of the intermolecular attractions in the liquid.

Entropy Entropy (S) is a measure of the unavailability of a system's energy to do work; in a closed system, an increase in entropy is accompanied by a decrease in energy availability. When a system undergoes a reversible change the entropy (S) changes by an amount equal to the energy (Q) transferred to the system by heat divided by the thermodynamic temperature (T) at which this occurs.

$$\Delta S \frac{\Delta Q}{T}$$

All real processes are to a certain extent irreversible changes and in any closed system an irreversible change is always accompanied by an increase in entropy.

Envelope conformation The conformation (of symmetry group C_S) of a five-membered ring in which four atoms are coplanar, and one atom (the flap) projects out of the plane.

Environment The aggregate, at a given moment, of all external conditions and influences to which a system is subjected (ISO, 1975). The term "system" covers all living organisms, including human beings.

Environmental chemistry chemical ecology.

The study of natural and man-made substances in the environment, including the detection, monitoring, transport, and chemical transformation of chemical substances in air, water, and soil.

Environmental health (Synonyms: environmental medicine, environmental hygiene) The health aspects of the human environment, including technical and administrative measures for improving the human environment from a health point of view (WHO, 1979).

Environmental quality standard (EQS) This regulatory value defines the maximum concentration of a potentially toxic substance which can be allowed in an environmental compartment, usually air or water, over a defined period. Synonym: ambient standard. See limit values.

Environmental sanitation Traditionally used to indicate activities concerned with the improvement of the basic environmental conditions affecting health, i.e., water supply, human and animal waste disposal, protection of food from biological contamination, and housing conditions, all of which are concerned with the quality of the human environment (WHO, 1979).

Environmental transformation Once emitted into the environment, a chemical substance may be transported in the biosphere and undergo various types of chemical changes (WHO, 1979).

Enzyme A macromolecule that functions as a biocatalyst by increasing the reaction rate, frequently containing or requiring one or more metal ions. In general, an enzyme catalyzes only one reaction type (reaction specificity) and operates on only a narrow range of substrates (substrate specificity). Substrate molecules are attacked at the same site (regiospecificity) and only one or preferentially one of the enantiomers of chiral substrate or of racemic mixtures is attacked (enantiospecificity).

Epidemiology The study of the distribution and determinants of health-related states or events in populations, and the application of this study to control of health problems (from Last, 1988).

Epimer A diastereoisomer that has the opposite configuration at only one of two or more tetrahedral "stereogenic" centres present in the respective molecular entity.

Epimerization Interconversion of epimers by reversal of the configuration at one of the "stereogenic" centres.

Epoxy resin Epoxy resins are thermosetting resins produced by copolymerising epoxide compounds with phenols (e.g. epichlorohydrin and bisphenol A). They contain ether linkages (-O-) and form a tight, cross-linked polymer network. Toughness, good adhesion, corrosive-chemical resistance, and good dielectric properties characterize epoxy resins. Most epoxy resins are two-part type which harden when blended.

EPR See electron paramagnetic resonance spectroscopy.

Equation of state Equation of state is an equation relating the pressure, volume, and temperature of a substance or system. Equation of state for ideal gas

$pV = nRT$

where p is pressure, V molar volume, T temperature, and R the molar gas constant (8.314 $JK^{-1}mol^{-1}$).

Equatorial 1. For the use of this term in the context of cyclohexane see axial, equatorial.

2. For the use of this term in the context of a bipyramidal structure see apical, basal, equatorial.

Equatorial rain forests An area located near the equator worldwide, covering seven percent of the earth's surface, characterized by high precipitation, absent or short dry season, high tree and vegetative growth rates, high humidity, high photosynthetic rates and extreme biodiversity. The equatorial rain forests produce much of the oxygen in the Earth's atmosphere and absorbs large quantities of carbon dioxide from the air. (See deforestation).

Equatorial The bonds in a chair cyclohexane that are oriented along the equator of the ring.

Equilibrium, chemical Reversible processes (processes which may be made to proceed in the forward or reverse direction by the (infinitesimal) change of one variable, ultimately reach a point where the rates in both directions are identical, so that the system gives the appearance of having a static composition at which the Gibbs energy, G, is a minimum. At equilibrium the sum of the chemical potentials of the reactants equals that of the products, so that

$G_r = G_r^0 + RT \ln K = 0$

$G_r^0 = -RT \ln K$

The equilibrium constant, K, is given by the mass-law effect.

Equilibrium expressions The expression giving the ratio between the products and reactants. The equilibrium expression is equal to the concentration of each product raised to its coefficient in a balanced chemical equation and multiplied together, divided by the concentration of the product of reactants to the power of their coefficients.

Equilibrium isotope effect See isotope effect.

Equilibrium-constant expressions These are algebraic equations that describe the concentration relationships that exist among reactants and products at equilibrium. Among other things, equilibrium-constant expressions permit calculation of the error resulting from the quantity of unreacted analyte that remains when a steady state has be reached.

Equinox Twice during the year, September 21 and March 21, the length of day and night are equal because the tilt of the Earth's axis (in relationship to the sun) is nullified and both the Northern and Southern Hemispheres receive equal quantities of sunlight. An animation showing this is here.

Equivalence point Compare with end point.

The equivalence point is the point in a titration when enough titrant has been added to react completely with the analyte.

Equivalent diameter (of a particle) The diameter of a spherical particle of the same density that, relative to a given phenomenon or property, would behave as the particle under investigation (ISO, 1979).

Erg Erg (erg) is a non-SI (cgs) unit of energy, equal to 10^{-7} J.

Erik christian clemmensen (1876-1941) Born in Odense in Denmark and was educated at the Royal Polytechnic Institute in Copenhagen. He emigrated to the United States in 1900 and worked as a research chemist for the pharmaceutical company, Parke, Davis & Co. He was awarded a doctorate by the University of Copenhagen for his development of the Clemmensen reduction. He co-founded the Commonwealth Chemical Corporation in 1914, which merged with the Mathieson Alkali Works in 1923. After a fire in 1929, the company was taken over by Monsanto, where Clemmensen worked on organic phosphates and thiophosphates. He founded the Clemmensen

Chemical Corporation in 1933. His name is associated with the Clemmensen reduction of ketones (1913).

Ernst gottfried fischer (1754-1831) Born in Hoheneiche near Saalfeld. After studying theology and mathematics at the University of Halle, he was a teacher in Berlin before becoming Professor of Physics at Berlin in 1810. He translated Claude-Louis Berthollet's* Recherches sur les lois de l'affinitié in 1802, thereby giving publicity to Richter's views. He proposed a system of equivalents based on sulfuric acid = 100. *The Berthel(l)ots will appear in this feature soon.

Ernst otto beckmann (1853-1923) Born in Solingen, he began his career as an pharmacist's assistant in Arolsen, Burg (Wupper), Leipzig and Cologne. After working for Fresenius in Wiesbaden, Beckmann studied chemistry and pharmacy at Leipzig. He taught pharmacy and chemistry at the Technische Hochschule in Brunswick, before moving back to Leipzig as an assistant to Wislicenus. He was briefly a Professor of Physical Chemistry at Leipzig and Giessen in turn before moving to Erlangen. He then became Director of the Laboratory of Applied Chemistry at Leipzig, and finally, Director of the new Kaiser Wilhelm Institute for Physical Chemistry and Electrochemistry in Berlin-Dahlem. His name is associated with the Beckmann thermometer, used to measure freezing and boiling point depressions, and the Beckmann rearrangement of oximes to amides. He was also interested in the analysis of foodstuffs and the chemistry of camphor.

Ernst otto fischer (b.1918) Born in Munich, the son of Karl Tobias Fischer (1871-1953), a physics professor at the Technische Hochschule. He took his doctorate in organometallic chemistry there under Hieber in 1952. Soon afterwards, he began his study of ferrocene which won him the Nobel Prize (with Geoffrey Wilkinson) in 1973. He taught at the University and Technische Hochschule in Munich.

Erythema In medicine, this term is applied to redness of the skin due to blood vessel distension.

ESE See electron spin-echo spectroscopy.

ESR Acronym for electron spin resonance. See electron paramagnetic resonance spectroscopy.

Ester A molecule containing a carbonyl group adjacent to an oxygen (RCOOR'). Also a functional group.

Esters Esters are organic compounds formed by reaction between alcohol and acids. Esters formed from carboxylic acids have the general formula RCOOR'. Triesters, molecules containing three ester groups, occur in nature as oils and fats.

Etectic point Eutectic temperature; eutectic composition.

The composition and the melting point of a eutectic mixture. For example, the eutectic point of a mixture of NaCl and water occurs at 23.3% NaCl (by mass) and -21.1°C. That means that the lowest possible temperature at which a liquid NaCl solution can exist is -21.1°C; below the eutectic point the solution will freeze into a mixture of ice and salt crystals.

Eter An ester is a compound formed from an acid and an alcohol. In esters of carboxylic acids, the -COOH group and the -OH group lose a water and become a -COO- linkage:

R-COOH + R'-OH = R-COO-R' + H_2O

where R and R' represent organic groups.

Ethane A hydrocarbon found in the atmosphere whose primary fate is to react with free radicals such as chlorine and nitrous oxide. This reaction would prevent these radicals from reacting with ozone which is needed in the stratosphere to block ultraviolet light from reaching the surface of the earth.

Ethanol Also known as ethyl alcohol or alcohol, (C_2H_5OH) is an oxygenated hydrocarbon that can be burned as a fuel, or blended into gasoline. Thought of as an alternative fuel and clean fuel, ethanol is a renewable energy source that is made from corn or other grains. If blended into gasoline, ethanol helps with the combustion process and therefore less unburned hydrocarbons form. Ethanol blended fuels competed with methyltertbutylether (MTBE), but because of the cost and the fact MTBE can be shipped through exisisting pipelines, MTBE was chosen. MTBE is now being banned because of its ability to travel in ground water and the bad taste it imparts to drinking water, because of this ethanol is once again being put back into the gas.

Ethers Ethers are organic compounds with formula R-O-R, where R is not equal to H. They may be derived from alcohols by elimination of water, but thc major method is catalytic hydration of olefins. They are volatile highly flammable compounds; when containing peroxides they can detonate on heating. The term ether is often used synonymously with ethyl ether.

Ethylbenzene C_8H_{10}, Colorless, flammable liquid with an aromatic odor, it is found in coal tar, petroleum, as well as in manufactured products like inks, insecticides, and paints. Ethylbenzene is also used to make styrene. It is during the manufacturing process that ethylbenzene becomes an airborne hazard. It takes about 3 days for ethylbenzene to breakdown into other chemicals. Exposure to ethylbenzene causes dizziness, throat and eye irritation, tightening of the chest, and a burning sensation in the eyes. The Occupational Safety and Health Administration (OSHA) has set an occupational exposure limit of 100 ppm of ethylbenzene for an 8-hour workday for a 40-hour workweek.

Eugen bamberger (1857-1932) Born in Berlin, he took his doctorate at Berlin University under Liebermann. He worked as an assistant to Rammelsberg in Berlin and taught in Munich before taking up a chair at ETH in Zurich. He carried· out extensive research in heterocyclic and alicylic chemistry, as well as diazo compounds. He prepared nitrosobenzene and synthesised isoquinolines. His name is associated with the Bamberger rearrangement of N-phenylhydroxylamines to p-aminophenols.

Euivalent. Compare with normality. 1. The amount of substance that gains or loses one mole of electrons in a redox reaction.

2. The amount of substances that releases or accepts one mole of hydrogen ions in a neutralization reaction.

3. The amount of electrolyte that carries one mole of positive or negative charge, for example, 1 mole of Ba^{2+}(aq) is 2 equivalents of Ba^{2+}(aq).

Eukaryotes Organisms whose cells have their genetic material packed in a membrane-surrounded, structurally discrete nucleus, and that have well-developed cell organelles. Eukaryotes include all organisms except archaea and eubacteria.

Euler number Euler number (Eu) is a dimensionless quantity used in fluid mechanics, defined by

$$Eu = \frac{\Delta p}{\rho . v^2}$$

where p is pressure, ρ is density, and v is velocity.

Eustatic changes A consistent change in the sea level that affect oceans, throughout the world.

Eutectic Eutectic is a solid solution consisting of two or more substances and having the lowest freezing point of any possible mixture of these components.

Eutectic point is the lowest temperature at which the eutectic mixture can exist in liquid phase. A liquid having the eutectic composition will freeze at a single temperature without change of composition.

Eutectic mixture A mixture of two or more substances with melting point lower than that for any other mixture of the same substances.

Evaporate To convert a liquid into a gas.

Evaporation The process of the change in the state of a liquid or solid to a gas or vapor. Vanishing of the surface of a liquid to the atmosphere.

Evapotranspiration The total water loss due to the transpiration of vegetation plus the evaporation from the soil; higher climatic temperatures result in a higher evapotranspiration rate.

Ewens-bassett number See oxidation number.

Excimer An excited dimer, "non-bonding" in the ground state. For example, a complex formed by the interaction of an excited molecular entity with a ground state counterpart of this entity.

Exciplex An electronically excited complex of definite stoichiometry, "non-bonding" in the ground state. For example, a complex formed by the interaction of an excited molecular entity with a ground state counterpart of a different structure.

Excited state Compare with ground state.

An atom or molecule which has absorbed energy is said to be in an excited state. Excited states tend to have short lifetimes; they lose energy either through collisions or by emitting photons to "relax" back down to their ground states.

Excitotoxin An excitotoxin is a toxic molecule that stimulates nerve cells so much that they are damaged or killed. Domoic acid and glutamate are examples of excitotoxins.

Excretion rate The amount or proportion of a substance that is excreted per unit time. It should be noted that according to this definition excretion does not include the passing of a substance through the intestine without absorption. When discussing the total amount of a substance in faeces (including the unabsorbed part), it is preferable to speak about faecal content of substance

Excretion The discharge or elimination of an absorbed or endogenous substance, or of a waste product, via some tissue of the body and its appearance in urine, faeces, or other products normally leaving the body. Excretion of chemical compounds from the body occurs mainly through the kidney and the gut. For volatile compounds, however, elimination by exhalation may be important. Excretion via perspiration and through hair and nails may also be of importance under special circumstances. Excretion via the gastrointestinal tract may take place by various routes such as the bile, the shedding of intestinal cells, and transport through the intestinal mucosa (WHO, 1979).

Exogenous Originating externally. In the context of metalloprotein ligands, exogenous describes ligands added from an external source, such as CO or O_2.

Exon A section of DNA which carries the coding sequence for a protein or part of it. Exons are separated by intervening, non-coding sequences (called introns). In eukaryotes most genes consist of a number of exons.

Exothermic Exothermic reaction; exothermic process. Compare with endothermic.

A process that releases heat. The enthalpy change for an exothermic process is negative. Examples of exothermic processes are combustion reactions and neutralization reactions.

Experiment An experiment is direct observation under controlled conditions. Most experiments involve carefully changing one variable and observing the effect on another variable (for example, changing temperature of a water sample and recording the change volume that results).

Experimental yield Actual yield. Compare with theoretical yield and percent yield.

The measured amount of product produced in a chemical reaction.

Explosion The sudden and violent release of mechanical, chemical or nuclear energy from a confined space which creates a heat wave that travels at subsonic speeds. Often used interchangeably with detonation.

Exponentiation Raising something to a power.

Exposure The amount of an environmental agent that has reached the individual (external dose) or has been absorbed into the individual (internal dose, absorbed dose).

Exposure assessment The quantification of the amount of exposure to a hazard for an individual or group (WHO, 1979).

Exposure control see emission or exposure control.

Exposure limit A general term implying the level of exposure that should not be exceeded (WHO, 1979).

Expression The cellular production of the protein encoded by a particular gene. The process includes transcriptionof DNA, processing of the resulting mRNA product and its translation into an active protein.

N.B. A recombinant gene inserted into a host cell by means of a vector is said to be expressed if the synthesis of the encoded polypeptide can be demonstrated. For the expression of metalloproteins usually other gene products will be required.

Extended X-ray absorption fine structure (EXAFS) EXAFS effects arise because of electron scattering by atoms surrounding a particular atom of interest as that special atom absorbs X-rays and emits electrons. The atom of interest absorbs photons at a characteristic wavelength and the emitted electrons, undergoing constructive or destructive interference as they are scattered by the surrounding atoms, modulate the absorption spectrum. The modulation frequency corresponds directly to the distance of the surrounding atoms while the amplitude is related to the type and number of atoms. EXAFS studies are a probe of the local structure.EXAFS can be applied to systems which have local structure, but not necessarily long-range structure, such as non-crystalline materials. In particular, bond lengths and local symmetry (coordination numbers) may be derived. The X-ray absorption spectrum may also show detailed structure below the absorption edge. This X-ray absorption near edge structure (XANES) arises from excitation of core electrons to high level vacant orbitals.

Extensive property A property that changes when the amount of matter in a sample changes. Examples are mass, volume, length, and charge.

External compensation (usage strongly discouraged)

The absence of optical activity in a racemate.

Extinction coefficient A measure of the rate of the reduction of transmitted light through a substance.

Extraneous residue limit An extraneous residue limit is, for a particular commodity, the maximum toxicologically acceptable concentration of a residue unavoidably arising from sources other than the use of a pesticide directly or indirectly for the production of that commodity (WHO, 1976).

Extrapolation The calculation, based on quantitative observations in exposed test species, of predicted dose-effect and dose-response relationships for a chemical in humans and other environmental biota (WHO, 1979).

Extrusion transformation A transformation in which an atom or group Y connected to two other atoms or groups X and Z is lost from a molecule, leading to a product in which X is bonded to Z,

The reverse of an extrusion is called insertion.

F

F orbital F-orbital.

An orbital with angular momentum quantum number = 2. The f orbitals generally have 3 nuclear nodes and rather complex shapes.

F-430 A tetrapyrrole structure containing nickel, a component of the enzyme methyl-coenzyme M reductase, which is involved in the formation of methane in methanogenic bacteria. The highly reduced macrocyclic structure, related to porphyrins and corrins, is termed a corphin.

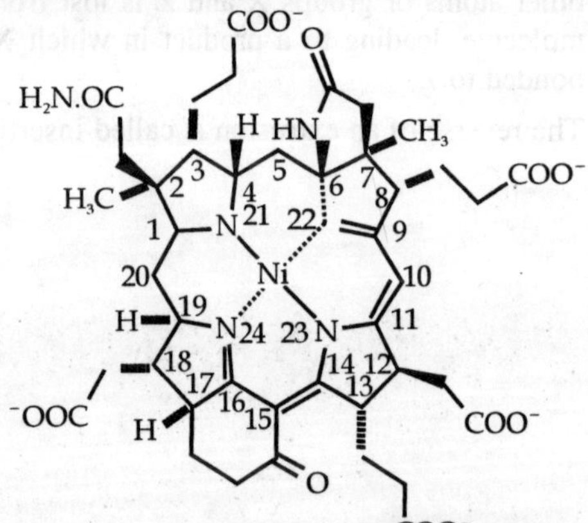

Fahrenheit scale of temperature The Fahrenheit scale of temperature, named after G.D. Fahrenheit (1686-1736) who invented it, has the temperature of boiling water as 212 degrees and the temperature of melting ice as 32 degrees. The Fahrenheit scale is no longer in scientific use.

To convert from the Fahrenheit scale to the Celsius scale the formula is

deg C = 5 (deg F - 32) / 9.

Farad Farad (F) is the SI derived unit of electric capacitance. The farad is the capacitance of an electric capacitor between the two plates of which there appears a difference of electric potential of one volt when it is charged by a quantity of electricity equal to one coulomb (F = C/V). The unit is named after the British scientist M. Faraday (1791-1867).

Faraday cup A Faraday cup is a metal cup that is placed in the path of the ion beam. It is attached to an electrometer, which measures the ion-beam current. Since a Faraday cup can only be used in an analog mode it is less sensitive than other detectors that are capable of operating in pulse-counting mode.

Faraday's constant The Faraday Constant, F, is the electric charge carried by one mole of electrons or singly ionised ions (i.e. the product of Avogadro constant and the charge on an electron or ion, disregarding sign).

It has the value 9.648670 X 10^4 coulombs per mole.

This number of coulombs is sometimes treated as a unit of electric charge called the Faraday.

Faraday's laws of electrolysis Faraday's Laws of Electrolysis govern the electrolysis of aqueous solutions and state

Faraday's First Laws of Electrolysis

The mass of a substance liberated at each electrode is directly proportional to the quantity of electricity which passes (i.e. to the current strength and to the time), and

Faraday's Second Laws of Electrolysis

The masses of the elements liberated by the same quantity of electricity are directly proportional to their chemical equivalents.

Fast-atom bombardment (FAB) mass spectroscopy A method in which ions are produced in a mass spectrometer from nonvolatile or thermally fragile organic molecules by bombarding the compound in the condensed phase with energy-rich neutral particles.

Fat A fat is a triester of glycerol, and on hydrolysis with sodium hydroxide gives glycerol and Soap (i.e. a mixture of the sodium salts of the fatty acids).

Fatty acids Fatty acids are aliphatic monocarboxylic acids characterized by a terminal carboxyl group (R-COOH). Natural fatty acids commonly have a chain of 4 to 28 carbons (usually unbranched and even-numbered), which may be saturated or unsaturated. The most important of saturated fatty acids are butyric (C4), lauric (C12), palmitic (C16), and stearic (C18). The most common unsaturated acids are oleic, linoleic, and linolenic (all C18).

The physical properties of fatty acids are determined by chain length, degree of unsaturation, and chain branching. Short-chain acids are pungent liquids, soluble in water. As chain length increases, melting points are raised and water-solubility decreases. Unsaturation and chain branching tend to lower melting points.

Fauna Animals or animal life of a particular region or a particular time.

Feedback cycle The increase of the input into a system, which produces a negative or a positive output in a system of interlocking components; the return of a part or the effects of the output to the input. A proposed positive feedback cycle involves the outgassing of carbon dioxide from the planet's oceans: Since CO_2 solubility in water DECREASES with temperature and since CO_2 is a greenhouse gas, a positive (and possibly runaway) feedback cycle might involve the following: increased greenhouse warming of the atmosphere by IR absorbing CO_2 in the gas phase—and the subsequent heating of the oceans—could increase the amount of carbon dioxide released into the atmosphere from the oceans. This might thereby increase greenhouse warming, which would, in turn, again heat the atmosphere and the ocean, etc. Though the process is no doubt more complicated than this, a runaway greenhouse effect is usually given as the reason for Venus' high surface temperatures.

Feedstock The Feedstock for the manufacture of petroleum products and petrochemicals is crude oil extracted from known reserves.

Felix hoffmann (1868-1946) Born in Ludwigsburg, Württemberg. Studied pharmacy and chemistry (under Eugen Bamberger) at Munich. He worked at Bayer, where he synthesised acetylsalicylic acid and proposed it as an alternative to sodium salicylate as an antirheumatic drug in 1897. After his retirement in 1928, he studied art history in Italy and Switzerland, where he died.

FeMo-cofactor An inorganic clusterthat is found in the FeMo protein of the molybdenum-nitrogenase and is essential for the catalytic reduction of N_2 to ammonia. This cluster contains Fe, Mo and S in a 7:1:9 ratio.

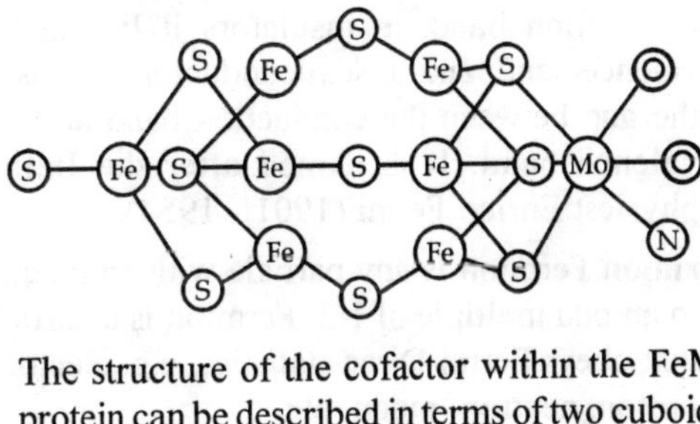

The structure of the cofactor within the FeMo protein can be described in terms of two cuboidal subunits, Fe_4S_3 and $MoFe_3S_3$ bridged by three S^{2-} ions and "anchored" to the protein by a histidine bound via an imidazole group to the Mo atom and by a cysteine bound via a

deprotonated SH group to an Fe atom of the Fe_4S_3 subunit. The Mo atom at the periphery of the molecule is six-coordinate and in addition to the three sulfido ligands and the histidine imidazole is also bound to two oxygen atoms from an (R)-homocitrate molecule.

Femto (f) Prefix used in the SI system meaning "multiply by 10^{-15}". For example 22 fg means 22×10^{-15} g.

Fenton reaction $Fe^{2+} + H_2O_2 \longrightarrow Fe^{3+} + OH\cdot + OH^-$. This is the iron-salt-dependent decomposition of dihydrogen peroxide, generating the highly reactive hydroxyl radical, possibly via an oxoiron(IV) intermediate. Addition of a reducing agent such as ascorbate leads to acycle which increases the damage to biological molecules.

Fermentation A class of biochemical reactions that break down complex organic molecules (such as carbohydrates) into simpler materials (such as ethanol, carbon dioxide, and water). Fermentation reactions are catalyzed by enzymes.

Fermentation of carbohydrates The fermentation of carbohydrates is a form of anaerobic respiration occurring in certain microorganism (e.g. yeasts), comprises a series of biochemical reactions by which sugars (i.e. saccharides), including glucose and sucrose, are converted to ethanol and carbon dioxide.

Fermi level Fermi level is the highest energy of occupied states in a solid at zero temperature. The Fermi level in conductors lies in the conduction band, in insulators it lies in the valence band, and in semiconductors it falls in the gap between the conduction band and the valence band. It is named after the Italian physicst Enrico Fermi (1901 - 1954).

Fermion Fermion is any particle with spin equal to an odd multiple of 1/2. Fermion is a particle that obeys Fermi-Dirac statistics, e.g. electron, proton, neutron, muon, etc.

Ferredoxin A protein containing more than one iron and acid-labile sulfur, that displays electron-transferactivity but not classical enzyme function.

Ferric compound Ferric compound are compounds of iron in its +3 oxidation state. Examples include iron (III) chloride, FeCl3, which was previously called ferric chloride.

Ferric Ferric ion.

Deprecated. 1. the iron(III) ion, Fe^{3+}.

2. A compound that contains iron in the +3 oxidation state.

Ferriheme An iron(III) porphyrin coordination complex.

Ferrites Ferrites are ceramic materials of nominal formula $MO.Fe_2O_3$, where M is a divalent metal (Co, Mn, NI, or Zn). The ferrites show either ferrimagnetism or ferromagnetism, but are not electrical conductors, and they are used in high-frequency circuits as magnetic cores, in rectifiers on memory and record tapes, and various related uses in radio, television, radar, computers, and automatic control systems.

Ferritin An iron storage protein consisting of a shell of 24 protein subunits, encapsulating up to 4500 iron atoms in the form of a hydrated iron(III) oxide.

Ferrochelatase An enzyme which catalyzes the insertion of iron into protoporphyrin IX to form heme. The mammalian enzyme contains an iron-sulfur cluster.

Ferroheme An iron(II) porphyrin coordination complex.

Ferroin A blood-red complex of Fe^{2+} ion with 1,10-phenanthroline, used as a redox indicator. Ferroin changes from red to pale blue when oxidized.

Ferromagnetism Ferromagnetism is a type of magnetism in which the magnetic moments of atoms in a solid are aligned within domains which can in turn be aligned with each other by a weak magnetic field. The total magnetic moment of a sample of the substance is the vector sum of the magnetic moments of the component domains. In an unmagnetized piece of ferromagnetic material the magnetic moments of the domains themselves are not aligned; when an external field is applied those domains that are aligned with the field increase in size at the expense of the others. Ferromagnetic materials can retain their magnetization when the external field is removed, as long as the temperature is below a critical value, the Curie temperature. They are characterized by a large positive magnetic susceptibility.

Ferrous ferrous ion.Deprecated. 1. the iron(II) ion, Fe2+.

2. A compound that contains iron in the +2 oxidation state.

Ferrous sulphate, Fe(II)SO4, is the salt formed when iron is dissolved in sulphuric acid.

Fertilisers Fertilisers are the minerals which are added to soil to increase its productivity in crop production. The major components of inorganic fertilisers are

Nitrogen

in the form of ammonia, nitrate and urea,

Phosphorus

in the form of phosphate or superphosphate,

Potassium

in the form of one of its salts) and

Sulphur

in the form of sulphate salts.

Trace levels of other minerals are also often included in fertilisers, either as trace nutrients for crops or for uptake by animals grazing on the crops.

Fetus (foetus) In medicine, this term is applied to the young of mammals when fully developed in the womb. In humans, this stage is reached after about 3 months of pregnancy. Prior to this, the developing mammal is in embryo stage.

Fibrosis Formation of fibrous tissue as a reaction or as a repair process; may be result of treatment and/or disease.

Fick's law Fick's law is the statement that the flux J of a diffusing substance is proportional to the concentration gradient, i.e.,

Fiducial group See helicity and torsion angle.

Field effect An experimentally observable effect symbolized by F (on reaction rates, etc.) of intramolecular coulombic interaction between the centre of interest and a remote unipole or dipole, by direct action through space rather than through bonds. The magnitude of the field effect (or "direct effect") depends on the unipolar charge/dipole moment, orientation of dipole, shortest distance between the centre of interest and the remote unipole or dipole, and on the effective dielectric constant. An approach to its calculation was described by Kirkwood and Westheimer (1938) and has been elaborated in more recent years.

Filling orbital The orbital of an element that is only partially filled with electrons when an atom is at a neutral state electrically.

Filter A filter is a device for separating solids from liquids or an insoluble substance from a soluble one.

The simplest laboratory filter for liquids is a fluted glass funnel in which a cone of special paper, known as filter paper, is placed. The liquid to be filtered is passed through the paper filter, whereupon the insoluble solids are retained on the filter.

Grouch crucibles, which are special containers with a porous base of sintered glass, are also used as filters.

Filter pump A filter pump is a simple laboratory vacuum pump in which air is removed from a system by a jet of water forced through a narrow nozzle. The lowest pressure possible is the vapour pressure or water.

Filtration Filtration is a method of separating a solid from a liquid or an insoluble substance from a soluble one. The mixture is poured through a filter; the liquid that goes through the filter and is called the filtrate, and the solid that remains on the filter is called the residue.

Fingerprint Numerical representation of a compound or library which describes in a computationally simple fashion a set of attributes (descriptors), such as atom connectives, 3-D structure or physical properties.

Fingerprint region Region of an IR spectrum below 1,500 cm^{-1}. The fingerprint region of the

IR spectrum is often complex and difficult to interpret.

Firn Granular snow that has endured a summer without melting and will transform into glacial ice.

First ionization energy (IE,IP) first ionization potential. Compare with second ionization energy, adiabatic ionization energy, vertical ionization energy, electronega-tivity, and electron affinity.

The energy needed to remove an electron from an isolated, neutral atom.

First ionization potential (first ionization energy) The minimum amount of energy needed to remove the outermost (highest energy) electron from a neutral atom in the gaseous state.

First law First law of thermodynamics.

The first law states that energy cannot be created or destroyed. Many equivalent statements are possible, including: Internal energy changes depend only on the initial and final states of the system, not on the path taken. The work done during an adiabatic process depends only on the initial and final states of the system, and not on the path taken. The internal energy change for any cyclic process is zero.

First order reaction Compare with zero order reaction and second order reaction.

The sum of concentration exponents in the rate law for a first order reaction is one. Many radioactive decays are first order reactions.

Fischer projection (or fischer-tollens projection) A projection formula in which vertically drawn bonds are considered to lie below the projection plane and horizontal bonds to lie above that plane. Thus for the molecule Cabcd.

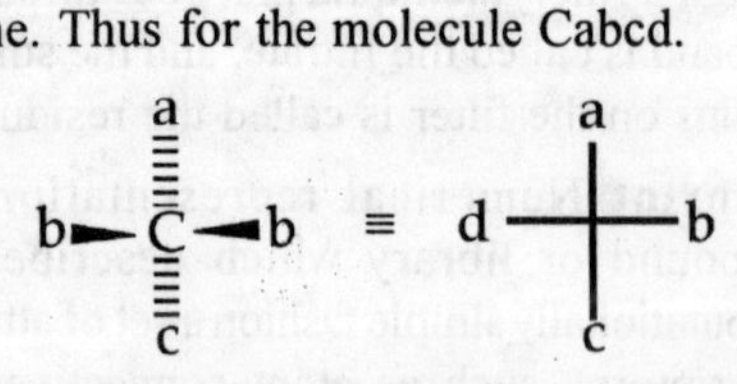

Fischer-rosanoff convention (or rosanoff convention) An arbitrary convention according to which (+)-glyceraldehyde, now known to be (R)-2,3-dihydroxypropanal, was named D-glyceraldehyde (with the enantiomer L-glyceraldehyde and its racemate DL-glyceraldehyde) and taken to have the absolute configuration represented by the Fischer projection formula.

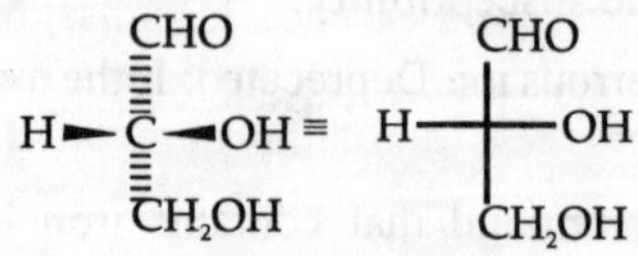

D-glyceraldehyde

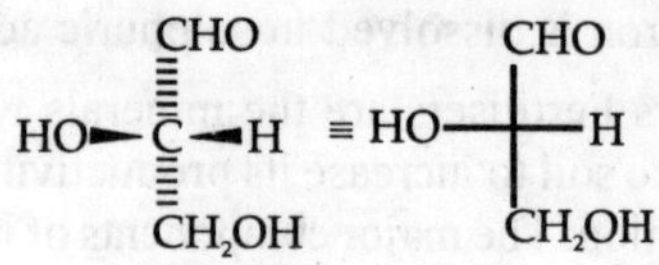

L-glyceraldehyde

Fission Nuclear fission.

A nuclear reaction in which an atomic nucleus breaks into smaller nuclei of comparable mass, releasing a large amount of energy.

Flash point Compare with auto-ignition temperature.

The temperature when vapor pressure of a substance becomes high enough to allow the air/vapor layer over the substance to be ignited. Ether and acetone have flash points below room temperature, which makes them very dangerous.

Flash vacuum pyrolysis (FVP) Thermal reaction of a molecule by exposing it to a short thermal shock at high temperature, usually in the gas phase.

Flocculation Flocculation is a process of aggregating into larger clumps. Finely divided precipitates, where the particles are small enough to pass through the pores in a filter, need to undergo flocculation to form larger particles before filtration.

Flora A plant or plant life of a specific region or particular period.

Flow cytometry Technique for characterizing or separating particles such as beads or cells, usually on the basis of their relative fluorescence.

Fluidic system Device for synthesis or screening in which fluids such as reagents or assay buffers maybe directed to specified locations by the opening and closing of valves in a stationary network of tubes and wells.

Fluorescence Fluorescence is emission of light from compounds which have been illuminated, due to the return of the compound from an excited state to the ground state.

Fluorides Fluorides are the salts of hydrofluoric acid.

Fluorination Fluorination is a chemical reaction in which a fluorine atom is introduced into a molecule.

Fluorocarbons Fluorocarbons are compounds obtained by replacing some or all the hydrogen atoms of hydrocarbons with fluorine atoms. Their high stability to temperature makes them suitable for a variety of uses, including aerosol propellants, oils, polymers, etc..

They are often known as freons. Their use in aerosols is one cause of the depletion of zone layer.

Fluorous synthesis Approach for solution phase synthesis which takes advantage of the ability of highly fluorinated groups to partition out of aqueous and most organic solutions into a third phase consisting of a fluorinated solvent.

Fluorspar Fluorite or fluorspar (i.e. calcium fluoride, CaF2) is a mineral that occurs in Derbyshire, crystallised in cubes or octahedral or in compact masses like marble. It is also called Derbyshire spar, or Blue Jhon, when the crystals are coloured blue or purple. The colourless transparent crystals show a bluish tinge when light falls on them, and this property is called fluorescence.

Fluorspar was described by Agricola in 1530AD, as fluor (from the Latin fluo, I flow), since it melts at a red heat, approximately 1330 degC.

Fluorspar is used chiefly as a flux material in the smelting of iron and steel. It is also used as a source of fluorine and hydrofluoric acid and in the manufacture of ceramics and in the optical-glass industry.

Flux The rate of continuous change, flow or movement of liquid, particles or energy. The rate of discharge of a liquid, removal of energy or particle depositing from one body to another.

Fluxional A chemical species is said to be fluxional if it undergoes rapid degenerate rearrangements (generally detectable by methods which allow the observation of the behaviour of individual nuclei in a rearranged chemical species, e.g. NMR, X-ray). The term is also used to designate positional change among ligands of complex compounds and organometallics. In these cases, the change is not necessarily degenerate.

Foam Foams are dispersions of gases in liquids or solids. The gas globule may be of any size, from colloidal to macroscopic, as in soap bubbles. Bakers' bread and sponge rubber are examples of solid foams. Typical liquid foams are those used in fire-fighting, shaving creams, etc. Foams made by mechanical incorporation of air are widely used in the food industry (e.g. whipped cream, egg white, ice cream, etc.). Foams can be stabilized by surfactants.

Foehn A warm, dry wind on the lee side of a mountain range that owes its relatively high temperatures largely to adiabatic heating during descent down mountain slopes. In Zurich Switzerland, this means a warm but clear view of the Alps.

Fog The state in which particles of very small water droplets are in contact or close to the earth's surface, generally reducing horizontal visibility.

Folate coenzymes A group of heterocyclic compounds that are based on the 4-(2-amino-3,4-dihydro-4-oxopteridin-6-ylmethylamino)benzoic acid (pteroic acid) and conjugated with one or more L-glutamate units. Folate derivatives are important in DNA synthesis and erythrocyte formation. Folate deficiency leads to anemia.

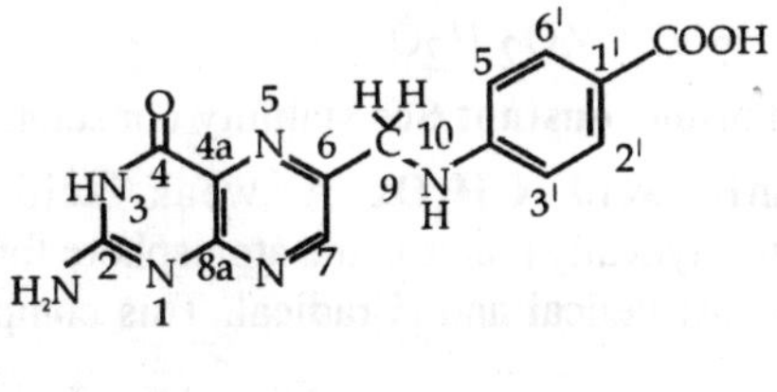

Folic acid Folic acid is a member of the vitamin B group of vitamins.

Follow up study (synonym: cohort study) A study in which individuals or populations, selected on the basis of whether they have been exposed to risk, received a specified preventive or therapeutic procedure, or possess a certain characteristic, are followed to assess the outcome of exposure, the procedure, or effect of the characteristic, e.g., occurrence of disease.

Food additive Any non-nutritive substances added intentionally to food, generally in small quantities, to improve its appearance, flavour, texture or storage properties, with the exception of sub- stances which are added to food exclusively for their nutritive properties, but including animal feed adjuncts which may result in residues in human food and components of packaging materials which may find their way into human food, and other contaminants.

Food chain The sequence of transfer of matter and energy in the form of food from organism to organism in ascending or descending trophic levels (WHO, 1979).

Food web A network of food chains (WHO, 1979).

Force An entity that when applied to a mass causes it to accelerate. Sir Isaac Newton's Second Law of mation states: the magnitude of a force = mass * acceleration.

Force-field calculations See molecular mechanics calculation.

Formaldehyde CH_2O A colorless, pungent, carcinogenic, and toxic gas which can irritate membranes. It is used as formalin in its aqueous state (37 %). Formaldehyde is reactive due to its carbonyl functional group.

Formation Formation reaction.

A reaction that forms one mole of a compound from its elements in their most stable forms. For example, the formation reaction for water is $H_2(g) + ½O_2\ H_2O$.

Formation constant See stability constant.

Formic acid CH_2O_2 A weak acid that photolytically reacts in the atmosphere forming COOH radical and H radical. This compound often reacts with the hydroxyl radical and atomic oxygen to form formic acid and a hydroperoxy radical. This is a sink for the hydroxyl radical and it also helps drive other reactions with the addition of hydroperoxy radicals to the atmosphere.

Formula A formula is a way of representing a chemical compound using symbols for the atoms present. Subscripts are used to denote the numbers of each type of atom.

Formula unit Compare with empirical formula.One formula weight of a compound.

Formula weight Formula mass. Compare with molecular weight and empirical formula.

The formula weight is the sum of the atomic weights of the atoms in an empirical formula. Formula weights are usually written in atomic mass units (u).

Fossil fuels There are coal, crude oils, oil shales, tar sands and natural gases such as butane, ethane, methane which occur naturally from the decomposition of plant and sea and land organisms over millions of years. These natural resources contain stored energy from the sun which is released upon combustion. These fuels also release various types of pollutants such as sulfur dioxide, carbon dioxide and nitrogen dioxide when burned.

Fractional crystallization Fractional crystallization is a method of separating a mixture of soluble solids by dissolving them in a suitable hot solvent and then lowering the temperature slowly. The least soluble component will crystallize out first, leaving the other components in solution. By controlling the temperature, it is sometimes possible to remove each component in turn.

Fractional distillation Compare with distillation.

A technique for separation of liquid mixtures by distillation that uses a tower attached to a flask containing the mixture to perform multiple distillations. Vapor moving up the column condenses on packing material inside the column, trickles down the column, and again vaporises. The more volatile component can then be drawn off at the top of the component, while the less volatile component remains at the bottom.

Fractionating column A fractionating column is added before the condenser in fractional distillation and contains glass rings or balls which provide a large surface area for condensation and re-evaporation.

Fractionation factor, isotopic The ratio $(x_1/x_2)_A/(x_1/x_2)_B$, where x is the abundance, expressed as the atom fraction of the isotope distinguished by the subscript numeral, when the two isotopes are equilibrated between two different chemical species A and B (or between specific sites A and B in the same or different chemical species). The term is most commonly met in connection with deuterium solvent isotope effects, when the fractionation factor expresses the ratio

$= (x_D/x_H)_{solute}/(x_D/x_H)_{solvent}$

for the exchangeable hydrogen atoms in the chemical species (or sites) concerned. The concept is also applicable to transition states.

Fragmentation 1. The heterolytic cleavage of a molecule according to the general reaction

$a\text{-}b\text{-}c\text{-}d\text{-}X\ (a\text{-}b)^+ + c{=}d + X^-$

where a-b is an electrofuge and X is a nucleofuge (which may emerge from the reaction in combined form), and the middle group affords the unsaturated fragment c=d. For example,

$Ph_3C\text{-}CO_2H + H^+\ Ph_3C^+ + C{=}O + H_2O$

GROB (1969).

2. The breakdown of a radical into a diamagnetic molecule or ion and a smaller radical, e.g.,

$(CH_3)_3C\text{-}O\cdot\ (CH_3)_2C{=}O + H_3C\cdot$

$(ArBr)\cdot^-\ Ar\cdot + Br^-$ (solution)

3. The breakdown of a radical ion in a mass spectrometer or in solution, forming an ion of lower molar mass and a radical, e.g.,

$((CH_3)_3C\text{-}OH)\cdot^+\ (CH_3)_2C{=}OH^+ + H_3C\cdot$ (mass spectrometer)

Franz josef emil fischer (1877-1947) Of the Fischer-Tropsch process, was born in Freiburg im Breisgau and studied under Elbs at Giessen. After a varied academic career, during which he worked with Wilhelm Ostwald and Emil Fischer, he became director of the Kaiser Wilhelm Institute for Coal Research in the Ruhr in 1913.

Franz; gottwalt fischer (1902-1960) Born in Florence, he took his doctorate under Wieland at Freiburg. He developed the hydrogenation of ozonides in the late 1920s and taught at the Universities of Munich, Freiburg and Würzburg.

Frasch process The Frasch Process is a method of obtaining sulphur from underground deposits using a tube consisting of three concentric pipes. Superheated steam is passed down the outer pipe to melt the sulphur, which is then forced up through the intermediate pipe by compressed air from the inner tube. The steam in the outer casing keeps the sulphur molten in the pipe.

Frederick eugene bailey, junior (b.1927) Born in Brooklyn, New York, he studied chemistry at Amherst and Yale, where he took his Ph.D., before joining Union Carbide. He has developed the polymerisation of ethylene oxide, and the production of high molecular weight poly(ethylene oxide).

Free electron Electron which is not attached to a nucleus.

Free energy Energy that is actually available to do useful work. A decrease in free energy accompanies any spontaneous process. Free energy does not change for systems that are at equilibrium.

Free radical A Free radical is an atom, or group of atoms, which normally exist only in combination with other atoms, but which may exist independently for short periods during the course of a chemical reaction or for longer periods under special conditions.

Examples of free radical formation include the case where a chlorine molecule is dissociated by a photon of light into two chlorine radicals (i.e. unpaired chlorine atoms), which are highly reactive

light

Cl2 ⇒ 2 Cl*

These radicals are then available to initiate and propagate other chemical reactions.

Free radical chain reaction The free radical chain reaction is the mechanism of many self-sustaining chemical reactions. Chain reactions are exothermic (i.e. they release large amounts of heat), but they may require a high temperature, or other special conditions, to start them. In the case of combustion, the initial heat is required for the bond breaking which is required to produce the free radicals, and initiate the chain reaction. However, when this initiation has been achieved, the chain reaction can proceed with the release of energy in sufficient quantity to continue to propagate the reaction.

For example, the combustion of alkanes which results in their complete oxidation to carbon dioxide and water is a free radical chain reaction. Although this reaction is exothermic, a high temperature is required for initiation. This initial heat is required for the bond breaking which produces the free radicals that initialise the reaction. Thereafter, the large amounts of heat released during the course of the reaction sustains the reaction. Alkanes can be oxidised completely, by combustion, to give carbon dioxide and water. For example, methane burns in air.

$$CH4 + 2O2 \Rightarrow CO2 + 2H2O + energy$$

Free rotation, hindered rotation, restricted rotation In a stereochemical context the rotation about a bond is called "free" when the rotational barrier is so low that different conformations are not perceptible as different chemical species on the time scale of the experiment.

The inhibition of rotation of groups about a bond due to the presence of a sufficiently large rotational barrier to make the phenomenon observable on the time scale of the experiment is termed hindered rotation or restricted rotation.

Freeze drying Freeze drying is the process used in dehydrating food, blood plasma and other heat sensitive substances. The product is deep frozen and the ice trapped in it removed by reducing the pressure and causing it to sublime. The water vapour is then removed, leaving an undamaged dry product.

Freezing Freezing is the process by which a liquid changes from liquid to solid. It is the opposite of melting.

Freezing point (mp) standard melting point; normal melting point; melting point.

The temperature at which the vapor pressure of a liquid is equal to the vapor pressure of the corresponding solid form. The liquid and solid forms can coexist at equilibrium at the freezing point. The standard melting point is the melting point at standard pressure.

Freezing point depression (T_{fp}) The freezing point of a solution is always lower than the freezing point of the pure solvent. The freezing point depression is roughly proportional to the molality of solute particles in the solution. Freezing point depression is an example of a colligative property of a solution.

Freezing point See Melting point

Freons (also see CFCs) Stable liquids or gases usually produced for solvents, aerosol propellants, refrigeration, air conditioner or styrofoam purposes. These high vapor pressure liquids/gases are chemically stable and have long atmospheric lifetimes; therefore, they can eventually become well-mixed in the troposphere and ultimately diffuse into the stratosphere. Upon encountering high energy UV light in the upper stratosphere these species are photodissociated and ultimately release all of their atoms there. The free radical chlorines (or bromines in the case of some halons) catalytically destroy ozone (see chlorine) in the stratosphere.

Friedrich bayer (originally Beyer) (1825-1880) Born in Barmen, the grandson of a weaver from Nördlingen. He served an apprenticeship as a chemical merchant, then set up his own business. Seeing the potential of the new synthetic dyes, he joined forces with a friend Friedrich Westkott and formed Friedrich Bayer & Cie in Barmen.

Initially, the new firm had only one worker, but a chemist was soon hired. A new factory was built near Elberfeld in 1867. By the time of Bayer's death, his firm had become a major dye manufacturer.

Friedrich hoffmann (1660-1742) Born in Halle, the son of a physician of the same name (d. 1675, confusingly his father's name was sometimes spelt Hofmann). Visited England and met Robert Boyle. Professor of medicine at Halle for forty-eight years. Carried out important work on spa waters, including the Epsom spa. Associated with Hoffmann's drops (solution of ether in ethanol).

Fritz arndt (1885-1969) Born in Hamburg and no relation to Kurt Arndt. Studied chemistry at the universities of Geneva, Freiburg im Breisgau and Berlin; he took his doctorate under Howitz at Freiburg. After a varied academic career in Freiburg (as an assistant to Gattermann), Hamburg, Griefswald, Kiel, Breslau and Constantiople, he became a professor at Breslau. Forced to leave Germany in 1933, he went briefly to Oxford before returning to Istanbul. In 1955, he retired to Hamburg. Although his positions were often in inorganic chemistry, he carried out research on azo compounds. His name is associated with the Arndt-Eistert synthesis of diazoketones.

Fritz hoffmann (1868-1920) Born in Basle. Married Adèle La Roche in 1895 and joined her surname to his. Founded the Swiss pharmaceutical company Hoffmann-La Roche in October 1896.

Fritz hofmann (1866-1956) Born in Kölleda, Thuringia (now in Saxony-Anhalt), apprenticed to a pharmacist in Göttingen (other sources say Berlin) and studied chemistry in Berlin and Rostock (under August Michaelis). After working as an assistant to Ludwig Claisen in Aachen, he joined Bayer. Developed synthetic rubbers, notably 'methyl rubber', between 1907 and 1918. Director of the Coal Research Institute at Breslau (1918-1934). Died in Hanover.

Frontal affinity chromatography Method for screening mixtures of compounds for affinity against an immobilized target.

Frontal wedging The lifting of air resulting when cool air acts as a barrier over which warmer, lighter air will rise.

Frontier orbitals The highest-energy occupied molecular orbital (HOMO) (filled or partly filled) and lowest-energy unoccupied molecular orbital (LUMO) (completely or partly vacant) of a molecular entity. Examination of the mixing of frontier molecular orbitals of reacting molecular entities affords an approach to the interpretation of reaction behaviour; this constitutes a simplified perturbation molecular orbital theory of chemical behaviour.

Froude number Froude number (Fr) is a dimensionless quantity used in fluid mechanics, defined by

$$Fr = \frac{v}{(l.g)^{\frac{1}{2}}}$$

where v is velocity, l is length, and g is acceleration due to gravity.

Fuel cells Fuel cells are electrochemical cells in which the chemical energy of a fuel is converted directly into electrical energy.

Fuel oil Fuel oil is a higher boiling fraction from the distillation of crude oil.

Fugacity Fugacity (f) is a thermodynamic function used in place of partial pressure in reactions involving real gases and mixtures. For a component of a mixture, it is defined by

$d\mu = RT(\ln f)$

where μ is the chemical potential.

The fugacity of a gas is equal to the pressure if the gas is ideal. The fugacity of a liquid or solid is the fugacity of the vapour with which it is in equilibrium. The ratio of the fugacity to the fugacity in some standard state is the activity.

Fully combinatorial Containing, or designed to contain, all possible combinations of building blocks. Pool/split libraries are generally fully combinatorial while parallel synthesis libraries may not be.

Fundamental particles The fundamental particles are the subatomic particles in the atomic particles, from which the atoms of all elements are made. The protons and neutrons exists in the nucleus, which is the central core of the atom, and the electrons travels in the orbitals about the nucleus of the atom. These atomic particles (i.e. protons, and neutrons) are themselves composed of the sub-atomic particles (i.e. the Quarks and Leptons), which are also called the fundamental particles.

Fungicide A chemical used to kill fungi. See pesticide.

Fur The iron uptake regulating protein present in prokaryotes, which binds simultaneously Fe and DNA thereby preventing the biosynthesis of enzymes for the production of scavenger chelates (siderophores).

G

Gaia hypothesis An hypothesis proposed during the early 1970s by James Lovelock which states that all living organisms have the ability to affect their surroundings such as the atmosphere, lithosphere, and climate to maximize its biological success. The hypothesis connects the evolution and survival of a species to the evolution and conditions of its environment.

Gal Gal is a non-SI unit of acceleration, equal to 1 cm/s^2. Also called galileo.

Gallon Gallon (US) is a unit of volume equal to 3.785412 L.

Gallon (UK, Imperial) is a unit of volume equal to 4.546090 L.

Gamma radiation Gamma radiation is electromagnetic radiation of extremely short wavelength. Gamma radiation ranges in energy from about 10^{-15} to 10^{-10} J (10 keV to 10 MeV) (wavelength less than about 1 pm). Gamma rays are emitted by excited atomic nuclei during the process of passing to a lower excitation state.

Gamma rays are extremely penetrating and are absorbed by dense materials like lead and uranium. Exposure to gamma radiation may be lethal.

Gamma ray Extremely short wavelength and intensely high-energy electromagnetic radiation. Gamma rays originate from an atom's nucleus and normally accompany alpha and beta particles as part of the emissions of the radioactive decay of an atom and always accompany nuclear fission. Because gamma rays are energy and not matter, they are very penetrating and can cause damage to animal and plant tissues. Gamma rays are absorbed by extremely dense materials like lead (Pb) and depleted uranium (U).

Gamma(γ) band Identical to Soret band.

Gas Gases; vapor.

Matter in a form that has low density, is easily compressible and expandable, and expands spontaneously when placed in a larger container. Molecules in a gas move freely and are relatively far apart. "Vapor" often refers to a gas made of a substance that is usually encountered as a liquid or solid; for example, gaseous H_2O is called "water vapor".

Gas-phase acidity The negative of the Gibbs energy (G_r^o) change for the reaction

$A\text{-}H\ A^- + H^+$

in the gas phase.

Gas-phase basicity The negative of the Gibbs energy (G_r^o) change associated with the reaction

$B + H^+\ BH^+$

in the gas phase. Also called absolute or intrinsic basicity.

Gauche conformation A type of staggered conformation in which two big groups are next to each other

Gauche Synonymous with a synclinal alignment of groups attached to adjacent atoms. See also torsion angle.

Gauche effect 1. The stabilisation of the gauche (synclinal) conformation in a two carbon unit bonded vicinally to electronegative elements e.g. 1,2-difluoroethane.

2. The destabilisation of the gauche (synclinal) conformation in a two carbon unit bonded vicinally to large, soft and polarisable elements such as sulfur and bromine.

Gauss Gauss (G) is a non-SI unit of magnetic flux density (B). 1 G = 10^{-4} T.

Gaussian system of units Gaussian system of units is a hybrid system used in electromagnetic theory, which combines features of both the electrostatic cgs subsystem (esu) and electromagnetic cgs subsystem (emu). With three base units, it uses em units in magnetism and es units in electrostatics. This involves using the constant c (the velocity of light in vacuum) to interrelate these sets of units.

Gavage Dose given by intragastric intubation (WHO, 1979).

Gay-lussac's law See Charles' law

Geiger counter Instrument that measures radiation output.

Gel Compare with colloid.

A gell is a sol in which the solid particles fuse or entangle to produce a rigid or semirigid mixture. For example, gelatin dissolved in water produces a sol of protein molecules. When the gelatin is cooked, the protein chains entangle and crosslink, forming a gel which is a mesh of solid protein with trapped pockets of liquid inside. Fruit jellies are also gels

Gel phase Description applied to certain solid supports which display properties intermediate between solid and liquid phases, e.g. in the apparent mobility of the support as determined by nuclear magnetic resonance (NMR) spectroscopy.

Geminate pair Pair of molecular entities in close proximity in solution within a solvent cage and resulting from reaction (e.g. bond scission, electron transfer, group transfer) of a precursor that constitutes a single kinetic entity.

Geminate recombination The reaction with each other of two transient molecular entities produced from a common precursor in solution. If reaction occurs before any separation by diffusion has occurred, this is termed "primary geminate recombination". If the mutually reactive entities have been separated, and come together by diffusion, this is termed "secondary geminate recombination".

Gene Structurally, a basic unit of hereditary material; an ordered sequenceof nucleotide bases that encodes one polypeptide chain (via mRNA). The gene includes, however, regions preceding and followingthe coding region (leader and trailer) as well as (in eukaryotes) intervening sequences (introns) between individual coding segments (exons). Functionally, the gene is defined by the cis-trans test that determines whether independent mutations of the same phenotype occur within a single gene or in several genes involved in the same function.

General acid catalysis The catalysis of a chemical reaction by a series of Brønsted acids (which may include the solvated hydrogen ion) so that the rate of the catalysed part of the reaction is given by k_{HA}(HA) multiplied by some function of substrate concentrations. (The acids HA are unchanged by the overall reaction.) General catalysis by acids can be experimentally distinguished from specific catalysis by hydrogen cations (hydrons) by observation of the rate of reaction as a function of buffer concentration. See also catalysis, catalytic coefficient, intramolecular catalysis, pseudo-catalysis, specific catalysis.

General base catalysis The catalysis of a chemical reaction by a series of Brønsted bases (which may include the lyate ion) so that the rate of the catalysed part of the reaction is given by k_B(B) multiplied by some function of substrate concentration.

General circulation models Three dimensional computer simulations of climate and weather which, through calculations, are used to investigate gravity wave fields. The models may help to predict atmospheric chemistry distribution and meteorological trends for the future.

Generic structure The general structural formula of a library, consisting of the scaffold plus an indication of the position of attachment of the various residues.

Genetic algorithm Method for library design by evaluating the fit of a parent library to some desired property (e.g. the level of activity in a biological assay or the computationally determined diversity of the compound set) as measured by a fitness function. The design of more optimal daughter libraries is then carried out by a heuristic process with similarities to genetic selection in that it employs replication, mutation, deletions, etc. over a number of generations.

Geochemistry Geological chemistry.

The study of materials and chemical reactions in rocks, minerals, magma, seawater, and soil.

Geologic time scale A time scale in which the earth and its atmosphere are dated. It is broken into two division; Relative Time: Determines the age relative to other surrounding objects; Absolute Time: Obtained with radiometric dating of the object.

Geometric isomer Geometric isomers are molecules that have the same molecular formula and bond connections, but distinctly different shapes.

Geometric isomerism (usage strongly discouraged) Obsolete synonym for cis-trans isomerism.

Gertrude maude walsh (1886-1954) Born in Over near Manchester, the daughter of a coal salesman, and studied chemistry at Manchester University, where she worked under Chaim Weizmann (1874-1952). She then joined the research group of Robert Robinson (1886-1975) and married him in 1912. She continued her research alongside her husband and did some important work on plant pigments and penicillin. Her name is associated (with her husband) with the Piloty-Robinson reaction (1910, 1918).

G-factor See electron paramagnetic resonance spectroscopy.

Gibbs phase rule Gibbs phase rule is the relationship used to determine the number of state variables, usually chosen from among temperature, pressure, and species composition in each phase, which must be specified to fix the thermodynamic state of a system in equilibrium:

$$F = C - P + 2$$

where C is the number of components in a mixture, P is the number of phases, and F is the degrees of freedom, i.e., the number of intensive variables that can be changed independently without affecting the number of phases.

Gigayears Time measured in billions of years; a common measure of geologic time.

Glacier A slow-moving mass of ice formed in higher latitudes and elevations. When snowfall is greater than melting and the increasing amounts of snow become compacted and pressurized it forms firn (see above) and ultimately a glacier. As the glacier moves, it carries rocks and soil, and can form u-shaped valleys over geologic time. Glaciers can move at variable rates.

Glass transition temperature Glass transition temperature is the temperature at which an amorphous polymer is transformed, in a reversible way, from a viscous or rubbery condition to a hard and relatively brittle one.

Glauber's salt Glauber's salt is sodium sulfate decahydrate ($Na_2SO_4 \times 10H_2O$). Loses water of hydration at 100 °C. Energy storage capacity is more than seven times that of water.

Global climate model A computer program that is used by atmospheric scientists to study and predict world-wide trends in weather patterns and the forces and physical laws that affect climate. The formulae are complicated and require skilled mathematicians, atmospheric scientists, and powerful computers for execution. Five regimes that influence climate are often taken into consideration: atmosphere, hydrosphere, biosphere, cryosphere, and pedosphere. Atmospheric and environmental physical laws as well as principles of thermodynamics and hydrodynamics are all used as part of the extensive mathematical parameters for the program. The results of these models are used as reference for other models, policy decisions, and by meteorologists.

Global warming The overall increase of atmospheric temperature due to a buildup of greenhouse gases.

Glutamate Ionic salts of glutamic acid used as flavor enhancers in many foods. Glutamate is usually manufactured by acid hydrolysis of vegetable proteins. Besides being a basic building block of proteins, glutamate functions as a neurotransmitter that helps neurons grow new connections; as such, glutamate plays an important role in learning and memory. At high concentrations, glutamate can function as an excitotoxin.

Glutamic acid Glu; -aminoglutaric acid. Compare with glutamate and glutamine.

A naturally occurring amino acid that has a carboxylic acid group on its side chain. In proteins, glutamic acid residues can occur with the side-chain carboxylic acid group converted to an amide; the residue is then called glutamine.

Glutamate receptors Glutamate receptors are protein molecules that helps gate the flow of ions across a nerve cell's membrane. They play a role in the formation of new connections between nerve cells (and so, in learning and memory). The receptors are normally activated by aspartate and glutamate. In amnesic shellfish poisoning, domoic acid acts as an excitotoxin that very strongly activates some of these receptors, preventing their proper functioning.

Glutamine Gln. The amide of the amino acid glutamic acid. Glutamic acid often occurs as glutamine when built into proteins.

Glycerides Glycerides are esters of glycerol (propane-1,2,3-triol) with fatty acids, widely distributed in nature. They are by long-established custom subdivided into triglycerides, 1,2- or 1,3-diglycerides, and 1- or 2-monoglycerides, according to the number and positions of acyl groups.

Glycerol ($HOCH_2CH(OH)CH_2OH$) Glycerol is a small molecule with three alcohol groups. It is a basic building block of fats and oils.

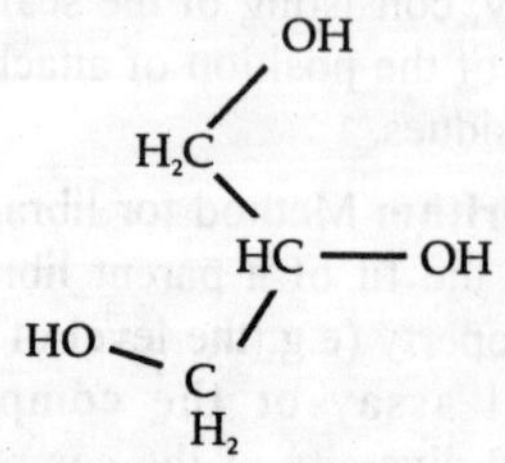

Glycine (NH_2CH_2COOH) Gly; aminoacetic acid.

A naturally occurring aliphatic amino acid, found in large quantities in gelatin.

Glycolic aldehyde ($C_2H_4O_2$) Also called hydroxyacetaldehyde, is formed during the oxidation of ethane and isoprene, and also by the emission of biomass fires. It is the simplest form of sugar. Glycol aldehydes have also been found in trace amounts in dust clouds near the Milky Way of outer space.

Gold drugs Gold coordinationcompounds used in the treatment of rheumatoid arthritis, examples being auranofin, (tetraacetylthioglucosato-S)(triethylphosphane)gold(I), and myocrysin, disodium thiomalonatogold(I).

Gondwanaland A hypothetical supercontinent comprising approximately the present continents of the southern hemisphere.

Good agricultural practice in the use of pesticides Good agricultural practice in the use of pesticide is the officially recommended or authorized usage of pesticides under practical conditions at any stage of production, storage, transport, distribution, and processing of food and other agricultural commodities, bearing in mind the variations in requirements within and between regions and taking into account the minimum quantities necessary to achieve adequate control, the pesticide being applied in such a manner as to leave residues that are the smallest amounts practicable and that are toxicologically acceptable (WHO, 1976).

GPCR G - protein coupled receptors form a large super - family of proteins composed of three major classes and more than 30 subfamilies. They are integral membrane proteins characterized by seven membrane- spanning (transmembrane; TM) regions. They are involved with signal transduction across cell membranes. Many medically and pharmacologically important proteins are included in this super-family: e.g., Acetylcholine receptors, Dopamine receptors, and Opioid receptors.

Graded effect An effect that can usually be measured on a graded scale of intensity or severity and its magnitude related directly to the dose (WHO, 1978a).

Graham's law The rate of diffusion of a gas is inversely proportional to the square root of its molecular weight.

Grain Grain (gr) is a non-SI unit of mass, equal to 64.79891 mg.

Gram A metric unit of mass, equal to 1/1000 of a kilogram. Kilograms are the base SI units for mass, not grams.

Graphite An amorphous form of carbon, made of carbon atoms bound hexagonally in sheets (like chickenwire).

Grashof number Grashof number (Gr) is a dimensionless quantity used in fluid mechanics, defined by

$$Gr = l^3 g\alpha\Delta T\rho^2/\eta^2$$

where T is temperature, ρ is density, l is length, η is viscosity, α is cubic expansion coefficient, and g is acceleration of gravity.

Gravimetry Gravimetry is the quantitative measurement of an analyte by weighing a pure, solid form of the analyte. Since gravimetric analysis is an absolute measurement, it is the principal method for analyzing and preparing primary standards.

A typical experimental procedure to determine an unknown concentration of an analyte in solution is as follows:

- quantitatively precipitate the analyte from solution
- collect the precipitate by filtering and wash it to remove impurities
- dry the solid in an oven to remove solvent
- weigh the solid on an analytical balance
- calculate the analyte concentration in the original solution based on the weight of the precipitate.

Gravitational constant Gravitational constant (G) is the universal constant in the equation for the gravitational force between two particles

$$F = Gm_1m_2/r^2$$

where r is the distance between the particles and m_1 and m_2 are their masses.

Gravity A natural force exerted on a body toward the earth's surface or between two bodies. Gravity has a strong effect on how matter interacts. Gravitational attraction depends on the masses of the bodies involved and the distance between them. Centrifugal force caused by the earth's rotation on its axis tends to lessen the force of gravity.

Gray Gray (Gy) is the SI derived unit of absorbed dose of radiation. The gray is the absorbed dose when the energy per unit mass imparted to matter by ionizing radiation is one joule per kilogram (Gy = J/kg). The unit is named after the British scientist Louis Harold Gray (1905-1965).

Greenhouse effect The phenomenon in which outgoing infrared radiation that would normally exit from a planet's atmosphere but instead, is trapped or reflected because of the presence of the atmosphere and its components is called the greenhouse effect. It has been calculated that this effect is necessary to maintain the earth's climate and surface temperature and, more importantly, the liquid state of water in the majority of the earth's biosphere; however, the best scientific estimates to date suggest that increasing amounts of greenhouse gases are resulting in higher temperatures worldwide. This could result in melting of icecaps that would raise the sea level and cause devastating floods in coastal areas, more extremes in rainfall and intensity, and the distribution of species in the biosphere.

Greenhouse gases Those atmospheric components that absorb strongly in the infrared region of the spectrum. Infrared radiation is reflected and emitted by the earth's surface as heat and causes a fairly large warming effect when trapped by these gases in the atmosphere. In order of abundance and importance as greenhouse gases are water vapor, carbon dioxide, ozone, nitrous oxide, methane, and chlorofluorocarbons (CFCs). Absorption by water vapor, the most common greenhouse gas, explains why many humid or cloudy days feel much hotter than dry, clear days of the same air temperature. Because of these gases, only about 5% of the radiation escapes from the atmosphere while more than 90% is radiated back to the surface of the earth.

Greenhouse warming See global warming

Grignard reagents Grignard reagents are organomagnesium halides, RMgX, having a carbon- magnesium bond (or their equilibrium mixtures in solution with $R_2Mg + MgX_2$).

Gross error Compare with systematic error, random error and mistake.

Gross errors are undetected mistakes that cause a measurement to be very much farther from the mean measurement than other measurements.

Ground state Compare with excited state.

The lowest energy state for an atom or molecule. When an atom is in its ground state, its electrons fill the lowest energy orbitals completely before they begin to occupy higher energy orbitals, and they fill subshells in accordance with Hund's rule (usually!)

Group The vertical columns (major classes or divisions) into which elements are arranged in the periodic table of elements. There are three common numbering systems for these groups:

The new IUPAC system, which numbers each column with Arabic numbers from 1 (one) through 18 (eighteen). To reduce confusion caused by the other two systems, this is the system that is used in articles on this web site.

The old IUPAC system, which labeled columns with Roman numerals followed by either the letter 'A' or 'B'. Columns were numbed such that columns one through seven were numbered 'IA' through 'VIIA', columns 8 through 10 were labeled 'VIIIA', columns 11 through 17 were numbered 'IB' through 'VIIB' and column 18 was numbered 'VIII'.

The CAS system, which also used Roman numerals followed by an 'A' or 'B'. This method, however, labeled columns 1 and 2 as 'IA' and 'IIA', columns 3 through 7 as 'IIIB' through 'VIB', column 8 through 10 as 'VIII', columns 11 and 12 as 'IB' and 'IIB' and columns 13 through 18 as 'IIIA' through 'VIIIA'.

Because of the confusion the old IUPAC and the CAS system created, the IUPAC adopted their new system.

Elements are arranged in these groups according to whose proprieties are similar. All elements in Group 1 for instance are alkali metals. They have only one electron in the outer shell (valence electron) and as a result are highly reactive. Elements in Group 17 are the halogens. They all have seven electrons in the outer orbital (two in level *s* and five in level *p)*. They are also very reactive because they have seven electrons in the outer shell and will readily accept an electron in order to reach the ion configuration with the ideal number of eight electrons in the outer shell. Elements Group 18 have a complete outer shell with eight electrons. These noble gases are highly stable and do not react to form compounds under normal conditions.

Guanylate cyclase An enzyme catalyzing the conversion of guanosine 5'-triphosphate to cyclicguanosine 3',5'-monophosphate, which is involved in cellular regulation processes. One member of this class is a heme-containing enzyme involved in processes regulated by nitrogen monoxide.

Guest An organic or inorganic ion or molecule that occupies a cavity, cleft or pocket within the molecular structure of a host molecular entity and forms a complex with it or that is trapped in a cavity within the crystal structure of a host.

Guideline level A guideline level is the maximum concentration of a pesticide residue that might occur after the officially recommended or authorized use of a pesticide for which no acceptable daily intake or temporary acceptable daily intake is established and that need not to be exceeded if good practices are followed. It is expressed in milligrams for the residue per kilogram of food (WHO, 1976).

Guides to air quality Sets of concentrations and exposure times that are associated with specific effects of varying degrees of air pollution on man, animals, vegetation, and on the environment in general (WHO, 1979).

Guides to environmental quality Sets of levels and exposure times that are associated with the specific effects of varying levels of environmental factors on man, animals, vegetation, and the environment in general (WHO, 1979).

Guinea pig maximization test This is a skin test for screening possible contact allergens. It is concidered to be a useful model for predicting moderate and strong sensitizers in humans.

Gulf stream The Gulf Stream is a powerful current in the Atlantic Ocean. Its movement is related to density distribution system of waters and currents. Beginning in the Gulf of Mexico, it flows along the east coast of North America and across the Atlantic to Europe. The Gulf Stream carries warm water from the tropics and warms the climate of the land near where it passes.

H

Haber-weiss reaction The Haber-Weiss cycle consists of the following two reactions : H_2O_2 + OH $\rightarrow$ H_2O + O_2^- + H^+ and H_2O_2 + O_2^- $\rightarrow$ O_2 + OH^- + OH·

The second reaction achieved notoriety as a possible source of hydroxyl radicals. However, it has a negligible rate constant. It is believed that iron(III) complexes can catalyze this reaction: first Fe(III) is reduced by superoxide, followed by oxidation by dihydrogenperoxide.

Hadley circulation, Proposed by George Hadley in 1735, it is the tropical convection system that describes the atmospheric thermal circulation of air from the equator (0*f* latitude) poleward to 30*f* both north and south of the equator. When saturated air at the intertropical convergence zone (ITCZ), located approximately at the equator, is heated, convergence and convection causes it to rise into the upper atmosphere, creating an area of low pressure at the earthÃs surface. As the pocket of saturated air at the ITCZ rises, it cools, releasing large amounts of precipitation. Rather than remaining in the tropics, the air then flows horizontally towards the higher latitudes. At approximately 30*f* both north and south of the equator, the pocket of air in the upper atmosphere has sufficiently mixed and cooled enough to return to the surface. To complete the circle, the area of low pressure at the equator pulls air in from the higher latitudes.

Half cell A half-cell is an electrode in contact with a solution of ions, and it forms one half of a cell. Gas half cells have a gold or platinum plate in a solution with gas bubbled over the metal plate. The hydrogen half-cell is used as a reference half cell and is assigned the electrode potential of zero. Common half cells include the zinc and copper half cells.

Half-chair The conformation of a six-membered ring structure in which four contiguous atoms are in a plane and the other two atoms lie on opposite sides of the plane is called a half-chair.

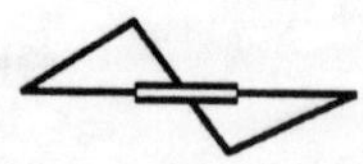

Half-life For a given reaction the half life $t_{1/2}$ of a reactant is the time required for its concentration to reach a value that is the arithmetic mean of its initial and final (equilibrium) value.

In nuclear chemistry, (radioactive) half life is defined, for a simple radioactive decay process, as the time required for the activity to decrease to half its value by that process.

$N = N_0/2$

Halides Halides are compounds which contain one of the halogen elements in chemical combination with another element. The halides of typical metals are ionic. Metals form halides in which the chemical bonding is largely covalent.

The halide salts are the fluorides, chlorides, bromides or iodides.

Halocarbon Halocarbon is a compound containing no elements other than carbon, one or more halogens, and sometimes hydrogen. The simplest are compounds such as tetrachloromethane (CCl_4), tetrabromomethane (CBr_4), etc. The lower members of the various homologous series are used as refrigerants, propellant gases, fireextinguishing agents, and blowing agents for urethane foams. When polymerized, they yield plastics characterized by extreme chemical resistance, high electrical resistivity, and good heat resistance.

Halochromism Halochromism means the colour change which occurs on addition of acid (or base, or a salt) to a solution of a compound. A chemical reaction (e.g. ion formation) transforms a colourless compound into a coloured one.

Halocline in The oceans, a well-defined vertical gradient of salinity.

Haloperoxidase A peroxidase which catalyzes the oxidative transformation of halides to XO^- (X being Cl, Br or I), or organic halogen compounds. Most are hemeproteins, but some bromoperoxidases from algae are vanadium-containing enzymes .

Hammett acidity function See acidity function.

Hammett equation (or hammett relation) The equation in the form

$lg(k/k_o)$ = or

$lg(K/K_o)$ = applied to the influence of meta- or para-substituents X on the reactivity of the functional group Y in the benzene derivative m- or p-XC_6H_4Y. k or K is the rate or equilibrium constant, respectively, for the given reaction of m- or p-XC_6H_4Y; k_o or K_o refers to the reaction of C_6H_5Y, i.e. X = H; is the substituent constant characteristic of m- or p-X; is the reaction constant characteristic of the given reaction of Y. The equation is often encountered in a form with lg k_o or lg K_o written as a separate term on the right hand side, e.g.

lg k = + lg k_o

or

lg K = + lg K_o

It then signifies the intercept corresponding to X = H in a regression of lg k or lg K on .

Hammond principle (or Hammond postulate) The hypothesis that, when a transition state leading to an unstable reaction intermediate (or product) has nearly the same energy as that intermediate, the two are interconverted with only a small reorganization of molecular structure. Essentially the same idea is sometimes referred to as "Leffler's assumption", namely, that the transition state bears the greater resemblance to the less stable species (reactant or reaction intermediate/product). Many text books and physical organic chemists, however, express the idea in Leffler's form, but attribute it to Hammond.

As a corollary, it follows that a factor stabilising a reaction intermediate will also stabilize the transition state leading to that intermediate.

Handedness This term has been used in two ways, either chirality or chirality sense.

Hans fischer (1881-1945) Born in Frankfurt-Höchst. He took his Ph.D. under Zincke at Marburg, and studied medicine at the University of Munich. After undertaking research with Emil Fischer and a spell at the University of Innsbruck, he became Professor of Organic Chemistry in Munich in 1921. An expert on porphyrins, he won the Nobel Prize for the synthesis of hæmogoblin in 1930. Fischer committed suicide at the end of the Second World War. His father Eugen Fischer (1854-1917) was an industrial chemist and a director of the dye firm, Kalle.

Hans meerwein (1879-1965) Born in Hamburg, the son of a architect, he studied chemistry at the Fresenius chemistry school in Wiesbaden before going to Bonn. After taking his doctorate at the polytechnic in Berlin-Charlottenburg, he returned to Bonn. He later held Chairs at Königsberg and Marburg. Meerwein's main interest was carbonium ion chemistry and carbonium ion polymerisation. The chemical giant I.G. Farben was very interested in his research and funded his work generously. In

1914, during his study of the pinacol-pinacolone rearrangement, he generalised Egor Vagner's mechanism for the conversion of ?-pinene into bornyl chloride (Wagner-Meerwein rearrangement). Meerwein's name is also associated with the Meerwein methylation, the Meerwein arylation (1939) and Meerwein-Ponndorf-Verley reduction (1925-6).

Hans stobbe (1860-1938) Born in Tiegenhof, near Danzig, and was educated at Heidelburg, Munich, Strassburg and Leipzig, finally switching to chemistry under the influence of Johannes Wislicenus. He became Professor of Organic Chemistry at Leipzig in 1904, after the death of Wislicenus. Stobbe was not only one of the founders of physical organic chemistry, but also a pioneer of polymer chemistry, and he investigated the polymerisation of styrene as early as 1909. Between 1928 and 1938, Stobbe was an editor of "Poggendorff" (properly speaking the Biographisch-Literarisches Handwörterbuch). His name is associated with the Stobbe condensation of diethyl succinate (1893).

Hansch constant A measure of the capability of a solute for hydrophobic (lipophilic) interaction based on the partition coefficient P for distribution of the solute between octan-1-ol and water. The most general way of applying P in correlation analysis, QSAR, etc. is as log P, but the behaviour of substituted benzene derivatives may be quantified by a substituent constant scale, , which is defined in a way analogous to the Hammett scale. There are various scales, depending on the substrate series used as reference..

Hapten A molecule (usually a small organic molecule) which can be bound to an antigenic determinant/epitope. Usually they are too small to give a response of their own. They become antigenic if they are coupled to a suitable macromolecule, such as a protein.

Hapto The hapto symbol, η (Greek eta), with numerical superscript, provides a topological description for the bonding of hydrocarbons or other π-electron systems to metals, by indicating the connectivity between the ligand and the central atom. The symbol is prefixed to the ligand name, or to that portion of the ligand name most appropriate. The right superscript numerical index indicates the number of coordinating atoms in the ligand which bind to the metal. Examples :

$(PtCl_2(C_2H_4)(NH_3))$ amminedichloro(η^2-ethene)platinum

$(Fe(\eta^5\text{-}C_5H_5)_2)$ bis(η^5-cyclopentadienyl)iron (ferrocene)

Hard acid A Lewis acid with an acceptor centre of low polarizability. It preferentially associates with hard bases rather than with soft bases, in a qualitative sense (sometimes called "HSAB rule"). Conversely a soft acid possesses an acceptor centre of high polarizability and exhibits the reverse preferencefor a partner for coordination.

Hard base A Lewis base with a donor centre of low polarizability; the converse applies to soft bases. See also hard acid.

Hard drug A nonmetabolizable compound, characterized either by high lipid solubility and accumulation in adipose tissues and organelles, or by high water solubility. In the lay press the term refers to a powerful drug of abuse such as cocaine or heroin.

Hardness Hardness is the resistance of a material to deformation of an indenter of specific size and shape under known load. This definition applies to all types of hardness scales except Mohs scale, which is a based on the concept of scratch hardness and is used chiefly for minerals. The most generally used hardness scales are Brinell (for cast iron), Rockwell (for sheet metal and heat-treated steel), Knoop (for metals).

Hardness of water The hardness of water results from dissolved salts of calcium and magnesium, which are introduced into the water in the environment when it passes through limestone areas. The nature of the materials dissolved will depend on the geology of the region. However, calcium and magnesium salts are the source of the hardness of water. Total hardness and calcium hardness can be distinguished chemically.

The hardness may be classified as temporary hardness which is removed on boiling, or permanent hardness which is due to dissolved salts.

Harmonic oscillator model for a diatomic molecule We can model the bond in a molecule as a spring connecting two atoms and use the harmonic oscillator expression to describe the potential energy for the periodic vibration of the atoms. The potential energy, V(x), of a particle moving in one dimension is given by:

$$V(x) = \frac{1}{2}kx^2$$

where k is the force constant as above and the constant of integration is zero. We can make this expression more useful by changing x to $R-R_e$, where R is the internuclear distance (the distance between atoms) and R_e is the equilibrium internuclear distance (the bond length):

$$V(R)=\frac{1}{2}k(R-R_e)^2$$

Hartley-huggins band of ozone absorption This spectral band in the UV is responsible for the absorption and filtering of solar ultraviolet radiation. Photons with a wavelength longer than 210 nm are weakly absorbed by oxygen in the atmosphere. Ozone then becomes the major UV absorber, with a band of 210-380 nm. Compare to ozone's Chappuis band in the visible wavelength region.

Hartree (E_h) Compare with atomic unit.

The atomic unit of energy, equal to 4.359 743 81 × 10^{-18} J ± 0.000 000 34 × 10^{-18} J .

Hazard A source of danger: a qualitative term expressing the potential that an environmental agent can harm health (WHO, 1988).

Hazard identification The identification of the substance of concern, its adverse effects, target populations, and conditions of exposure (WHO, 1988).

Hazardous material (*HazMat*) Any material or substance, which even in normal use, poses a risk to health, safety, property or the environment. Hazardous materials are broken down into nine primary classes by the United Nations, the USDOT, IATA and other regulatory bodies for the purposes of the placarding and identification the hazards of shipments.

Haze An atmospheric aerosol of sufficient concentration to be visible. The particles are so small that they cannot be seen individually but are still effective in visual range restriction. The Blue Ridge Mountains of the Eastern United States have a haze caused by the natural emissions of hydrocarbons which are oxidized and form aerosols.

HDPE (*high density polyethylene*) The structure is similar to PVC except that there is no chlorine (Cl) associated with the molecule. The tight structure makes it very dense.

Health A state of complete physical, mental and social well-being, and not merely the absence of disease or infirmity (WHO, 1978b).

Healthy worker effect A phenomenon observed initially in studies of occupational diseases: workers usually exhibit lower overall death rates than the general population, due to the fact that the severely ill and disabled are ordinarily excluded from employment. Death rates in the general population may be inappropriate for comparison if this effect is not taken into account.

Heat Compare with work, energy, enthalpy, and temperature.

Heat is a transfer of energy that occurs when objects with different temperatures are placed into contact. Heat is a process, not a property of a material.

Heat capacity Heat capacity is defined in general as dQ/dT, where dQ is the amount of heat that must be added to a system to increase its temperature by a small amount dT. The heat capacity at constant pressure is $C_p = (H/ T)_p$; that at constant volume is $C_V = (E/ T)_V$, where H is enthalpy, E is internal energy, p is pressure,

V is volume, and T is temperature. An upper case C normally indicates the molar heat capacity, while a lower case c is used for the specific (per unit mass) heat capacity.

Heat capacity of activation, Cp^o (SI unit: J mol-1 K-1)

A quantity related to the temperature coefficient of H (enthalpy of activation) and S (entropy of activation) according to the equations:

$C_p = (H/T)_p = T(S/T)_p$

If the rate constant is expressible in the form ln k = a/T + b + c ln T + dT, then

Cp = (c - 1) R + 2dRT

Heat flux The amount of heat that is transferred across a surface of unit area in a unit of time.

Heat of combustion The heat of combustion of a substance is the heat evolved when one gram-molecule of the substance combines with oxygen. When a substance burns in air, combustion is the process of reacting a substance with oxygen. In everyday life, we make use of the combustion of coal or gas in air as a source of heat. Chemical reactions accompanied by an evolution of heat are called exothermic reactions.

Heat of formation The heat of formation of a substance is the heat evolved when one gram-molecule of the substance is produced from its elements.

Heat of fusion The heat required to convert a solid into a liquid with no temperature change. This is also called the latent heat of fusion or melting.

Heat of neutralization The heat of neutralisation is the quantity of heat liberated when one mole of a strong acid is neutralised by one mole of strong base.

The heat Of neutralisation is independent of the nature of the acid or base. For this reason, it is assumed to be the heat released due to the recombination of the hydrogen ion from the acid and hydroxide ion from the base to form water.

H(+) + OH(-) $\Rightarrow$ H2O

The heat of neutralisation is the amount of heat evolved when one gram equivalent of an acid is neutralised by one gram equivalent of a base to give one gram equivalent of a neutral salt. Thus, in dilute solution, an acid or base is considered to be completely ionized, and the neutralisarion reaction is

Na(+) + OH(-) + H(+) + Cl(-) Na(+) + Cl(-) + H2O

The value of H should always be - 57.26 kj per mole for neutralisation of any strong acid and strong base.

Heat of reaction The heat of reaction is a difference between the intrinsic energy in the products of a chemical reaction and the intrinsic energy in the reactants, and it is either adsorbed or released during the course of the chemical reaction.

Heat of solution The heat of solution is the amount of heat evolved or absorbed when one gram molecule of a compound is dissolved in a large quantity of water.

Heat of vaporization The heat required to convert a substance from the liquid to the gaseous state with no temperature change. This is also called the latent heat of vaporization.

Heavy water (D2O) Water that contains 2H, rather than 1H. Heavy water is about 11% denser than ordinary water.

Hectare Hectare (ha) is a unit of area equal to 10^4 m^2. The unit is still used in agriculture.

Heinrich August Wilhelm; ferdinard fischer (1842-1916). Born in Rödermühle in the Harz and educated at Göttingen and Berlin, taking his Ph.D. at Jena in 1869. He was a pioneering chemical technologist, specialising in fuel technology, and from 1894 taught at Göttingen University. Fischer founded the Gesellschaft für angewandte Chemie in 1886, which became the Verein deutscher Chemiker ten years later. His brother Hermann Fischer (1840-1915) was a well-known engineer.

Heinrich caro (1834-1910) Famous dye chemist and colourist, born in Posen. Studied chemistry and dyeing in Berlin at the Royal Trades Institute and the university. Worked at Troost's calico printing works in Mülheim, Roberts, Dale in Manchester and finally BASF in Ludwgishafen. Associated with Caro's acid (persulfuric acid) and Caro's reagent.

Heisenberg uncertainty principle The Heisenberg Uncertainty Principle describes the uncertainty with which the velocity and position of an electron in an orbital can be known. Only the wave function for the electron is known with certainty, and the electron density at any point is the square of the wave function at that point. Thus, the probability of finding an electron at any point is proportional to the electron density at that point.

Helicity The chirality of a helical, propeller or screw-shaped molecular entity. A right-handed helix is described as P (or plus), a left-handed one as M (or minus).

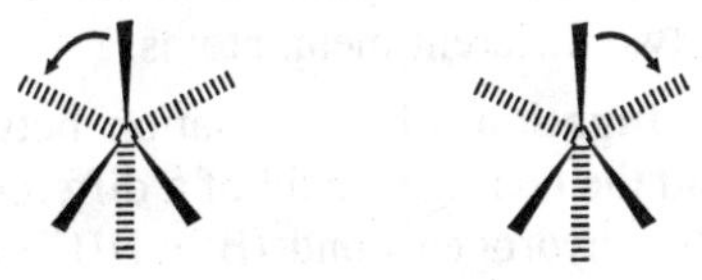

M or A P or A

(M)-hexahelicene

The application of this system to the description of conformations considers the torsion angle between two specified (fiducial) groups that are attached to the atoms linked by that bond. The sign of the smaller torsion angle between the fiducial groups defines the chirality sense of the helix.

Helium He.Element 2, atomic weight 4.0026. A colorless, odorless, inert gas, first discovered in the emission spectrum of the sun in 1868.

Helix A particular rigid left- or right-handed arrangement of a polymeric chain, characterised by the number of strands,the number (n) of units per turn and its pitch (p), the distance the helix rises along its axis per full turn. Examples of single-stranded helices are the protein helices: α -helix:n = 3.6, p = 540 pm; 3_{10}-helix: n = 3.0, p = 600 pm; -helix: n = 4.4, p = 520 pm.

Hellmut johannes fischer (1902-1976) Took his Ph.D. under Stock at Berlin in 1926. He worked for Siemens & Halske at the huge Siemens complex in Berlin. He carried out research on inorganic analysis and microanalysis, particularly the use of diphenyl thiocarbazone (dithizone), first synthesised by Emil Fischer.

Helmholtz free energy (A) Arbeitfunktion.

A thermodynamic property that can be used to predict whether a process will occur spontaneously at constant volume and temperature. Helmholtz free energy A is defined as A = U - TS where U, T and S are the internal energy, temperature, and entropy. Changes in A correspond to changes in free energy for processes occuring at constant temperature and volume. The sign of DeltaA is negative for spontaneous processes and zero for processes at equilibrium.

Heme A near-planar coordination complex obtained from iron and the dianionic form of porphyrin. Derivatives are known with substituents at various positions on the ring named a, b, c, d etc. Heme b, derived from protoporphyrin IX, is the most frequently occurring heme.

Hemerythrin A dioxygen-carrying protein from marine invertebrates, containing an oxo-bridged dinuclear iron center.

Hemileptic See homoleptic.

Hemochromatosis A genetic condition of massive iron overload leading to cirrhosis and/or other tissue damage, attributable to iron.

Hemocyanin A dioxygen-carrying protein (from invertebrates, e.g arthropods and molluscs), containing dinuclear type 3 copper sites.

Hemoglobin A dioxygen-carrying heme protein of red blood cells, generally consisting of two alpha and two beta subunits, each containing one molecule of protoporphyrin IX.

Henderson-hasselbach equation An equation of the form

$$pH = pK_a - lg((HA)/(A^-))$$

for the calculation of the pH of solutions where the ratio $(HA)/(A^-)$ is known.

Henry Henry (H) is the SI derived unit of inductance equal to the inductance of a closed circuit in which an e.m.f. of one volt is produced when the electric current in the circuit varies uniformly at the rate of one ampere per second (H = V'"s/A). The unit is named after the US physicst Joseph Henry (1797-1878).

Henry's law Henry's law constant.

Henry's law predicts that the solubility (C) of a gas or volatile substance in a liquid is proportional to the partial pressure (P) of the substance over the liquid:

P = k C

where k is called the Henry's law constant and is characteristic of the solvent and the solute.

Hepatotoxic The adjective applied to anything which is harmful to the liver.

Heptane Heptane, C7H26, is the seventh member of the alkane series of hydrocarbons. It is a liquid which is obtained from petroleum. Its relative density is 0.684, its melting point is -90.6 degC, and its boiling point is 98.4 degC.

Herbert brereton baker (1862-1935) Born in Blackburn, he studied chemistry under Dixon at Balliol College, Oxford. After teaching at Dulwich College, he went back to Oxford as Dr Lees' Reader at Christ Church. He later became Professor of Chemistry at Imperial College. Baker is most famous for his research on the catalytic effect of water on reactions, for instance distilling phosphorous in rigorously dried oxygen without any reaction taking place..

Herbicide The descriptor applied to a chemical used to kill plants. See pesticide.

Hertz Hertz (Hz) is the SI derived unit, with a special name, for frequency. The hertz is the frequency of a periodic phenomenon of which the period is one second ($Hz = 1\ s^{-1}$). It is named after the German scientist Heinrich Hertza (1857 - 1894).

Hess's law Law of constant heat summation; Hess's law of heat summation.

The heat released or absorbed by a process is the same no matter how many steps the process takes. For example, given a reaction A B, Hess's law says that H for the reaction is the same whether the reaction is written as A C B or as A B. This is the same as writing that H(A B) = H(A C) + H(C B).

Hess's law of heat summation Hess's Law of Heat Summation states that the internal energy of a substance is independent of the process by which it was made.

Thus, if a chemical reaction takes place in stages, the algebraic sum of the amount of heat evolved in each separate stage is equal to the total amount of heat that would be evolved if the reaction took place in one stage.

Heterobimetallic complex A metal complex having two different metal atoms.

Heteroconjugation +1. Association between a base and the conjugate acid of a different base through a hydrogen bond ($B'\cdots HB^+$ or $A'H \cdots A^-$). The term has its origin in the conjugate acid-base pair and is in no way related to conjugation of orbitals. Heteroassociation is a more appropriate term.

+2. Some authors refer to conjugated systems containing a heteroatom, e.g. pyridine, as "heteroconjugated systems". This usage is discouraged since it inappropriately suggests an analogy to homoconjugation 2., and conflicts with the currently accepted definition of that term.

Heterocyclic compounds Heterocyclic compounds are cyclic compounds having as ring members atoms of at least two different elements, e.g., quinoline, 1,2-thiazole.

Heterocyclic Heterocycle; heterocyclic ring.

An organic group or molecule containing rings with at least one noncarbon atom on the ring.

Heterogeneous A mixture or material consisting of more than one substance. The earth's atmosphere is heterogeneous in that it typically consists of 20.7% oxygen, 78% nitrogen, 0.93%

argon and 0.03% CO_2 with other gasses making up the remaining percentage. Heterogeneous is Latin for "different kinds".

Heterogeneous mixture Heterogeneous. Compare with homogeneous mixture, solution, element, and compound.

A sample of matter consisting of more than one pure substance and more than one phase. Blood, protoplasm, milk, chocolate, smoke, and chicken soup are examples of heterogeneous mixtures.

Heteroleptic Transition metal or Main Group compounds having more than one type of ligand. See also homoleptic.

Heterolysis The cleavage of a bond (heterolytic cleavage or heterolytic fission) so that both bonding electrons remain with one of the two fragments between which the bond is broken.

Heterolytic bond-dissociation energy The energy required to break a given bond of some specific compound by heterolysis. For the dissociation of a neutral molecule AB in the gas phase into A^+ and B^- the heterolytic bond-dissociation energy $D(A^+B^-)$ is the sum of the bond dissociation energy, D(A-B), and the adiabatic ionization energy of the radical A· minus the electron affinity of the radical B·.

Heterotopic See stereoheterotopic.

Heterotrophic organisms Organisms that are not able to synthesize cell components from carbon dioxide as sole carbon source. Heterotrophic organisms use preformed oxidizable organic substrates such as glucose as carbon and energy sources, while energy is gained through chemical processes (chemoheterotrophy) or through light sources (photoheterotrophy).

Heterovalent hyperconjugation See hyperconjugation.

Heusler alloys Heusler alloys are alloys of manganese, copper, aluminum, nickel, and sometimes other metals which find important uses as permanent magnets.

Hexafluoroethane (C_2F_6) An odorless halocarbon that meets all requirements of a greenhouse gas: unreactive to OH... and ozone, does not photodissociate from either UV or visible light, is insoluble in water, and absorbs in the infrared spectrum.

Hexanone $C_6H_{12}O$ is a volatile organic compound (VOC) which are a large family of carbon-containing compounds which are emitted into the atmosphere from a variety of industrial processes. 2-hexanone may have the potential to contribute directly to global warming by absorbing infrared radiation from the earth's surface. In general the more complex a VOC, the greater its ability to absorb infrared radiation. 2-hexane may also contribute indirectly to global warming through the change in concentration of ozone, which is a potent greenhouse gas.

High density polyethylene High density polyethylene, HDPE, is an important plastic which is manufactured industrially by the polymerisation of ethylene.

High resolution electron microscopy (HREM) High resolution electron microscopy (HREM) is a variation of TEM that uses phase-contrast microscopy to achieve a resolution of approximately 0.2 nm, which is sufficient to provide atomic-scale resolution. The phase-contrast method uses the phase difference of multiple diffracted beams in addition to the intensity of the diffracted beams. The lattice image that is reconstructed is valid only when the electron beam has a spatial resolution of approximately 0.2 nm and the sample is thin enough that electrons are not scattered multiple times (less than approximately 10 nm). Under these conditions the reconstructed lattice image represents the true structure.

High spin complex High-spin complex.

A metal-ligand complex with the same number of unpaired electrons as the uncomplexed metal ion. When a weak ligand complexes the metal ion, the crystal field splitting is small and the electrons can still occupy all of the d orbitals without pairing.

High-throughput screening Process for rapid assessment of the activity of samples from a combinatorial library or other compound collection, often by running parallel assays in plates of 96 or more wells. A screening rate of

100 000 assays per day has been termed "ultra high-throughput screening" (UHTS).

Hildebrand parameter A parameter measuring the cohesion of a solvent (energy required to create a cavity in the solvent).

Hindered rotation See free rotation, hindered rotation, restricted rotation.

HiPIP Formerly-used abbreviation for High-Potential Iron-sulfur Protein, now classed as a ferredoxin. An electron transfer protein from photosynthetic and other bacteria, containing a (4Fe-4S) cluster which undergoes oxidation-reduction between the $(4Fe\text{-}4S)^{2+}$ and $(4Fe\text{-}4S)^{3+}$ states.

His son boris aleksandrovich arbuzov Was also a Professor of Chemistry at the University of Kazan and worked on petroleum chemistry, an important research area in Kazan as there are many oilfields nearby.

His son hermann otto laurenz fischer (1888-1960) Born in Würzburg, studied under Knorr at Jena and became a well-known carbohydrate chemist. After teaching at Berlin, he went to the University of Basel in 1934, Banting Institute at the University of Toronto in 1937 and the University of California at Berkeley in 1948. His name is associated with the Grosheintz-Fischer-Reissert aldehyde synthesis and the Macdonald-Fischer degradation of aldoses.

His son ludwig eduard heinrich anschütz (1889-1954) Was born in Bonn and studied chemistry at Bonn and Munich before taking his doctorate under K. von Auwers at Marburg. After teaching in Bonn, Berlin and Marburg, he became Professor of Organic Chemistry at the German Technische Hochschule in Brünn, Czechoslovakia. After the Second World War, he was eventually called to a Chair at Würzburg. His main interest was organophosphorus chemistry, especially the reaction of phosphorus chlorides with phenolic compounds (pyrocatechol, salicylic acid).

Hit Library component whose activity exceeds a predefined, statistically relevant threshold.

Hit explosion Process of establishing structure-activity relationships around a hit by preparing new libraries or series of analogues using related building blocks and/or scaffolds to those employed in the preparation of that hit.

Hofmann rule "The principal alkene formed in the decomposition of quaternary ammonium hydroxides that contain different primary alkyl groups is always ethylene, if an ethyl group is present." Originally given in this limited form by A.W. Hofmann, the rule has since been extended and modified as follows: "When two or more alkenes can be produced in a -elimination reaction, the alkene having the smallest number of alkyl groups attached to the double bond carbon atoms will be the predominant product." This orientation described by the Hofmann rule is observed in elimination reactions of quaternary ammonium salts and tertiary sulfonium salts, and in certain other cases.

Hofmann voltmeter A Hofmann Voltmeter is an apparatus for the volumetric analysis of gases produced during electrolysis. For example, the hydrogen and oxygen produced during electrolysis can be collected in separate graduated tubes, and volumes produced compared.

Holoenzyme An enzyme containing its characteristic prosthetic group(s) and/or metal(s).

Holography Holography is a technique for creating a three-dimensional image of an object by recording the interference pattern between a light beam diffracted from the object and a reference beam. The image can be reconstructed from this pattern by a suitable optical system.

HOMO 1. An acronym for highest occupied molecular orbital..

2. A prefix (consisting of lower case letters, homo,) used to indicate a higher homologue of a compound.

Homoaromatic Whereas in an aromatic molecule there is continuous overlap of p-orbitals over a cyclic array of atoms, in a homoaromatic molecule there is a formal discontinuity in this overlap resulting from the presence of a single sp^3 hybridized atom at one or several positions within the ring; p-orbital overlap apparently bridges these sp^3 centres, and features associated with aromaticity are manifest in the properties of the compound. Pronounced homoaromaticity is not normally associated with neutral molecules, but mainly with species bearing an electrical charge, e.g., the "homotropylium" cation, $C_8H_9^+$

In bis, tris, (etc.) homoaromatic species, two, three, (etc.) single sp^3 centres separately interrupt the pi-electron system.

Homoconjugation 1. Association between a base and its conjugate acid through a hydrogen bond ($B \cdots HB^+$ or $AH \cdots A^-$). Homoassociation is a more appropriate term for this phenomenon.

2. The orbital overlap of two pi systems separated by a non-conjugating group, such as CH_2.

Homogeneous A substance or material that contains only one kind of compound or one element. Homogeneous is Latin for "the same kind". An example of a homogeneous substance would be pure water, which only contains the compound H_2O or pure table salt that only contains the compound NaCl.

Homogeneous mixture Solution. Compare with heterogeneous mixture, element and compound.

A sample of matter consisting of more than one pure substance with properties that do not vary within the sample.

Homoleptic Transition metal or Main Group compounds having only one type of ligand are said to be homoleptic, e.g. $TaMe_5$. See also heteroleptic.

Homologous series A Homologous Series is a family of organic compounds which have 1.a common general formula,

2. similar methods of preparation,

3. similar chemical properties,

4. a regular trend in physical properties, and

5. a molecular weight difference of 14 between adjacent members of the series, due to each member of the series containing an extra -CH2- group (i.e. a methylene group).

Homolysis The cleavage of a bond (homolytic cleavage or homolytic fission) so that each of the molecular fragments between which the bond is broken retains one of the bonding electrons.

Homomorphic Superposable ligands are called homomorphic.

Homotopic Atoms or groups of a molecule which are related by an n-fold rotation axis (n = 2, 3, etc) are called homotopic. For example, chiral tartaric acid (C_2 axis), chloroform (C_3 axis) and cyclohexaamylose (?-cyclodextrin, C_6 axis) have respectively two homotopic carboxyl groups, three homotopic chlorine atoms and six homotopic D-glucose residues.

Hooke's law When a load is applied to any elastic body is deformed or strained, then the resulting stress (the tendency of the body to resume its normal condition) is proportional to the strain. Stress is measured in units of force per unit area, strain is the extent of the deformation.

Hormone A molecule produced by endocrine glands that controls specific biological processes like growth and metabolism.

Horse latitudes This is the area between 30*f* and 35*f* latitude where the dense cool air descends and the winds become weak. According to some traditions sailors relying solely on wind power threw their horses overboard in hopes that speed would increase due to the resulting lighter load of the ship. Other traditions state that the horses were thrown overboard to avoid depleting their provisions and still another states that this is simply the region where the horses perished.

Horse power Horse power is obsolete non-SI of power introduced by James Watt in 1782 to allow to describe the power of steam machinery. It was equal to the work effort of a horse needed to raise vertically 528 cubic feet of water to one metre in one minute (HP = 735.498750 W).

Host A molecular entity that forms complexes with organic or inorganic guests, or a chemical species that can accommodate guests within cavities of its crystal structure. Examples include cryptands and crowns (where there are ion-dipole attractions between heteroatoms and positive ions), hydrogen-bonded molecules that form "clathrates" (e.g. hydroquinone and water), and host molecules of inclusion compounds (e.g. urea or thiourea). van der Waals forces and hydrophobic interactions bind the guest to the host molecule in clathrates and inclusion compounds.

HPLC (High-performance liquid chromatography is used to determine the amount of a compound in various media such as the air, water or blood. Samples of the desired media (air, water, etc.) are dissolved in solution and then the components are separated by injecting the solution through a mobile phase solution and then a stationary phase solution. As the solution passes through the stationary phase, the components are separated out due to different partitioning behaviors between the components in the mobile phase and the stationary phase. This allows each component to be analyzed separately as they pass through the detector.

Hückel (4n + 2) rule Monocyclic planar (or almost planar) systems of trigonally (or sometimes digonally) hybridized atoms that contain (4n + 2) electrons (where n is a non-negative integer) will exhibit aromatic character. The rule is generally limited to n = 0-5.

This rule is derived from the Hückel MO calculation on planar monocyclic conjugated hydrocarbons $(CH)_m$ where m is an integer equal to or greater than 3 according to which (4n + 2) electrons are contained in a closed-shell system.

Human ecology The interrelations between man and his physical, biological, socio-economic, and cultural environment, including the interrelations between himself and other individuals or groups of other species, and the interrelationship between himself and his entire environment. Because of the broad approach, human ecology studies are almost always multidisciplinary (WHO, 1988).

Humectant A substance that absorbs or retains moisture, added to a product to keep it from drying out.

Humic matter is part of humus, and is a mixture of amorphous, polydispersed colloidal, yellow to brown-black organic substances. They are hydrophyllic, and high in molecular weights, which range from a few hundreds to thousands of daltons. They have originated from the decomposed, non-humic, organic fraction by "new formation," called humification.

Humidity The moisture content of air. Relative humidity is the ratio of the amount of water vapor actually present in the air to the greatest amount possible at that same temperature.

Hund's rule Rule of maximum multiplicity.

A rule of thumb stating that subshells fill so that the number of unpaired spins is maximized, or "spread them out and line them up."

Hund's rule of maximum multiplicity The Hund's Rule of Maximum Multiplicity states that when electrons occupy the orbitals about a nucleus, and two or more orbitals are at the same energy level, each orbital is filled singly, before any is filled doubly. Ignition Temperature

Ignition temperature is the temperature to which a substance must be heated before it will burn in air.

Hyaluronic acid A viscous polysaccharide found in the connective tissue space and the synovial fluid of movable joints and the humors of the eye; it protects tissue.

Hybrid orbitals Orbitals formed from mixing together atomic orbitals, like the sp^x orbitals, which result from mixing s and p orbitals.

Hybridization Linear combination of atomic orbitals on an atom. Hybrid orbitals are often used in organic chemistry to describe the bonding molecules containing tetrahedral (sp^3), trigonal (sp^2), and digonal (sp) atoms.

Hydrate Compare with addition compound.

A hydrate is an addition compound that contains water in weak chemical combination with another compound. For example, crystals of $CuSO_4 \cdot 5 H_2O$ (copper sulfate pentahydrate) are made of regularly repeating units, each containing 5 molecules of water weakly bound to a copper(II) ion and a sulfate ion.

Hydration Addition of water or of the elements of water (i.e. H and OH) to a molecular entity. For example, hydration of ethene:

$CH_2=CH_2 + H_2O\ CH_3CH_2OH$

The term is also used in a more restricted sense for the process:

A (gas) A (aqueous solution)

cf. the use of the term in inorganic/physical chemistry to describe the state of the ions of an electrolyte in aqueous solution. Ahrland (1979). See also aquation, solvation.

Hydrazine (NH2NH2) A colorless, fuming, corrosive liquid that is a powerful reducing agent. NH2NH2 is used in jet and rocket fuels, and as an intermediate in the manufacture of agricultural, textile, photographic, and industrial chemicals.

Hydride ion (H^-) hydride.

A -1 ion formed from hydrogen. Hydride ions and hydride ionic compounds react instantly and sometimes violently with water.

Hydrocarbons Chemicals containing only carbon and hydrogen. These are of prime economic importance because they encompass the constituents of the major fossil fuels, petroleum and natural gas, as well as plastics, waxes, and oils. In urban pollution, these components—along with NO_x and sunlight—contribute to the formation of tropospheric ozone.

Hydrochlorofluorocarbons HCFCs Chemical species slated to replace CFCs in the near future in most Western nations. When the normal chlorofluorocarbons (see above) reach the stratosphere and are photo-decomposed, their released chlorinc radicals destroy the natural ozone that acts as our umbrella to shield the Earth from ultraviolet radiation (see chlorine and ozone). With one or more hydrogen-carbon bonds, HCFCs are still useful as replacement for CFCs in most applications. However, this bonding structure makes HCFCs much more chemicallyy unstable—as compared to CFCs—and therefore subject to hydroxyl radical and ozone attack early in their gas phase career in the atmosphere. In fact they react in the troposphere instead of the stratosphere. Their atmospheric lifetime is shorter than CFCs and they, therefore, have a smaller chance of reaching the stratosphere where their chlorine could be released by destructive photolysis and enter the catalytic ozone destruction cycle.

Hydrofluoric acid (HF) , A clear, corrosive liquid that has an extremely pungent odor and forms dense white vapor clouds if released. Both the liquid and the vapor can cause severe burns to all parts of the body, and medical treatment is required for all exposures. In fact, HF burns covering only 2% of the body can be fatal. If this chemical comes in contact with glass, concrete and other silicon-containing materials, it yields silicon tetrafluoride gas. If it comes into contact with common metals, it will produce hydrogen fluoride, which poses a fire and explosion hazard. In general, this chemical is used to produce environmentally safer fluorocarbon products, chemical derivatives, and fluoropolymers. Environmentally speaking, hydrofluoric acid is much too reactive to ever reach the upper atmosphere, and would not interfere with the ozone layer.

Hydrogen Hydrogen, H2, is the simplest element. It is the first element in the periodic table, and it is placed in Group I of the periodic table. It has three Isotopes

Atomic Number : 1

Atomic Mass : 1.008

Melting Point : -259 degC

Boiling Point : -253 degC

Density : 0.09

Deuterium, D2, (which is also known as Heavy Hydrogen) is an isotope of hydrogen. The atomic number of deuterium is 1, its atomic mass is 2.0144, and its boiling point is 23.6 deg K

Hydrogen bond The hydrogen bond is a form of association between an electronegative atom and a hydrogen atom attached to a second, relatively electronegative atom. It is best considered as an electrostatic interaction, heightened by the small size of hydrogen, which permits proximity of the interacting dipoles or charges. Both electronegative atoms are usually (but not necessarily) from the first row of the Periodic Table, i.e., N, O, or F. Hydrogen bonds may be intermolecular or intramolecular. With a few exceptions, usually involving fluorine, the associated energies are less than 20-25 kJ mol^{-1} (5-6 kcal mol^{-1}).

Hydrogen bonding Strong type of intermolecular dipole-dipole atttraction. Occurs between hydrogen and F, O or N.

Hydrogen chloride HCl Ahydrous hydrochloric acid. Its molecular weight is 36.47 g/mol; it is a colorless, corrosive gas. Hydrogen chloride is a covalent bonded nonflammable gas and ionizes almost completely when dissolved in water. When dissolved in water, hydrogen chloride forms a strong acid, hydrochloric acid. In the stratosphere, HCl acts as a reservoir species temporarily removing chlorine radicals from a catalytic ozone destruction cycle. (See ozone and chlorine.)

Hydrogen peroxide H_2O_2 Tis compounds molecular weight is 34.02g/mol; it is a colorless, rather unstable oxidant with a bitter taste and caustic to the skin.. Hydrogen peroxide will decompose, liberating oxygen. Pure hydrogen peroxide is stable, but the slightest impurity will enhance decomposition, often violently. Concentrated solutions of hydrogen peroxide are highly corrosive and toxic. H_2O_2 is used as bleach, deodorizer, and in the manufacturing of rocket fuel. The hydrogen peroxide in your bathroom is approximately 3 percent in water. In the atmosphere this is probably one of the oxidizers for sulfur dioxide in cloud water droplets that produces sulfuric acid, a major component in acid rain.

Hydrogen sulfide H_2S Hdrogen sulfide is a reducing species, which is relatively unstable but survives for rather a long time owing to its slow reaction with atmospheric oxygen. This sulfur-containing gas is a major participant in gas to particle conversion in the atmosphere. Many sulfur-containing gases are reactive and thus are rapidly converted to sulfuric acid.

Hydrogenase An enzyme, dihydrogen:acceptor oxidoreductase, that catalyzes the formation or oxidation of H_2. Hydrogenases are of various types. One class ((Fe)-hydrogenases) contains only iron-sulfur clusters. The other major class ((NiFe)-hydrogenases) has a nickel-containing center and iron-sulfur clusters; a variation of the latter type ((NiFeSe)-hydrogenases) contains selenocysteine.

Hydrolase An enzyme of EC class 3, also known as a hydro-lyase, that catalyzes the hydrolysis of a substrate.

Hydrologic cycle The recycling of water on earth. There are five steps: Condensation occurs when water vapor in the atmosphere turns into a liquid and forms clouds. The clouds release precipitation in the form of rain, snow, or sleet when the moisture in the clouds becomes too heavy. Infiltration occurs when the water seeps into the ground, The infiltration rate depends on the permeability of the ground. When there is too much water on the surface to infiltrate into the ground, it becomes runoff and ends up in lakes, streams, or oceans. Evapotranspiration is a two part process driven by the sun. Evaporation occurs when water is turned into a vapor from the ground and transpiration is water loss by plants, The water vapor rises and starts the process over again.

Hydrolysis Hydrolysis is a chemical reaction in which water reacts with another substance to form two or more new substances. This involves ionization of the water molecule as well as splitting of the compound hydrolyzed, e.g.

$$CH_3COOC_2H_5 + H_2O \rightleftharpoons CH_3COOH + C_2H_5OH$$

Examples are conversion of starch to glucose by water in presence of suitable catalysts and reaction of the ions of a dissolved salt to form various products, such as acids, complex ions, etc.

Hydrometeors Any condensed form of water—larger than a single water molecule—that is falling or suspended in the atmosphere; some examples of hydrometeors are snow, fog, cloud or rain. When these particles reach the size of 10-20 micrometers, available water is depleted and a stable cloud is formed.

Hydrometer An instrument for measuring the specific gravity of liquids. A hydrometer is a weight with a vertical scale attached. When placed into a liquid, the hydrometer bobs upright, and sinks to a certain level. The specific gravity or solution composition can be read from the liquid level on the vertical scale. Hydrometers are often calibrated in degrees Baumé.

Hydron General name for the ion H^+ either in natural abundance, or where it is not desired todistinguish between the isotopes, as opposed to proton for $^1H^+$, deuteron for $^2H^+$ and triton for $^3H^+$.

Hydronium ion (H3O+) Hydronium.The H3O+ ion, formed by capture of a hydrogen ion by a water molecule. A strong covalent bond is formed between the hydrogen ion and water oxygen; all hydrogen ions in aqueous solution are bound inside hydronium ions.

Hydroperoxy radical HO_2 Tis radical readily reacts with nitrogen oxides and hydrocarbons in the atmosphere; in urban atmospheres, the major source of the hydroperoxy radical is formaldehyde photolysis.

Hydrophilic Hydrophilic is having a strong tendency to bind or absorb water, which results in swelling and formation of reversible gels. This property is characteristic of carbohydrate.

Hydrophobic Hydrophobicity; hydrophobic group.

A nonpolar molecule or group that has little affinity for water. Hydrophobic groups on molecules in solution tend to turn in on themselves or clump together with other hydrophobic groups because they are unable to disrupt the network of strong hydrogen bonds in the water around them.

Hydrophobic interaction Hydrophobic interaction is the tendency of hydrocarbons (or of lipophilic hydrocarbon-like groups in solutes) to form intermolecular aggregates in an aqueous medium, and analogous intramolecular interactions. The name arises from the attribution of the phenomenon to the apparent repulsion between water and hydrocarbons. Use of the misleading alternative term hydrophobic bond is discouraged.

Hydrosol A dispersion of solid particles is a liquid system.

Hydrothermal chemistry Aqueous chemistry under high temperature conditions. In biospheric chemistry, this refers to water whose chemical composition has been altered by deep, hot rock and which tends to come to the surface at fault lines. This creates boiling springs and geysers.

Hydroxide (OH-) Hydroxide ion. 1. The OH- ion.

2. Compounds containing the OH- ion.

Hydroxyl Compare with hydroxide. 1. An -OH group within a molecule.

2. A free radical formed by abstraction of a hydrogen atom from water.

Hydroxyl radical OH Aradical consisting of one hydrogen atom and one oxygen atom which does not normally exist in a stable form; this radical readily reacts with methane and carbon monoxide. The source of the hydroxyl radical in the atmosphere occurs primarily through the 1. photolysis of hydrogen peroxide, heat, and light and

2. the attack on water of an excited oxygen radical (created by the photolysis of ozone).

Hygroscopic Able to absorb moisture from air. For example, sodium hydroxide pellets are so hygroscopic that they dissolve in the water they absorb from the air.

Hygroscopically By absorbing moisture from air.

Hygroscopicity The ability of a substance to absorb moisture from air. For example, sodium hydroxide pellets are so hygroscopic that they dissolve in the water they absorb from the air.

Hyperconjugation Weak interaction (electron donation) between sigma bonds with p orbitals. Hyperconjugation explains why alkyl substituents stabilize carbocations.

Hyperfine See electron paramagnetic resonance spectroscopy.

Hypergolic mixture Hypergolic fuel; hypergolic.

An oxidizing agent and a fuel that react or ignite instantly and spontaneously on contact.

top"Methylhydrazine and top"dinitrogen tetroxide is a hypergolic mixture used as a propellant in the space shuttle's orbital maneuvering engines.

Hypertonic Compare with osmotic pressure. Describes a solution which has higher osmotic pressure than some other solution (usually, higher osmotic pressure than cell or body fluids). Freshwater fish die if placed in seawater because the seawater is hypertonic, and causes water to leave the cells in fish's body.

Hypothesis Hypotheses. Compare with theory. A hypothesis is a conjecture designed to guide experimentation. Hypotheses are extremely useful in problem solving, and are essential in developing new theories.

Hypotonic Compare with osmotic pressure. Describes a solution which has lower osmotic pressure than some other solution (usually, lower osmotic pressure than cell or body fluids). Washing your contact lenses with distilled water rather than saline is painful because distilled water is hypotonic; it causes water to move into cells, and they swell and burst.

Hypsochromic shift Shift of a spectral band to higher frequency or shorter wavelength upon substitution or change in medium. It is informally referred to as blue shift.

Hypsometric diagram A graph that shows in any way the relative amounts of Earth's surface at different elevations with regard to sea level.

Hypsometry Hypsometry is a technique of mapping the elevation contour of both land and the ocean floor using tinting to illustrate a change in contour.

I

Ideal gas Ideal gas is a gas in which there is complete absence of cohesive forces between the component molecules; the behavior of such a gas can be predicted accurately by the ideal gas equation through all ranges of temperature and pressure. The concept is theoretical, since no actual gas meets the ideal requirement.

Ideal gas law constant (R) ideal gas constant; universal gas constant.

A constant R equal to PV/(nT) for ideal gases, where the pressure, volume, moles, and temperature of the gas are P, V, n, and T, respectively. The value and units of R depend on the units of P, V, and T. Commonly used values and units of R include: 82.055 cm^3 atm K^{-1} mol^{-1}; 0.082055 L atm mol^{-1} K^{-1}; 8.31434 J mol^{-1} K^{-1}; 1.9872 cal K^{-1} mol^{-1}; 8314.34 L Pa mol^{-1} K^{-1}; 8.31434 Pa m^3 mol^{-1} K^{-1}.

Ideal gas law The generalized ideal gas law is derived from a combination of the laws of Boyle and Charles. Ideal gas law is the equation of state

$$pV = RT$$

which defines an ideal gas, where p is pressure, V molar volume, T temperature, and R the molar gas constant (8.314 $JK^{-1mol-1}$).

Ideal solution Ideal solution is a solution in which solvent-solvent and solvent-solute interactions are identical, so that properties such as volume and enthalpy are exactly additive. Ideal solutions follow Raoult's law, which states that the vapor pressure p_i of component i is $p_i = x_i\, p_i^*$, where x_i is the mole fraction of component i and p_i^* the vapor pressure of the pure substance i.

Identity A math property which states: A+0=A and A*1=A.

Identity reaction A chemical reaction whose products are chemically identical with the reactants, for example the bimolecular self exchange reaction of CH3I with I^-.

Idiosyncrasy The increased individual sensitivity of an organism to the effect of certain substances

IDLH(Immediately dangerous to life and health). In the event of the accidental exposure to a chemical, this is the concentration below which an individual could escape within 30 minutes without experiencing any escape-impairing or irreversible health effects.

Imaging A medical diagnostic technique by which useful organ images are obtained from the radiation emitted by radionuclides that are introduced into organs, or from radiation absorbed by atomic nuclei within the organs. Typical examples are imaging obtained by recording the radiation emitted by a radionuclide such as ^{99m}Tc, and the ^1H-NMR imaging obtained by whole body nuclear magnetic resonance measurements.

Imbalance The situation in which reaction parameters that characterize different bond forming or bond breaking processes in the same reaction have developed to different extents as the transition state is approached along some arbitrarily defined reaction coordinate. For example, in the nitroalkane anomaly, the Brønsted exponent for proton removal is smaller than the Brønsted for the nitroalkane, because of imbalance between the amount of bond breaking and resonance delocalization in the transition state. Imbalance is common in reactions such as elimination, addition and other complex reactions that involve proton (hydron) transfer.

Immiscible Immiscibility. Compare with miscible and partial miscibility.

Two liquids are considered "immiscible" or unmixable if shaking equal volumes of the liquids together results in a meniscus visible between two layers of liquid. If the liquids are completely immiscible, the volumes of the liquid layers are the same as the volumes of liquids orginally added to the mixture.

Immune response The immune response is a general reaction of the body to substances that are foreign or treated as foreign. It may take various forms: antibody production,cell-mediated immunity, immunological tolerance, or allergy.

Immunogold A method for visualizing proteins in electron microscopy within a cell using gold particles attached to an antibody that binds specifically to that protein.

Impedance symbol: Z, unit: ohm (Ω)

Impedance is the degree to which an electronic component impedes the flow of current. In general it is a frequency-dependent quantity.

The impedance of a resistor is also called its resistance. The impedance of capacitors and inductors is also called their reactance.

Implosion Implosion is the inward collapse of a vessel, usually when the pressure of the gas in the container is less than atmospheric pressure.

In situ scaffold formation Process whereby a scaffold is formed during library production which contains residues of at least two building blocks.

In vitro A term applied to any study carried out in isolation from the living organism in an experimental system ('in a test tube').

In vivo The term used in contrast with 'in vitro' describing any study carried out within the living organism.

Incidence The number of instances of illness commencing, or of persons falling ill, during a given period in a specific population. Incidence is usually expressed as a rate, the denominator being the average number of persons in the specified population during a defined period or the estimated number of persons at the mid-point of that period. The basic distinction betweenincidenceandprevalenceis that whereas incidence refers only to new cases, prevalence refers to all cases, irrespective of whether they are new or old. When the terms incidence and prevalence are used, it should be stated clearly whether the data represent the numbers of instances of the disease recorded or the numbers of persons ill (WHO, 1966).

Inclusion compound (Or inclusion complex) A complex in which one component (the host) forms a cavity or, in the case of a crystal, a crystal lattice containing spaces in the shape of long tunnels or channels in which molecular entities of a second chemical species (the guest) are located. There is no covalent bonding between guest and host, the attraction being generally due to van der Waals forces. If the spaces in the host lattice are enclosed on all sides so that the guest species is "trapped" as in a cage, such compounds are known as "clathrates" or "cage" compounds".

Incomplete combustion A combustion reaction or process that does not convert all of the fuel's carbon and hydrogen into carbon dioxide and water, respectively. For example, incomplete combustion of carbon produces carbon monoxide.

Incomplete octet. 1. An atom with less than eight electrons in its valence shell.

2. An atom with less than eight total bonding and nonbonding electrons in a Lewis structure, for example, B in BH_3 has an incomplete octet.

Independent variable Compare with dependent variable.

An independent variable that can be set to a known value in an experiment. Several independent variables may be controlled in an experiment. For example, in an experiment where the vapor pressure of a liquid is measured at several different temperatures, temperature is the independent variable and vapor pressure is the dependent variable.

Index of refraction For a non-absorbing medium, index of refraction (n) is the ratio of the velocity of electromagnetic radiation in vacuo to the phase velocity of radiation of a specified frequency in the medium.

Indicator diagram PV diagram.

A plot of pressure vs. volume. Lines or curves on the indicator diagram represent processes. The areas under curves on the indicator diagram are equal to the work released by the process.

Indoor air quality A measure of the value or comfort of people with the air they breathe inside buildings and homes. "Suitable" indoor air quality can conferred on air in which there are no known contaminants at harmful levels as determined by appropriate authorities or—more likely—air with which 80% or more of the people exposed do not express dissatisfaction.

Induction period The initial slow phase of a chemical reaction which later accelerates. Induction periods are often observed with radical reactions, but they may also occur in other systems (for example before steady-state concentration of the reactants is reached).

Induction, stereochemical See asymmetric induction.

Inductive effect In strict definition, an experimentally observable effect (on rates of reaction, etc.) of the transmission of charge through a chain of atoms by electrostatic induction. A theoretical distinction may be made between the field effect, and the inductive effect as models for the Coulomb interaction between a given site within a molecular entity and a remote unipole or dipole within the same entity. The experimental distinction between the two effects has proved difficult, except for molecules of peculiar geometry, which may exhibit "reversed field effects". Ordinarily the inductive effect and the field effect are influenced in the same direction by structural changes in the molecule and the distinction between them is not clear. This situation has led many authors to include the field effect in the term "inductive effect". Thus the separation of values into inductive and resonance components does not imply the exclusive operation of a through-bonds route for the transmission of the non-conjugative part of the substituent effect. To indicate the all-inclusive use of the term inductive, the phrase "so-called inductive effect" is sometimes used. Certain modern theoretical approaches suggest that the "so-called inductive effect" reflects a field effect rather than through-bonds transmission. See also field effect, mesomeric effect, polar effect.

Inductomeric effect A molecular polarizability effect occurring by the inductive mechanism of electron displacement. The consideration of such an effect and the descriptive term have been regarded as obsolescent or even obsolete, but in recent years theoretical approaches have reintroduced substituent polarizability as a factor governing reactivity, etc. and its parametrization has been proposed.

Industrial nitrogen fixation Industrial nitrogen fixation is the conversion of atmospheric nitrogen into ammonia which is carried out using the Haber Process.

Inert A chemically inactive element, compound or material. Because of their extremely low chemical activity the noble gasses are considered to be inert substances. This term is also used for substances that are added to mixtures, primarily for their bulk and weight.

Inert electrodes Inert electrodes are electrodes which do not undergo chemical change during the course of an electrolysis experiment.

Inert gas Inert gases; noble gas; noble gases.

Any of the elements of Group 18, which includes top"helium, top"neon, top"argon, top"krypton, top"xenon, top"radon, and element 118. These elements are referred to as "inert" or "noble" because they do not easily form compounds with other elements.

Inert pair Inert pair effect.

Valence electrons in an s orbital penetrate to the nucleus better than electrons in p orbitals, and as a result they're more tightly bound to the nucleus and less able to participate in bond formation. A pair of such electrons is called an "inert pair". The inert pair effect explains why common ions of Pb are Pb^{4+} and Pb^{2+}, and not just Pb^{4+} as we might expect from the octet rule.

Inertia The tendency of a body to stay at rest or to continue to move at the same velocity, unless acted on by an outside force. A tractor trailer has more inertia than a bicycle. A bowling ball has more inertia than a tennis ball.

Infrared radiation (IR) infrared.

Electromagnetic radiation with wavelength longer than visible light but shorter than that of microwaves. Infrared radiation is produced by hot objects; absorption of infrared radiation causes chemical bonds to vibrate.

Infrared spectroscopy IR spectroscopy.

A technique for determining the structure (and sometimes concentration) of molecules by observing how infrared radiation is absorbed by a sample.

Inhibition The decrease in rate of reaction brought about by the addition of a substance (inhibitor), by virtue of its effect on the concentration of a reactant, catalyst or reaction intermediate. For example, molecular oxygen and p-benzoquinone can react as "inhibitors" in many reactions involving radicals as intermediates by virtue of their ability to act as scavengers toward these radicals.

If the rate of a reaction in the absence of inhibitor is v_0 and that in the presence of a certain amount of inhibitor is v, the degree of inhibition (i) is given by

$i = (v_0 - v)/v_0$.

Inhibitor A substance that decreases the rate of an enzyme catalyzed or other chemical reaction.

Inhibitors Agents that block or suppress the activity of enzymes such as proteases.

Initiation Initiation is the first step in a free radical mechanism that creates the initial free radicals necessary to start the sequence of reactions.

Inner transition metals The Inner Transition Metals are the series of elements

From Cerium (Atomic Number 58) to Lutetium (Atomic Number 71), which are called the Lanthanoids, and

From Thorium (Atomic Number 90) to Lawrencium (Atomic Number 103), which are called the Actinoids.

The inner transition elements are found between group 2 and the transition elements in the fifth row of the periodic table. The transition elements are also known as the f-block elements.

These two series make up the f-block elements in the periodic table, and their chemical properties of the elements derive from the filling of the f atomic sub-orbitals. The electronic configuration of these elements are characterised as having full outer orbitals and full second outermost orbitals, while the second outermost orbitals are incompletely filled. Thus, in the case of the first inner transition metals series, the electronic configuration of the outermost and second outermost orbitals is 4s2 3d10, while the third outermost orbitals (i.e. the 4f level) are incompletely filled.

Inner-sphere (Electron transfer) Historically an electron transfer between two metal centres sharing a ligand or atom in their respective coordination shells. The definition has more recently been extended to any situation in which the interaction between the donor and acceptor centres in the transition state is significant (>20 kJ mol^{-1}).

Inorganic chemistry Inorganic chemistry is the chemistry of all other substances other than organic compounds, although it is convenient to include in it such common carbon compounds as are frequently encountered e.g. carbon dioxide, carbonates or are essential to placing the element carbon in its correct relationship in the periodic system of classification.

Inorganic compound Inorganic. Compare with organic.

A compound that does not contain carbon chemically bound to hydrogen. Carbonates, bicarbonates, carbides, and carbon oxides are considered inorganic compounds, even though they contain carbon.

In-out isomerism Isomerism found in bicyclic systems having long enough bridges to allow the bridgehead exocyclic bond or lone pair of electrons to point either inside the structure or outside.

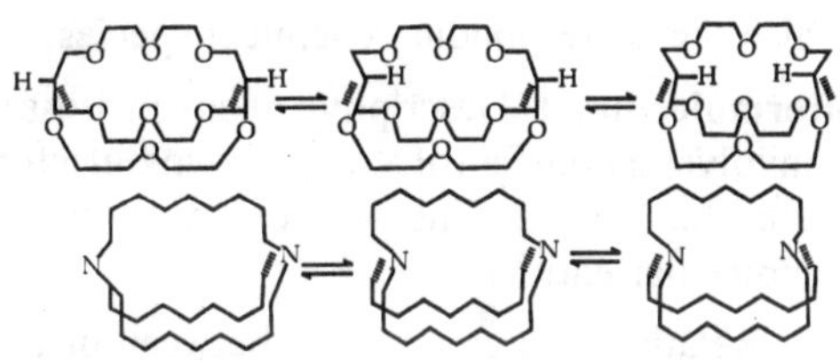

Insecticide A chemical used to kill insects. See pesticides.

Insertion reaction A chemical reaction or transformation of the general type X-Z + Y $\longrightarrow$ X-Y-Z in which the connecting atom or group Y replaces the bond joining the parts X and Z of the reactant XZ.

Insolation The solar radiation incident on a unit horizontal surface at the top of the atmosphere. It is sometimes referred to as solar irradiance. The latitudinal variation of insolation supplies the energy for the general circulation of the atmosphere. Insolation depends on the angle of incidence of the solar beam and on the solar constant.

Insoluble A solid is said to be Insoluble in a liquid if it does not dissolve in the liquid.

Instrumentation A typical fluorimeter contains an excitation source, sample cell, fluorescence detector. Molecules in solution are usually excited by uv light and the excitation source is usually a deuterium or xenon lamp. Broad-band excitation light from a lamp passes through a monochromator, which passes only a selected wavelength. The fluorescence is dispersed by another monochromator and detected by a photomultiplier tube. Scanning the excitation monochromator gives the excitation spectrum and scanning the fluorescence monochromator gives the fluorescence spectrum. Simple instruments sometimes use only a bandpass filter to select the excitation wavelength.

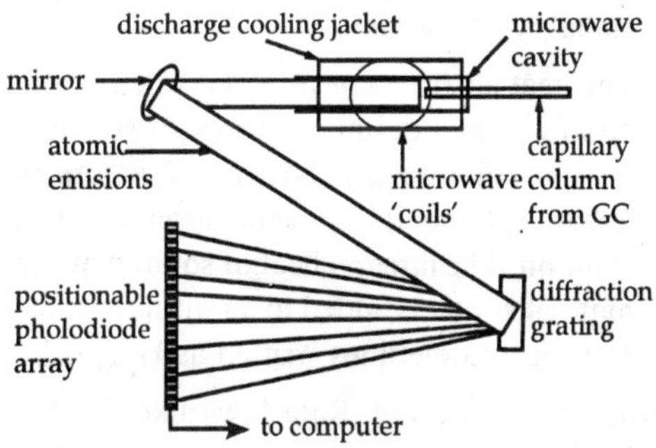

Gas Chromatography (GC) Gas chromatography is a chromatographic technique that can be used to separate organic compounds that are volatile. A gas chromatograph consists of a flowing mobile phase, an injection port, a separation column containing the stationary phase, a detector, and a data recording system. The organic compounds are separated due to differences in their partitioning behavior between the mobile gas phase and the stationary phase in the column.

Insulator Insulator is a material in which the highest occupied energy band (valence band) is completely filled with electrons, while the next higher band (conduction band) is empty. Solids with an energy gap of 5 eV or more are generally considered as insulators at room temperature. Their conductivity is less than 10^{-6} S/m and increases with temperature.

Intake The amount of a substance or material that is taken into the body, regardless of whether or not it is absorbed. The daily intake may be expressed as the amount taken in by a particular exposure route, e.g., ingestion or inhalation. The daily intake from food is the total amount of a given substance taken in during one day through the consumption of food. The daily intake by inhalation is calculated by multiplying the concentration of the substance (or agent) in air by the total amount of air inhaled during one day (24 hours). The total daily intake is the sum of the daily intake by an individual from food, drinking-water, and inhaled air (WHO, 1979).

Integral enthalpy of solution (H_{soln}) integral heat of solution; integrated heat of solution; integrated enthalpy of solution. Compare with enthalpy of solution.

The heat absorbed or released when a solute is dissolved in a definite amount of solvent. The heat of solution depends on the nature of the solute and on its concentration in the final solution. The integral heat of solution when one mole of solute is added to an infinite amount of solvent is sometimes written as $H_{, soln}$.

Integrated rate law Rate laws like d(A)/dt = -k(A) give instantaneous concentration changes. To find the change in concentration over time, the instantaneous changes must by added (integrated) over the desired time interval. The rate law d(A)/dt = -k(A) can be integrated from time zero to time t to obtain the integrated rate law $\ln((A)/(A)_o) = -kt$, where $(A)_o$ is the initial concentration of A.

Intensive property Intensive; intensive properties. Compare with extensive property.

A property that does not change when the amount of sample changes. Examples are density, pressure, temperature, color.

Intercalation compounds Compounds resulting from inclusion, usually without covalent bonding, of one kind of molecule (the guest molecule) in a matrix of another compound (the host compound), which has a layered structure. The host compound, with a rather rigid structure, may be macromolecular, crystalline, or amorphous.

Interface The area where two immiscible phases of a dispersion come into contact. It may involve the same or different states of matter. There are five possible types: liquid/liquid (oil/water), liquid/gas (water/air), liquid/solid (water/clay), solid/gas (smoke/air), solid/solid (rubber carbon black).

Interference Interfering. Compare with constructive interference and destructive interference.

The amplitudes of waves moving into the same region of space add to produce a single resultant wave. The resultang wave can have higher or lower amplitude than the component waves.

Intermediate A molecular entity with a lifetime appreciably longer than a molecular vibration (corresponding to a local potential energy minimum of depth greater than RT) that is formed (directly or indirectly) from the reactants and reacts further to give (either directly or indirectly) the products of a chemical reaction; also the corresponding chemical species.

Intermolecular 1. Descriptive of any process that involves a transfer (of atoms, groups, electrons, etc.) or interactions between two or more molecular entities.

2. Relating to a comparison between different molecular entities.

See also intramolecular.

Intermolecular force An attraction or repulsion between molecules. Intermolecular forces are much weaker than chemical bonds. Hydrogen bonds, dipole-dipole interactions, and London forces are examples of intermolecular forces.

Internal compensation (usage strongly discouraged) See meso-compound.

Internal energy (U, E) Compare with enthalpy and energy.

Internal energy (U) is defined so that changes in internal energy (U) are equal to the heat absorbed or released by a process running at constant volume. While changes in internal energy can be measured using calorimetry, absolute values of internal energy usually cannot be determined. Changes in internal energy are equal to the heat transferred plus the work done for any process.

International system of units International System of Units (SI) is the unit system adopted by the General Conference on Weights and Measures in 1960 and recommended for use in all scientific and technical fields. It consists of seven base units (meter, kilogram, second, ampere, kelvin, mole, candela), plus derived units and prefixes.

Interpretation of data The evaluation of all the facts available from a given investigation or study with a view to their significance for health.

Intertropical convergence zone (ITCZ) The location in the atmosphere along both sides of the equator at which the strong upward motion of air (Hadley cell) is characterized by heavy precipitation in convective thunderstorms and by relatively low surface pressure.

Intervention study An epidemiologic investigation designed to test a hypothesized cause-effect relationship by modifying a supposed causal factor in a population .

Intestinal reabsorption Absorption further down the intestinal tract of material that has been absorbed before and subsequently excreted into the intestinal tract, usually through the bile (WHO, 1979).

Intramolecular 1. Descriptive of any process that involves a transfer (of atoms, groups, electrons, etc.) or interactions between different parts of the same molecular entity.

2. Relating to a comparison between atoms or groups within the same molecular entity.

Intramolecular catalysis The acceleration of a chemical transformation at one site of a molecular entity through the involvement of another functional ("catalytic") group in the same molecular entity, without that group appearing to have undergone change in the reaction product. The use of the term should be restricted to cases for which analogous intermolecular catalysis by chemical species bearing that catalytic group is observable. Intramolecular catalysis can be detected and expressed in quantitative form by a comparison of the reaction rate with that of a comparable model compound in which the catalytic group is absent, or by measurement of the effective molarity of the catalytic group. See also effective molarity, neighbouring group participation.

Intramolecular forces Forces within molecules. Forces caused by the attraction and repulsion of charged particles.

Intravenous Into or inside the vein

Intrinsic barrier The Gibbs energy of activation (G) in the limiting case where $G^{o} = 0$, i.e. when the effect of thermodynamic driving force is eliminated. According to the Marcus equation, the intrinsic barrier is related to the reorganization energy, , of the reaction by the equation

$G = /4$

Intrinsic energy The intrinsic energy, U, (i.e. its internal energy) of a substance is the chemical energy of a substance (i.e. the amount of energy stored within the substance).

Introduction to chemistry Chemistry is the science of the nature, composition and properties of material substances, including their transformations and interconversions. It is involved with a study of all matter, including solids, liquids and gases.

The present body of knowledge represented by chemistry was built on a development of the experimental methodologies first used by the alchemists. Although many of their theories were incorrect, and their interest in the transformation of base metals into gold, and the preparation of magic potions has no place in modern chemistry, the experimental approach which they adopted allowed later experimenters to deduce of the structure of matter

Introduction to organic chemistry Organic chemistry is concerned with the properties of those compounds which are found in living organisms. By the beginning the nineteenth century, organic chemistry was firmly established as an experimental science, but the underlying theoretical principals were still to be discovered. There was a tacit assumption at that time that organic chemistry was a wholly separate branch of chemical science, largely governed by the workings of a mysterious agency called Vital Force, which was thought to be inherited by organic compounds from the living organisms from which ultimately they had been derived.

In 1828 Wohler was the first chemist to demonstrate the conversion of the inorganic compound (i.e. ammonium cyanate, NH4NCO) into an organic compound (i.e. Urea, NH2.CO.NH2). This experiment established that organic chemistry was just another branch of chemistry.

Organic compounds (e.g. wine, beer, dyes, perfumes, drugs, \ poisons, soaps, etc.) have been manufactured and used since early times. The first step in the development of organic chemistry was the preparation and description of individual organic compounds. Thus, distillation and other separation techniques were a major part of the early organic chemistry.

Alcohol was obtained by distillation of wine,

Acetic Acid was obtained by distillation of vinegar, and

Organic Acids were extracted by crystallisation from fruit and animal juices.

Although many compounds of carbon are still most conveniently isolated from plant and animal sources, many organic compounds are now easily synthesised in the laboratory. Petroleum and Coal are the two large reservoirs, from which many organic compounds may be obtained. The simple organic compounds isolated from these sources may then be used as building blocks for larger and more complicated compounds.

Inorganic Chemistry is the chemistry of all other substances. However, it is convenient to classify some common carbon compounds (e.g. carbon dioxide, carbon monoxide, sodium carbonates, etc.) as inorganic, so as to place carbon in its correct relationship to the other elements in the periodic system of classification. Inorganic chemistry is essentially the chemistry described by the periodic table. Z

By definition, carbon is always present in all organic compounds, and the vast majority of organic compounds also contain hydrogen. Compounds consisting of carbon and hydrogen only are called hydrocarbons. Many organic compounds contain oxygen, and some contain nitrogen, a halogens, sulphur, silicon and a few other elements. The classification of organic compounds will be illustrated for each functional group in this hypertext. Z

There are more than seven million chemical compounds known and more than three-quarters of them are carbon compounds. This is a remarkable proportion, particularly as very few other elements are included in these compounds. The main reason for the very large number of carbon compounds is the unique ability of the Carbon atom to combine with itself almost indefinitely, to form straight-chain or branched-chains or ring-systems, of all sizes and degrees of complexity.

Just as inorganic chemistry considers each element with regard to its position in the periodic table and then deals with the properties of the compounds of that element, organic chemistry is concerned with the properties of groups of organic compounds which have structural similarities (i.e. compounds which have certain functional groups in common).

Alcohols Aliphatic

Aldehydes Aliphatic

Alkanes

Alkenes

Alkynes

Amines

Amino Acids

Amides

Arenes

Aromatic Compounds

Carboxylic Acid

Esters

Halides

Halogens

Ketones Polymer

Intron An intervening section of DNA which occurs almost exclusively within a eukaryotic gene, but which is not translated to amino-acid sequences in the gene product. The introns are removed from the pre-mature mRNA through a process called splicing, which leaves the exons untouched, to form an active mRNA.

Inverse A math property which states: A+(-A)=0 and A*(1/A)=1

Inverse kinetic isotope effect See isotope effect.

Inversion An anomaly in the normal positive atmospheric lapse rate (change of temperature with increasing altitude). This usually refers to a thermal inversion, in which temperature of the atmosphere increases rather than decreases with height.

Inverted micelle The reversible formation of association colloids from surfactants in non-polar solvents leads to aggregates termed inverted (or inverse, reverse or reversed) micelles. Such association is often of the type

Monomer Dimer Trimer ... n-mer

and the phenomenon of critical micelle concentration (or an analogous effect) is consequently not observed.

In an inverted micelle the polar groups of the surfactants are concentrated in the interior and the lipophilic groups extend towards and into the non-polar solvent.

Iodine Iodine, I; is a dark violet (Greek, ioeides, violet) non-metallic halogen element belonging to Group VIIb (i.e. the Halogen Group of elements) of the periodic table.

Atomic Number : 53

Atomic Mass : 126.904

Melting Point : 113.5 degC

Boiling Point : 1834 degC

Density : 4.94

There is one stable isotope, iodine-127 and fourteen radioactive isotopes.

Ion An atom or molecule that has acquired a charge by either gaining or losing electrons. An atom or molecule with missing electrons has a net positive charge and is called a cation; one with extra electrons has a net negative charge and is called an anion.

Ion channel Ion channels enable ions to flow rapidly through membranes in a thermodynamically downhill direction after an electrical or chemical impulse. Their structures usually consist of 4-6 membrane-spanning domains. This number determines the size of the pore and thus the size of the ion to be transported.

Ion exchange Ion exchange resin; ion exchanger.

Ion exchange is a method of separating ions from a solution by reversibly binding them onto a resin that has charged sites on its surface. Ion exchangers are used to remove metal ions from drinking water.

Ion exchange resins Ion exchange resins are the polymers to which ionic groups have been attached. An ionic resin has negative ions built into its structure and therefore exchanges positive ions. A cationic resin has positive ions built in and exchanges negative ions.

Ion exchanger Ion-exchanger is a solid or liquid material containing ions that are exchangeable with other ions with a like charge that are present in a solution in which the material is insoluble. Ion-exchange resins consist of various copolymers having a cross-linked three-dimensional structure to which ionic groups have been attached.

Ion pumps Ion pumps enable ions to flow through membranes in a thermodynamically uphill direction by the use of an energy source such as ATP or light. They consist of sugar-containing hetero-peptide assemblies, which open and close upon the binding and subsequent hydrolysis of ATP, usually transporting more than one ion towards the outside or the inside of the membrane.

Ion-dipole forces Intermolecular force that exist between charged particles and partially charged molecules.

Ionic bond An ionic bond, (which is also called an electrostatic bond), is the attractive force between oppositely charged ions.

An ionic bond results from the transfer of one or more electrons from the outer shell of one atom to the outer shell of another atom. This type of bond is usually formed between elements whose positions in the periodic table lie just before or just after the Nobel Gases.

Ionic compound Salt. Compare with covalent compound and ionic bond.

A compound made of distinguishable cations and anions, held together by electrostatic forces.

Ionic dissociation Ionize; ionization.

When ionic substances dissolve, their ions are surrounded by solvent molecules and separated from each other. This phenomena is also called ionization.

Ionic equation Complete ionic equation . Compare with net ionic equation and molecular equation.

An ionic equation is a balanced chemical equation in which strong electrolytes are written as dissociated ions. For example, $Ag^+(aq)$ + $NO_3^-(aq)$ + $Na^+(aq)$ + $Cl^-(aq)$ = $AgCl(s)$ + $Na^+(aq)$ + $NO_3^-(aq)$ is an ionic equation; $AgNO_3(aq)$ + $NaCl(aq)$ = $AgCl(s)$ + $NaNO_3(aq)$ is not.

Ionic radius The radius exhibited by an ion in an ionic crystal where the ions are packed together to a point where their outermost electronic orbitals are in contact with each other.

Ionic strength Ionic strength (μ or I) is a measure of the total concentration of ions in a solution, defined by

$$\mu = \frac{1}{2}\sum c_i z_i^2$$

where z_i is the charge of ionic species i and c_i is its concentration.

Ionization The generation of one or more ions. It may occur, e.g. by loss of an electron from a neutral molecular entity, by the unimolecular heterolysis of such an entity into two or more ions, or by a heterolytic substitution reaction involving neutral molecules, such as

$CH_3CO_2H + H_2O$ $H_3O^+ + CH_3CO_2^-$

$Ph_3CCl + AlCl_3$ $Ph_3C^+ + AlCl_4^-$ (electrophile-assisted)

Ph_3CCl Ph_3C^+ Cl^-(ion pair, in benzene)

The loss of an electron from a singly, doubly, etc. charged cation is called second, third, etc. ionization. This terminology is used especially in mass spectroscopy. See also dissociation, ionization energy.

Ionization energy, E_i (SI unit kJ mol^{-1} or J per molecule) The minimum energy required to remove an electron from an isolated molecular entity (in its vibrational ground state) in the gaseous phase. If the resulting molecular entity is considered to be in its vibrational ground state, one refers to the energy as the "adiabatic ionization energy". If the molecular entity produced possesses the vibrational energy determined by the Franck-Condon principle (according to which the electron ejection takes place without an accompanying change in molecular geometry), the energy is called the "vertical ionization energy". The name ionization energy is preferred to the somewhat misleading earlier name "ionization potential". See also ionization.

Ionizes When a substance breaks into its ionic components.

Ionizing power A term to denote the tendency of a particular solvent to promote ionization of an uncharged or, less often, charged solute. The term has been used both in a kinetic and in a thermodynamic context.

Ionophore A compound which can carry specific ions through membranes of cells or organelles.

Ipso-attack The attachment of an entering group to a position in an aromatic compound already carrying a substituent group (other than hydrogen). The entering group may displace that substituent group but may also itself be expelled or migrate to a different position in a subsequent step. The term "ipso-substitution" is not used, since it is synonymous with substitution.

IR spectrosco py An instrumental technique that measures IR light absorption by molecules. Can be used to determine functional groups in an unknown molecule.

IR thermography Infrared thermography Screening technique where the heat of reaction of a multitude of samples is simultaneously measured. Has been applied in particular to the screening of libraries of potential catalysts.

Iron-responsive element A specific base sequence in certain messenger RNAs that code for various proteins of iron metabolism, which allows regulation at translational level by the iron-responsive protein.

Iron-responsive protein (IRP) A protein that responds to the level of iron in the cell, and regulates the biosynthesis of proteins of iron metabolism, by binding to the iron-responsive element on messenger RNA.

Iron-sulfur cluster A unit comprising two or more iron atoms and bridging sulfide ligands in an iron-sulfur protein. The recommended designation of a cluster consists of the iron and sulfide content, in square brackets, for example (2Fe-2S), (3Fe-4S). The possible oxidation levels are indicated by the net charge excluding the ligands, for example a $(4Fe\text{-}4S)^{2+}$; $(4Fe\text{-}4S)^{1+}$ (or $(4Fe\text{-}4S)^{2+;1+}$)cluster.

Iron-sulfur proteins Proteins in which non-heme iron is coordinated with cysteine sulfur and usually also with inorganic sulfur. Divided into three major categories: rubredoxins; "simple iron-sulfur proteins", containing only iron-sulfur clusters, and "complex iron-sulfur proteins", containing additional active redox centers such as flavin, molybdenum or heme. In most iron-sulfur proteins the clusters function as electron transfer groups, but in others they have other functions such as catalysis of hydratase/ dehydratase reactions, maintenance of protein structure, or regulation of activity.

IRP See iron-responsive protein.

Irradiance The total radiant flux received on a unit area of a given real or imaginary surface. Also called the radiant flux density.

Irritant Applied to any substance causing inflammation following immediate, prolonged or repeated contact with skin or mucous membranes.

Isaac adams, junior (1836-1911) Born in Boston and educated at Bowdoin College, Harvard Medical School and the École de Médicine in Paris. He then pursued a dual career as a physician and an analyst in Boston. He developed a commercial nickel-plating process in 1869.

Ischemia Local deficiency of blood supply and hence dioxygen to an organ or tissue owing to constriction of the blood vessels or to obstruction.

Isobacteriochlorin 2,3,7,8-Tetrahydroporphryin. A reduced porphyrin with two pairs of con-fused saturated carbon atoms (C-2, C-3 and C-7, C-8) in two of the pyrrole rings.

Isobar Compare with isotope. 1. A contour line that corresponds to values measured at identical pressures. For example, curves on a plot of gas volumes measured at different temperatures in an open container are isobars.

2. Nuclides that have the same isotopic mass but different atomic number.

Isobaric Having constant pressure.

Isobars Isobars are nuclide having the same mass number but different atomic numbers. ^{54}Cr and ^{54}Fe, ^{112}Cd and ^{112}Sn are isobars.

Isochore A contour line that corresponds to values measured at identical volumes. For example, a curve on a plot of gas pressure measured at different temperatures in a rigid container is an isochore.

Isochoric Having constant volume.

Isodesmic reaction A reaction (actual or hypothetical) in which the types of bonds that are made in forming the products are the same as those which are broken in the reactants, e.g.

$PhCOOH + p\text{-}ClC_6H_4COO^- \; PhCOO^- + p\text{-}ClC_6H_4COOH$

$ClCH{=}CH_2 + ClCH_2CH_2Cl \; CH_2{=}CH_2 + Cl_2CHCH_2Cl$

Such processes have advantages for theoretical treatment. The Hammett equation as applied to equilibria (cf. (a)) essentially deals with isodesmic processes.

Isoelectric point Isoelectric point is the pH of a solution or dispersion at which the net charge on the macromolecules or colloidal particles is zero. In electrophoresis there is no motion of the particles in an electric field at the isoelectric point.

Isoelectronic Two or more molecular entities are described as isoelectronic if they have the same number of valence electrons and the same structure, i.e. number and connectivity of atoms, but differ in some of the elements involved. Thus

CO, N_2 and NO^+ are isoelectronic

$CH_2=C=O$ and $CH_2=N=N$ are isoelectronic

CH_3COCH_3 and $CH_3N=NCH_3$ have the same number of electrons, but have different structures, hence they are not described as isoelectronic.

Isoentropic A reaction series is said to be isoentropic if the individual reactions of the series have the same standard entropy of activation.

Isoenzymes Multiple forms of enzymes arising from genetically determined differences in primary structure. The term does not apply to those derived by modification of the same primary sequence.

Isoequilibrium relationship A relationship analogous to the isokinetic relationship but applied to equilibrium data. The equation defining the isoequilibrium temperature is

$_rH$ - $_rS$ = constant

where H and S are enthalpy and entropy of reaction, respectively. See also isokinetic relationship.

Isokinetic relationship When a series of structurally related substrates undergo the same general reaction or when the reaction conditions for a single substrate are changed in a systematic way, the enthalpies and entropies of activation sometimes satisfy the relation

H - S = constant

where the parameter is independent of temperature. This equation (or some equivalent form) is said to represent an "isokinetic relationship". The temperature T = (at which all members of a series obeying the isokinetic relationship react at the same rate) is termed the "isokinetic temperature".

Supposed isokinetic relationships as established by direct correlation of H with S are often spurious and the calculated value of is meaningless, because errors in H lead to compensating errors in S. Satisfactory methods of establishing such relationships have been devised.

Isolated double bonds Double bonds separated by more than one carbon-carbon single bond.

Isoleucine $(CH_3)_2CHCH(NH_2)COOH)$

A naturally occuring amino acid with a nonpolar side chain.

Isolobal The term is used to compare molecular fragments with each other and with familiar species from organic chemistry. Two fragments are isolobal if the number, symmetry properties, approximate energy, and shape of the frontier orbitals and the number of electrons in them are similar.

Isomer One of several species (or molecular entities) that have the same atomic composition (molecular formula) but different line formulae or different stereochemical formulae and hence different physical and/or chemical properties.

Isomerase An enzyme of EC class 5, which catalyzes the isomerization of a substrate.

Isomeric Adjective derived from isomer.

Isomerism The relationship between isomers.

Isomerization A chemical reaction, the principal product of which is isomeric with the principal reactant. An intramolecular isomerization that involves the breaking or making of bonds is a special case of a molecular rearrangement.

Isomerization does not necessarily imply molecular rearrangement (e.g. in the case of the interconversion of conformational isomers).

Isomers Isomers are substances which have the same molecular formula but which have different structural formulae. Isomers are distinct chemical compounds, which have different physical and chemical properties. Structural isomerism are very common in organic chemistry.

Isometric, anisometric Two molecular entities that are superposable or can be made superposable by reflection of one of them in a mirror are isometric; otherwise they are anisometric.

Isomorphism Isomorphism is the existence of two or more substances that have the same crystal structure, so that they form solid solutions.

Isoprene C_5H_5 Acolorless volatile liquid which is insoluble in water, soluble in alcohol, and very reactive because of its low vapor pressure and double bonds. From an atmospheric chemistry point of view, isoprene is emitted from trees and plants, globally on the order of teragrams (10^{12} g) per year. This reactive, natural hydrocarbon may influence the oxidative/ reductive balance in the biosphere where its concentrations are significant, in regions such as rainforests and large temporal forests. Most isoprene is directly emitted into the atmosphere from plant surfaces and some no doubt condenses on particulate matter. Emission rates from trees are positively correlated with leaf temperature and therefore, to a degree, time of day. Ultimately the carbon in isoprene is oxidized to carbon monoxide and finally carbon dioxide

Isopropylbenzene (Cumene, C9H12) A colorless liquid with a gasoline odor, that is produced by petroleum refining, and combustion of petroleum products. It is a primary skin and eye irritant. Isopropylbenzene is used in paint thinner, additives to high octane aviation fuel, and the production of acetone. In the atmosphere it reacts with photochemically generated hydroxyl radicals. Its half life in the atmosphere is aproximately 49 hours.

Isosbestic point This term is usually employed with reference to a set of absorption spectra, plotted on the same chart for a set of solutions in which the sum of the concentrations of two principal absorbing components, A and B, is constant. The curves of absorbance against wavelength (or frequency) for such a set of mixtures often all intersect at one or more points, called isosbestic points.

Isosbestic points are commonly met when electronic spectra are taken (a) on a solution in which a chemical reaction is in progress (in which case the two absorbing components concerned are a reactant and a product, A + B), or (b) on a solution in which the two absorbing components are in equilibrium and their relative proportions are controlled by the concentration of some other component, typically the concentration of hydrogen ions, e.g., an acid-base indicator equilibrium.

$A\ B + H^{+}{}_{aq}$

The effect may also appear (c) in the spectra of a set of solutions of two unrelated non-interacting components having the same total concentration. In all these examples, A (and/or B) may be either a single chemical species or a mixture of chemical species present in invariant proportion.

If A and B are single chemical species, isosbestic points will appear at all wavelengths at which their molar absorption coefficients (formerly called extinction coefficients) are the same. (A more involved identity applies when A and B are mixtures of constant proportion.)

If absorption spectra of the types considered above intersect not at one or more isosbestic points but over progressively changing wavelength, this is prima facie evidence in case (a) for the formation of a reaction intermediate in substantial concentration (A C B), in case (b) for the involvement of a third absorbing species in the equilibrium, e.g.

$A\ B + H^{+}{}_{aq}\ C + 2\ H^{+}{}_{aq}$

or in case (c) for some interaction of A and B, e.g.,

A + B C

Isoselective relationship A relationship analogous to the isokinetic relationship, but applied to selectivity data of reactions. At the isoselective temperature, the selectivities of the series of reactions following the relationship are identical.

Isosteres Molecules or ions of similar size containing the same number of atoms and valence electrons.

Isosteric Having identical valence electron configurations.

Isotherm A contour line that corresponds to values measured at identical temperatures. For example, curves on a plot of gas pressure measured at different volumes in a constant temperature bath are isotherms.

Isothermal process Isothermal process is a thermodynamic process in which the temperature of the system does not change.

Isotone Compare with isotope.

One of a group of atoms or ions with nuclei that contain the same number of neutrons but different numbers of protons.

Isotones Isotones are nuclides having the same neutron number N but different atomic number Z.

Isotonic Isotonic solution.

Refers to solutions that have equal osmotic pressure.

Isotope The isotopes of an element are the atom of different atomic mass for that element. All atoms of an element must have the same atomic number (i.e. the same number of protons in its nucleus) However, they may have different atomic mass (i.e. due to numbers of neutrons in the nucleus). The isotopes of an element have different physical properties. However, because they all have the same number of electrons, their chemical properties are identical.

Isotopes are separated by the principal described in Graham's Law which uses diffusion as the method of separation.

By using a simple mass spectrometer, F.W.Aston discovered that all atoms of a particular element do not have the same mass. In the Mass Spectrometer, atoms are given a charge and accelerated in an electric field so that they move at high speeds. When this beam of charged atoms passes through an electric field and a magnetic field, all particles of the same mass are focused on a line. The particles strike a photographic plate or detector so that their positions are recorded. These atoms of different mass number but having the same atomic number are called Isotopes.

Hydrogen has three isotopes.

The first isotope (called Hydrogen) has a mass number of 1 and has a single proton in the nucleus.

The second isotope (called Deuterium) has a mass number of 2, and has one proton and one neutron in the nucleus.

The third isotope (called Tritium) has a mass number of 3, and has one proton and two neutrons in its nucleus.

Chlorine consists of two isotopes :

The first isotope has a mass number of 35 which has 18 neutrons and

The second isotope has a mass number of 37 which has 20 neutrons in the nucleus.

The observed atomic weight of chlorine is 35.46, which indicates that chlorine is a mixture of these two isotopes, and that the isotope with mass number 35 is the more abundant atom present. Indeed, the isotope of mass number 35 is approximately three times more abundant than the isotope of mass number 37.

Joule

The joule is the unit of energy in the SI system of units. It is the energy dissipated when a current of one ampere flows through a resistance of one ohm. Ketones

Ketones are organic compounds of carbon, hydrogen and oxygen, where the ketone functional group, >C=O is contained within the molecule. The are the compounds which form a homologous series which can be represented by the general formula R.CO.R', where R and R' are alkyl groups, and may be the same or different groups.

Ketones are formed by the oxidation of secondary alcohols.

(O)

R2CHOH ⇒ R.CO.R' + H2

Secondary

Ketone

Alcohol

The chemistry of the ketones is governed largely by the presence of the Carbonyl Group, >C=O. The polarity of this group means that the electron-deficient carbon atom of the carbonyl group is susceptible to attack by reagents which are rich in electrons.

The first member of the homologous series of ketones is acetone, CH3.CO.CH3. Other aliphatic ketones are named by naming the groups attached to the carbonyl group and adding the word ketone. Using systematic nomenclature, the longest chain carrying the carbonyl group, >C=O, is made the parent structure and compounds named by replacing the "e" of the parent alkane by the suffix "-one". The positions of substituents are indicated by numbers, where the carbonyl carbon atom is allocated the lowest possible number.

Isotope effect The effect on the rate or equilibrium constant of two reactions that differ only in the isotopic composition of one or more of their otherwise chemically identical components is referred to as a kinetic isotope effect (see isotope effect, kinetic) or a thermodynamic (or equilibrium) isotope effect (see isotope effect, thermodynamic), respectively.

Isotope effect, equilibrium See isotope effect, thermodynamic.

Isotope effect, heavy atom An isotope effect due to isotopes other than those of hydrogen.

Isotope effect, intramolecular A kinetic isotope effect observed when a single substrate, in which the isotopic atoms occupy equivalent reactive positions, reacts to produce a non-statistical distribution of isotopologue products. In such a case the isotope effect will favor the pathway with lower force constants for displacement of the isotopic nuclei in the transition state.

Isotope effect, inverse A kinetic isotope effect which $k^l/k^h < 1$, i.e. the heavier substrate reacts more rapidly than the lighter one, as opposed to the more usual "normal" isotope effect, in which $k^l/k^h > 1$. The isotope effect will normally be "normal" when the frequency differences between the isotopic transition states are smaller than in the reactants. Conversely, in inverse isotope effect can be taken as evidence for an increase in the corresponding force constants on passing from the reactant to the transition state.

Isotope effect, primary A kinetic isotope effect attributable to isotopic substitution of an atom to which a bond is made or broken in the rate-controlling step or in a pre-equilibrium step of a specified reaction is termed a primary isotope effect. The corresponding isotope effect on the equilibrium constant of a reaction in which one or more bonds to isotopic atoms are broken, is called a "primary equilibrium isotope effect". See also isotope effect, secondary.

Isotope effect, secondary A kinetic isotope effect that is attributable to isotopic substitution of an atom to which bonds are neither made nor broken in the rate-controlling step or in a pre-equilibrium step of a specified reaction, and is therefore not a primary isotope effect, is termed a secondary isotope effect. One speaks of, (etc.) secondary isotope effects, where , (etc.) denote the position of isotopic substitution relative to the reaction centre. The corresponding isotope effect on the equilibrium constant of such a reaction is called a "secondary equilibrium isotope effect".

Secondary isotope effects have been discussed in terms of the conventional electronic effects of physical organic chemistry, e.g. induction, hyperconjugation, hybridization, etc., since these properties are determined by the electron distribution, that depends on vibrationally averaged bond lengths and angles which vary slightly with isotopic substitution. While this usage is legitimate, the term "electronic isotope effect" should be avoided, because of the misleading implication that such an effect is electronic rather than vibrational in origin. See also isotope effect, steric.

Isotope effect, solvent A kinetic or equilibrium isotope effect resulting from change in the isotopic composition of the solvent.

Isotope effect, steric A secondary isotope effect attributed to the different vibrational amplitudes of isotopologues. For example, both the mean and mean-square amplitudes of vibrations associated with C-H bonds are greater than those of C-D bonds. The greater effective bulk of molecules containing the former may be manifested by a steric effect on a rate or equilibrium constant.

Isotope exchange A chemical reaction in which the reactant and product chemical species are chemically identical but have different isotopic composition. In such a reaction the isotope distribution tends towards equilibrium (as expressed by fractionation factors) as a result of transfers of isotopically different atoms or groups.

Isotopic abundance Compare with natural abundance. The fraction of atoms of a given isotope in a sample of an element.

Isotopic perturbation, method of NMR shift difference measurement of the isotope effect on a fast (degenerate) equilibrium between two, except for isotopic substitution, species which are equivalent. This can be used to distinguish a rapidly equilibrating mixture with time-averaged symmetry from a single structure with higher symmetry.

Isotopic scrambling The achievement, or the process of achieving, an equilibrium distribution of isotopes within a specified set of atoms in a chemical species or group of chemical species.

Isotopologue A molecular entity that differs only in isotopic composition (number of isotopic substitutions), e.g. CH_4, CH_3D, CH_2D_2.

Isotopomer Isomers having the same number of each isotopic atom but differing in their positions. The term is a contraction of "isotopic isomer".

Isotopomers can be either constitutional isomers (e.g. $CH_2DCH{=}O$ and $CH_3CD{=}O$) or isotopic stereoisomers (e.g. (R)- and (S)-CH_3CHDOH or (Z)- and (E)-$CH_3CH{=}CHD$). Iupac stereochemical terminology (1993).

Isotropy Lack of anisotropy; the property of molecules and materials of having identical physical properties in all directions.

Iterative deconvolution Multistep application of deconvolution where successively smaller sublibraries are prepared and tested to identify individual active members of a combinatorial library.

IUPAC (international union of pure and applied chemists) The standards body that among other things, makes recommendations regarding the names of newly discovered elements and forms other chemistry related standards (such as the labeling of groups on the periodic table).

J

J = -D(dc/dx) Where D is called the diffusion coefficient.

J value The coupling constant between two peaks in an NMR signal. Given in units of Hz.

Jacob volhard (1834-1910) Born in Darmstadt and he studied chemistry at Giessen under his fellow Darmstadter Justus von Liebig, whom he followed to Munich. After working in London (under Hofmann) and Marburg (under Kolbe), he became Professor of Organic Chemistry at Munich, briefly at Erlangen and finally in Halle. Volhard is best known for his development of volumetric analysis, notably the titration of silver with ammonium thiocyanate, but he also synthesised creatine in 1868. His name is associated with the Hell-Volhard-Zelinski bromination (1881-1887) and the Volhard-Erdmann cyclisation (1885).

Jet Stream The major atmospheric air currents centers over the Pacific and Atlantic Oceans, North America and over the Middle East. These systems arc fairly robust from period to period, and across different observational data sets. The most important jet stream from our perspective at Sam Houston State University, in Huntsville Texas, USA, is the Pacific Jet, which helps set the storm tracks over the west coast of North America—like most northern hemispheric weather—moving towards the east.

Johann carl wilhelm ferdinand tiemann (1848-1899) Born in Rübeland, in the Harz, and originally trained as a pharmacist at the polytechnic in Brunswick. He then switched to chemistry and worked as an assistant to August Wilhelm Hofmann (1818-1892), preparing TNT for the first time in 1870. Tiemann became a silent partner to Haarmann in Holzminden and Georges de Laire (1836-1908) in Paris, and carried out research on fragrances. He later elucidated the structure of the interrelated terpenes. His name is associated with the Reimer-Tiemann reaction (1876) and the Tiemann rearrangement of amide oxides (1891).

Johann Friedrich Wilhelm; Adolf (von) Baeyer (1835-1917). Born in Berlin, the son of a Prussian military surveyor who became a lieutenant-general. He studied chemistry at Heidelberg under Bunsen, then followed Kekulé to Ghent. he taught at the Gewerbeschule in Berlin before joining the newly Germanised University of Strassburg. Baeyer then succeeded Liebig at Munich where he remained for the rest of his career. He also acted as a consultant and research leader for Badische Anilin- & Soda-Fabrik. His major work was the structure and synthesis of indigo. He also discovered the phthalein dyes and synthesised oxonium compounds. His centric formula for benzene is well-known. Baeyer ennobled by Ludwig II in 1885, and was awarded the Nobel Prize in 1905 for his work on dyes and hydroaromatic compounds. His name is associated with the Baeyer-Drewson indigo synthesis from o-nitrobenzaldehyde, the Baeyer-Villiger oxidation of ketones to esters or lactones, the (discredited) Baeyer photosynthesis equation, the Baeyer strain theory of alicyclic compounds and the Baeyer test for olefins (decolourisation of glacial acetic acid and potassium permanganate).

John william baker (1898-1967) Born in London, he went up to Imperial College to study chemistry, but almost immediately became a chemist in the Royal Engineers in Mesopotamia. After returning to Imperial College where he did research under Sir Jocelyn Thorpe, he briefly became a schoolmaster. He then joined Ingold at the University of Leeds, where he remained for the rest of his career. Baker's name is associated with the Baker-Nathan effect, where the reactivity of (e.g.) alkyl chlorides is the opposite of that expected from the inductive effect. Baker and W. S. Nathan incorrectly assumed it was a result of hyperconjugation (a term coined by R. S. Mulliken in 1939), but nowadays it is usually explained by differential solvation. Baker published Hyperconjugation in 1953 and Electronic Theories of Organic Chemistry in 1958. He was an active member of the Chemical Society and the Congregational Church.

Joseph karl anton fischer (1901-1978) Born in Mainz, he took his doctorate in engineering under Ruff at the Technische Hochschule in Breslau. After working as an analytical chemist for Degussa and its associated companies for many years, he became Professor of Analytical Chemistry at the University of Frankfurt in 1951.

Joule Joule (J) is the SI derived unit of energy, work, and heat. The joule is the work done when the point of application of a force of one newton is displaced a distance of one metre in the direction of the force (J = N m). The unit is named after the British scientist James Prescott Joule (1818-1889).

Joule-thomson coefficient Joule-Thomson coefficient (μ) is a parameter which describes the temperature change when a gas expands adiabatically through a nozzle from a high pressure to a low pressure region. It is defined by

$$\mu = \frac{(\partial T)}{(\partial p)_H}$$

where H is enthalpy.

K

Kaiser test Analytical method for the determination of primary amines. Particularly useful for resin-bound analysis as the chromophoric product is released into solution allowing quantitation by colorimetry.

Kamlet-taft solvent parameters Parameters of the Kamlet-Taft solvatochromic relationship which measure separately the hydrogen bond donor, hydrogen bond acceptor, and dipolarity/ polarizability (*) properties of solvents as contributing to overall solvent polarity . Kamlet, Abboud and Taft (1981).

Karl andreas hofmann (1870-1940) Born in Ansbach, Bavaria, and studied chemistry under Adolf von Baeyer at Munich. Became professor of inorganic chemistry at the Technische Hochschule, Charlottenburg. Best known for his textbook of inorganic chemistry, first published in 1918.

Karl fischer (1901-1958) He introduced the eponymous method of water determination in 1935. Fischer worked for a German petroleum company (the DEA Group), and became the leader of its research laboratories.

Karl josef bayer (1847-1904) Born in Bielitz, Silesia, he initially trained as an architect (his father was an architect) before switching to the study of chemistry at Wiesbaden. After working in a steelworks, he took his doctorate under Bunsen at Heidelberg. After teaching briefly at the University of Brünn, he worked in various chemical factories in Russia. His name is associated with the Bayer process for the purification of bauxite.

Karl ludwig reimer (1845-1883) Born in Leipzig, but grew up in Berlin. He studied chemistry at Göttingen, Griefswald and Heidelberg, and took his doctorate under August Wilhem Hofmann (1818-1892) in 1871. He then entered industry, but continued his research. In 1876, he discovered a new synthesis of salicylaldehyde from phenol, which he developed in collaboration with his former Berlin colleague Ferdinand Tiemann (1848-1899). Reimer soon dropped out of this research, presumably to concentrate on the development of Haarmann and Reimer of Holzminden, founded in 1876 and now part of Bayer. His place was taken by Karl Ludwig Reimer (1856-1921).

Karl ludwig reimer (1856-1921) Was also the son of a Berlin publisher and may have been a relative. The younger Reimer also went into industry after taking his doctorate and eventually worked in the potash industry at Stassfurt.

Karl theophil fries (1875-1962) Born at Kiedrich near Wiesbaden on the Rhine, and was educated at Darmstadt Polytechnic and Marburg, where he took his doctorate under Theodor Zincke (1843-1928). After working for the chemical firm of Hoechst, he returned to Marburg in 1906 and became a professor. In 1918, he became Professor of Chemistry at Brunswick Polytechnic. When he retired in 1938, Fries returned to teaching at Marburg. He was very interested in polycyclic compounds, and fused ring heterocyclic compounds. His name is associated with the Fries rearrangement of phenolic esters (1908).

Karl-friedrich schmidt (1887-1971) Born in Heidelberg and studied chemistry in his home town and at the University of Munich, before taking his doctorate under Theodor Curtius (1857-1928) at Heidelberg. He became Professor of Organic Chemistry at the Swedish University at Åbo (or Turku) in Finland in 1922, returning to Heidelberg three years later. He then became a board director of the pharmaceutical firm Chemische Fabrik Knoll of Ludwigshafen, but set up his own laboratory in 1936.

Katabatic winds Strong winds associated with mountains and valleys. They occur when cold, dense air is produced by radiational cooling flows downhill due to the pull of gravity to a lower, warmer region such as depressions and valleys.

Kekulé structure (for aromatic compounds) A representation of an aromatic molecular entity (such as benzene), with fixed alternating single and double bonds, in which interactions between multiple bonds are assumed to be absent.

Kelvin Kelvin (K) is the SI base unit of thermodynamic temperature. The kelvin, unit of thermodynamic temperature, is the fraction 1/273.16 of the thermodynamic temperature of the triple point of water. The unit is named after the British scientist Sir. W. Thompson, Lord Kelvin (1824-1907).

Kelvin scale Scale for measuring temperature that sets zero degrees at the point of which molecular motion desists, that is, the point at which there is no heat. It is also known as the absolute scale.

Kenneth bailey (1909-1963) Born at Alsagers Bank, near Stoke-on-Trent, he studied chemistry at Birmingham and at Imperial College, where he studied under Astbury. He then worked briefly with Edwin Cohn at Harvard Medical School. He then moved to Cambridge, where he remained for the rest of his career. He studied the chemistry of proteins, mechanism of muscle contraction and of blood coagulation, notably myosin and ATPase.

Kenneth claude bailey (1896-1951) His birthplace is not recorded, but probably Dublin. After serving in the First World War, he won a classical studentship at Trinity College Dublin in 1921 at the height of the troubles, before taking a doctorate in chemistry at the University of Toulouse. He then became a fellow of TCD and subsequently Professor of Physical Chemistry at the University of Dublin. He had two major interests, the chemical writings of the elder Pliny and the inhibition of chemical reactions.

Ketones Ketones are compounds in which a carbonyl group is bonded to two carbon atoms: $R_1R_2C=O$ (neither R may be H). They are derived by oxidation of secondary alcohols. The simplest member of the series is acetone, $(CH_3)_2CO$.

Kilo-. (k) Prefix used in the SI system meaning "one thousand of". For example 1 km means "one thousand meters"; 2.8 kg could also be written "2.8×10^3 g" or "2800 g".

Kilogram (kg)

The kilogram (kg) is the base unit of mass in the SI system of units. The standard kilogram is a 1 kg corrosion resistant platinum/iridium cylinder, carefully preserved in the suburbs of Paris (with a backup copy kept in Gaithersburg, Maryland.) Efforts are underway to replace these artifacts by redefining the kilogram as the mass of a certain number of silicon atoms.

Kinetic ambiguity Synonymous with kinetic equivalence .

Kinetic control (of product composition) The term characterizes conditions (including reaction times) that lead to reaction products in a proportion governed by the relative rates of the parallel (forward) reactions in which the products are formed, rather than by the respective overall equilibrium constants. See also thermodynamic control .

Kinetic electrolyte effect (kinetic ionic-strength effect) The general effect of an added electrolyte (i.e. an effect other than, or in addition to, that due to its possible involvement as a reactant or catalyst) on the observed rate constant of a reaction in solution. At low concentrations (when only long-range coulombic forces need to be considered) the effect on a given reaction is determined only by the ionic strength of the solution and not by the chemical identity of the ions. For practical purposes, this concentration range is roughly the same as the region of validity of the Debye-Hückel limiting law for activity coefficients. At higher concentrations, the effect of an added electrolyte depends also on the chemical identity of the ions. Such specific action can usually be interpreted as the incursion of a reaction path involving an ion of the electrolyte as reactant or catalyst, in which case the action is not properly to be regarded just as a kinetic electrolyte effect.

Kinetic electrolyte effects are usually (too restrictively and therefore incorrectly) referred to as "kinetic salt effects".

A kinetic electrolyte effect ascribable solely to the influence of the ionic strength on activity coefficients of ionic reactants and transition states is called a "primary kinetic electrolyte effect".

A kinetic electrolyte effect arising from the influence of the ionic strength of the solution upon the pre-equilibrium concentration of an ionic species that is involved in a subsequent rate-limiting step of a reaction is called a "secondary kinetic electrolyte effect". A common case encountered in practice is the effect on the concentration of hydrogen ion (acting as catalyst) produced from the ionization of a weak acid in a buffer solution.

Kinetic energy Compare with potential energy.

The energy an object possesses by virtue of its motion. An object of mass m moving at velocity v has a kinetic energy of ½mv^2.

Kinetic equivalence Two reaction schemes are kinetically equivalent if they imply the same rate law .

For example, consider the two schemes (i) and (ii) for the formation of C from A:

(i) $A \underset{k_{-1}\,.\,OH^-}{\overset{k_1\,.\,OH^-}{\rightleftharpoons}} B \xrightarrow{k_2} C$

Providing that B does not accumulate as a reaction intermediate.

$$\frac{d[C]}{dt} = \frac{k_1k_2[A][OH^-]}{k_2 + k_{-1}[OH^-]} \quad (1)$$

(ii) $A \underset{k_{-1}\,.\,OH^-}{\overset{k_1\,.\,OH^-}{\rightleftharpoons}} B \xrightarrow[OH^-]{k_2} C$ Providing that B does not accumulate as a reaction intermediate

$$\frac{d[C]}{dt} = \frac{k_1k_2[A][OH^-]}{k_1 + k_2[OH^-]} \quad (2)$$

Both equations for d(C)/dt are of the form

$$\frac{d[C]}{dt} = \frac{r[A][OH^-]}{1 + s[OH^-]} \quad (3)$$

where r and s are constants (sometimes called "coefficients in the rate equation"). The equations are identical in their dependence on concentrations and do not distinguish whether OH^- catalyses the formation of B, and necessarily also its reversion to A, or is involved in its further transformation to C. The two schemes are therefore kinetically equivalent under conditions to which the stated provisos apply.

Kinetic isotope effect See isotope effect .

Kinetic product The product that forms the fastest. (This product has the lowest energy of activation.)

Kinetic resolution The achievement of partial or complete resolution by virtue of unequal rates of reaction of the enantiomers in a racemate with a chiral agent (reagent, catalyst, solvent, etc.).

Kinetic theory of gases The kinetic theory of gases explains the physical and thermodynamic properties of gases as molecules in relatively unconstrained random motion, which are free to fill completely the container in which they are contained. Brownian Movement and the observed diffusion of gases are vindications of this theory. The mathematical treatment of the random motion of molecules was published by Albert Einstein in 1905AD and predicts properties which have been verified experimentally.

Kinetics The study of the rates at which chemical reactions occur and the influence of physical and chemical conditions of these rates.

Kjeldhal flask Kjeldhal flask is a round bottom flask with a long wide neck that is used in the determination of nitrogen by Kjeldhal's method.

Kjeldhal's method Kjeldhal's method is an analytical method for determination of nitrogen in certain organic compounds. The method was developed by the Danish chemist Johan Kjeldahl (1849-1900).

It involves addition of a small amount of anhydrous potassium sulfate to the test compound, followed by heating mixture with concentrated sulfuric acid, often with a catalyst such as copper sulfate. As a result ammonia is formed. After alkalyzing the mixture with sodium hydroxyde, the ammonia is separated by distilation, collected in standard acid, and the nitrogen determined by back-titration.

Klaus H. hofmann (1911) Born in Karlsruhe, studied chemsitry at ETH, Zurich, under Ruzicka. Professor at University of Pittsburgh School of Medicine. Studied biotin and fatty acids containing cyclopropene and cyclopropane rings; also synthesised an ACTH-related polypeptide.

Knocking Knocking is the premature spontaneous explosion of the petrol mixture during the compression stroke in a petrol engine before the mixture is ignited by a spark. Knocking causes significant loss of power and may result in damage to the engine.

Knorr resin Amide-releasing, acid-cleavable solid support.

Knudsen number Knudsen number (Kn) is a dimensionless quantity used in fluid mechanics, defined by

$Kn = \lambda/l$

where λ is mean free path and l is length.

Kohlrausch's law If a salt is dissolved in water, the conductivity of the solution is the sum of the conductances of the anions and cations. The law, which depends on the independent migration of ions, was deduced experimentally by German chemist Friedrich Kohlrausch (1840-1910).

Konimeter Defined as an instrument utililized to determine the amount of dust particles in a sample of air.

Koppel-palm solvent parameters Parameters to measure separately the ability of a solvent to enter into non-specific solvent-solute interactions (permittivity and refractive index n_D) and specific solvent-solute interaction (solvent basicity or nucleophilicity B and solvent acidity or electrophilicity E) as contributing to overall solvent polarity.

Kosower Z-value See Z-value .

Kurt arndt (1873-1946) Born in Frankfurt on the Oder and studied chemistry at the Techische Hochschule in Berlin. He took his doctorate at the University of Basle, where he became an assistant to Kahlbaum. He spent the rest of his career at the Techische Hochschule in Berlin, mainly within the Institute of Applied Electrochemistry. He worked on a wide range of electrochemical processes, including the production of calcium, sodium perborate (thereby enabling the production of Persil) and silicon carbide, and the improvement of batteries.

K_w. Kw. Symbol for the autoprotolysis constant for water, equal to 1.01×10^{-14} at 25°C.

L

L, u Stereodescriptors of diastereoisomers with two chirality elements. When the two chirality elements are both R or both S the molecular entity is described as having an l relationship (l = like), when one chirality element is R and the other S the molecular entity is described as having a u relationship (u = unlike). The notation can be extended to molecules with more than two chirality elements whereby the l,u relationships of the chiral elements are considered in pairs.

Labile The term has loosely been used to describe a relatively unstable and transient chemical species or (less commonly) a relatively stable but reactive species. It must therefore not be used without explanation of the intended meaning.

Laccase A copper-containing enzyme, 1,4-benzenediol oxidase , found in higher plants and microorganisms. Laccases are multicopper oxidases of wide specificity that carry out one-electron oxidation of phenolic and related compounds, and reduce O_2 to water. The enzymes are polymeric and generally contain one each of type 1, type 2, type 3 copper centers per subunit, where the type 2 and type 3 are close together forming a trinuclear copper cluster.

Lachrymators Any substance that causes the eye to tear, redden, and or swell; eye irritant examples include formaldehyde, A-Brombenzyl cyanide, Chloroacetophenone, and Dichlorodimethyl ether .

Lactoferrin An iron-binding protein from milk, structurally similar to the transferrins.

Ladder synthesis Strategy for library assembly where a portion of compound is capped following incorporation of each building block, such that the final sample comprises a mixture of all possible truncated products. This may be designed such that approximately equimolar quantities of each truncated form are present as an approach to gain maximal diversity, or such that each truncate is present in a small amount relative to the fully elaborated product. In the latter case, analysis of the pattern of products serves to identify the parent and is termed ladder encoding.

Lambert's law The intensity of radiation passing through a material decays exponen-tially with path length b.

Laminar flow Laminar flow is smooth, uniform, non-turbulent flow of a gas or liquid in parallel layers, with little mixing between layers. It is characterized by small values of the Reynolds number.

Lanthanides Lanthanides (lanthanons, lanthanoids or rare-earth elements) are a series of fourteen elements in the periodic table, generally considered to range in proton number from cerium to lutetium inclusive. It was convenient to divide these elements into the cerium group or light earth: cerium (Ce), praseodymium (Pr), neodymium (Nd), promethium (Pm), samarium (Sm), europium (Eu); and the yttrium group or heavy earths: gadolinium (Gd), terbium (Tb), dysprosium (Dy), holmium (Ho), erbium (Er), thulium (Tm), ytterbium (Yb) i lutetium (Lu). The position of lanthanum is somewhat equivocal and, although not itself a lanthanides, it is often included with them for comparative purpose. The lanthanides are sometimes simply called the rare earths. Apart from unstable Pm the lanthanides are actually not rare. Cerium is the 26. most abundant of all elements, 5 times as abundant as Pb. All are silvery very reactive metals.

Lanthanide contraction An effect that causes sixth period elements with filled 4f subshells to be smaller than otherwise expected. The intervention of the lanthanides increases the effective nuclear charge, which offsets the size increase expected from filling the n=6 valence shell. As a consequence, sixth period transition metals are about the same size as their fifth period counterparts.

Lapse rate The theoretical rate of decrease of temperature with increasing height in the atmosphere. If heat is neither gained nor lost from the air parcel under consideration, then the lapse rate is said to be adiabatic. For air parcels of differing water content this value is different.

Laser Laser (light amplification by stimulated emission of radiation) is a light amplifier usually used to produce monochromatic coherent radiation in the infrared, visible, and ultraviolet regions of the electromagnetic spectrum.

Latent heat Latent Heat (L) is the quantity of heat absorbed or released when a substance changes its physical phase at constant temperature (e.g. from solid to liquid at the melting point or from liquid to gas at the boiling point).

Latent heat of evaporation of water The latent heat of evaporation of water is the amount of heat required to convert 1 gm. of water at 100 degreeC into steam at 1000C.

Latent heat of fusion of water The latent heat of fusion of water is the amount of heat required to convert 1 gm. of ice at 0 degrees C into water at the same temperature.

Latent period (synonym: latency) Delay between exposure to a disease-causing agent and the appearance of manifestations of the disease. After exposure to ionizing radiation, for instance, there is a latent period of five years, on average, before development of leukemia, and more than 20 years before development of certain other malignant conditions. The term "latent period" is often used as synonym with "induction period", that is, the period between exposure to a disease-causing agent and the appearance of manifestations of the disease. It has also been defined as the period from disease initiation to disease detection .

Lattice A regular array of ions or atoms.

Lattice constants Lattice constants are parameters specifying the dimensions of a unit cell in a crystal lattice, specifically the lengths of the cell edges and the angles between them.

Lattice energy The lattice energy of an ionic solid is the energy released when the requisite number of moles of each of the component ions is brought together from infinity to form 1 mole of the ionic crystalline solid.

This is also the energy required to remove the ions from their equilibrium positions in the crystal to infinity.

A negative value for the heat change due to the lattice energy indicates that the process is exothermic (i.e. heat is being released by the system). A positive value for the heat change due to the lattice energy indicates that the process is endothermic (i.e. heat is being absorbed by the system).

Law Natural law; scientific law.

Natural laws summarize patterns that recur in a large amount of data. Unlike human laws, natural laws don't forbid or permit; they describe.

Law of chemical equilibrium The law of chemical equilibrium (also called the law of mass action) states that the rate at which a substance reacts is proportional to its active mass (i.e. to its molar concentration).

Thus, the velocity of a chemical reaction is proportional to the product of the concentration of the reactants.

Law of combining volumes. Gay-Lussac's law.

When gases react, they do so in a definite proportion by volume, if the volumes are measured at the same pressure and temperature. For example, in the reaction $N_2(g) + 3 H_2(g) = 2 NH_3(g)$, 3 liters of hydrogen will react with 1 liter of nitrogen to give 2 liters of ammonia if all volumes are measured at the same temperature and pressure.

Law of conservation of energy The law of conservation of energy

states that in any chemical transformation, energy is conserved (i.e. it is neither created or destroyed).

Although that energy cannot be destroyed or created during the course of a chemical reaction, it may be altered from one form of energy to another form.

Law of conservation of mass There is no change in total mass during a chemical change. The demonstration of conservation of mass by Antoine Lavoisier in the late 18th century was a milestone in the development of modern chemistry.

Law of constant composition The law of constant composition

states that a pure compound has a fixed and invariable composition (i.e. the compound always contains the same elements, united together in the same fixed proportions by mass).

Law of definite composition States a pure compound has a fixed and invariable composition. It was discovered by a French chemist, Joseph Louis Prout (1754-1826) .

Law of definite proportions When two pure substances react to form a compound, they do so in a definite proportion by mass. For example, when water is formed from the reaction between hydrogen and oxygen, the 'definite proportion' is 1 g of H for every 8 g of O.

Law of mass action The law of mass action

states that the rate at which a chemical reaction proceeds is directly proportional to the product of the active masses of the reacting substances.

The active mass is defined as the concentration measured in terms of gram-molecular weights per unit volume, usually 1 liter. This law was formulated by Guldberg and Waage in 1864AD.

Law of multiple proportions When one element can combine with another to form more than one compound, the mass ratios of the elements in the compounds are simple whole-number ratios of each other. For example, in CO and in CO_2, the oxygen-to-carbon ratios are 16:12 and 32:12, respectively. Note that the second ratio is exactly twice the first, because there are exactly twice as many oxygens in CO_2 per carbon as there are in CO.

Law of octaves The law of octaves

states that each eight element had similar chemical properties.

The law of octaves was an early attempt to summarise the regularly repeating chemical properties of the elements with increasing atomic number.

LC50 This abbreviation is used for the exposure concentration of a toxic substance lethal to 50% of a test population.

LDPE (Low density polyethylene) This structure is similar to *high density polyethylene* except that it is a less rigid and less dense form of the molecule.

Le chatelier's principle Le Chatelier's principle predicts that when a stress is applied to an equilibrium mixture, the equilibrium will shift to relieve the stress. Stresses include temperature changes, pressure changes, and changes in the concentrations of species in the mixture. For example, increasing the concentration of a reactant drives the reaction forward; increasing the concentration of a product drives it backward.

Lead compounds as Pollutants There is anxiety that lead compounds are significant pollutants in the environment. The use of lead compounds as additives to fuels is being phased out.

Lead discovery The process of identifying active new chemical entities, which by subsequent modification may be transformed into a clinically useful drug.

Lead generation The term applied to strategies developed to identify compounds which possess a desired but non-optimized biological activity.

Lead optimization The synthetic modification of a biologically active compound, to fulfill stereo electronic, physicochemical, pharmacokinetic and toxicological clinical usefulness.

Least nuclear motion, principle of The hypothesis that, for given reactants, the reactions involving the smallest change in nuclear positions will have the lowest energy of activation . (It is also often simply referred to as principle of least motion.) .

Leaving group An atom or group (charged or uncharged) that becomes detached from an atom in what is considered to be the residual or main part of the substrate in a specified reaction.

For example, in the heterolytic solvolysis of benzyl bromide in acetic acid

$PhCH_2Br + AcOH\ PhCH_2OAc + HBr$

the leaving group is Br^- ; in the reaction

$MeS^- + PhCH_2N^+Me_3\ MeSCH_2Ph + NMe_3$

the leaving group is NMe_3; in the electrophilic nitration of benzene, it is H^+. The term has meaning only in relation to a specified reaction. The leaving group is not, in general, the same as the substituent group present in the substrate (e.g. bromo and trimethylammonio in the substrates of the first two examples above.)

A slightly different usage of the term prevails in the (non-mechanistic) naming of transformations, where the actual substituent group present in the substrate (and also in the product) is referred to as the leaving group.

Leffler's assumption See Hammond principle .

Left to right convention Arrangement of the structural formulae of the reactants so that the bonds to be made or broken form a linear array in which the electrons move from left to right..

Leghemoglobin A monomeric hemoglobin synthesised in the root nodules of leguminous plants that are host to nitrogen-fixing bacteria. Has a high affinity for dioxygen and serves as an oxygen supply for the bacteria.

Lepton The lepton is a fundamental subatomic particle.

Leucine $(CH_3)_2CHCH_2CH(NH_2)COOH$ A naturally occuring aliphatic amino acid with a nonpolar side chain.

Levelling effect The tendency of a solvent to make all Brønsted acids whose acidity exceeds a certain value appear equally acidic. It is due to the complete transfer to a protophilic solvent of a hydron from a dissolved acid stronger than the conjugate acid of the solvent. The only acid present to any significant extent in all such solutions is the lyonium ion . For example, the solvent water has a levelling effect on the acidities of $HClO_4$, HCl, and HI: aqueous solutions of these acids at the same (moderately low) concentrations have the same acidities. A corresponding levelling effect applies to strong bases in protogenic solvents.

Levorotatory Compare with dextrorotatory.

Having the property of rotating plane-polarized light counterclockwise.

Lewis acid A molecular entity (and the corresponding chemical species) that is an electron-pair acceptor and therefore able to react with a Lewis base to form a Lewis adduct , by sharing the electron pair furnished by the Lewis base. For example:

Me_3B (Lewis acid) + $:NH_3$ (Lewis base) $Me_3B^- -N^+H_3$ (Lewis adduct)

Lewis acidity The thermodynamic tendency of a substrate to act as a Lewis acid . Comparative measures of this property are provided by the equilibrium constants for Lewis adduct formation of a series of Lewis acids with a common reference Lewis base .

Lewis acids and bases The Bronsted-Lowry definition of acids and bases does not encompass all chemical compounds that exhibit acidic and basic properties. A more general definition is that of Lewis acids and bases:

A Lewis acid is an electron-pair acceptor. And

A Lewis base is an electron-pair donor.

These definitions are broader than the Bronsted-Lowry definition in that they include many compounds that do not have protons, but exhibit acid/base behavior. The Lewis definition encompasses the Bronsted-Lowry definition: In the reaction of H^+ and OH^-, H^+ is a Lewis acid because it accepts an electron pair from the OH^-. Since the OH^- donates an electron pair we call it a Lewis base.

As an example not described by the Bronsted-Lowry definition, Al^{3+} in water is a Lewis acid. It reacts with water to form an aqua complex: the Al^{3+} accepts the electron-pair from water molecules. In this example the water acts as a Lewis base.

Equilibria of Acids and Bases(h)

We will use the Bronsted-Lowry definitions for acids and bases:

Acids are species that donate a proton (H^+). and bases are species that accept a proton.

Lewis adduct The adduct formed between a Lewis acid and a Lewis base .

Lewis base A molecular entity (and the corresponding chemical species) able to provide a pair of electrons and thus capable of coordination to a Lewis acid , thereby producing a Lewis adduct .

Lewis basicity The thermodynamic tendency of a substance to act as a Lewis base . Comparative measures of this property are provided by the equilibrium constants for Lewis formula (electron dot or lewis structure) Molecular structure in which the valency electrons are shown as dots so placed between the bonded atoms that one pair of dots represents two electrons or one (single) covalent bond , e.g.

A double bond is represented by two pairs of dots, etc. Dots representing non-bonded outer-shell electrons are placed adjacent to the atoms with which they are associated, but not between the atoms. Formal charges (e.g. +, -, 2+, etc.) are attached to atoms to indicate the difference between the positive nuclear charge (atomic number) and the total number of electrons (including those in the inner shells), on the formal basis that bonding electrons are shared equally between atoms they join. (Bonding pairs of electrons are usually denoted by lines, representing covalent bonds, as in line formulae.)

Lewis structure Electron dot structure; dot structure.

A model pioneered by Gilbert N. Lewis and Irving Langmuir that represents the electronic structure of a molecule by writing the valence electrons of atoms as dots. Pairs of dots (or lines) wedged between atoms represent bonds; dots drawn elsewhere represent nonbonding electrons.

Lewis theory of acids Lewis introduced the idea that acids and bases had a reciprocal relationship and he introduced the concept of lewis conjugate acid-base pairs.

A lewis acid is an electron pair acceptor.

A lewis base is an electron pair donor.

Using this definition, ammonia is easily classified as a base.

NH3 + H3O(+) $\Rightarrow$ NH4(+) + H2O

Lewis Lewis Lewis Lewis

Base Acid Acid Base

In particular, water and the hydrated proton, H3O(+), are a lewis conjugate acid-base pair. Thus, the hydrated proton is able to donate a proton, (i.e. it is a Lewis Acid) and water is able to receive a proton by donating its lone pair of electrons (i.e. it is a lewis base).

H(+) + H2O $\Rightarrow$ H3O(+)

Lewis Lewis

Base Acid

It should be noted that the hydrogen ion exist in aqueous solution as hydrated proton.

Libraries from libraries **Strategy for accelerating library production, whereby an existing library is subjected to a relatively minor modification in order to generate a new library, thus avoiding the majority of chemical development and rehearsal required for a new library.**

Library equivalent The number of samples which equals the number of compounds in the library. Particularly applied to libraries in which individual beads are encoded, where one library equivalent is the number of beads which equals the number of compounds in the library.

Lifetime (mean lifetime) The lifetime of a chemical species which decays in a first-order process is the time needed for a concentration of this species to decrease to 1/e of its original value. Statistically, it represents the mean life expectancy of an excited species. In a reacting system in which the decrease in concentration of a particular chemical species is governed by a first-order rate law , it is equal to the reciprocal of the sum of the (pseudo)unimolecular rate constants of all processes which cause the decay. When the term is used for processes which are not first order, the lifetime depends on the initial concentration of the species, or of a quencher, and should be called apparent lifetime instead. See also chemical relaxation , half-life , rate of reaction .

Ligand 1. In inorganic chemistry, a molecule or ion that binds to a metal cation to form a complex.

2. In biochemistry, a molecule that binds to a receptor, having a biological effect.

Ligands complexed to a metal ion will raise the energy of some of its d orbitals and lower the energy of others. The difference in energy is called the crystal field splitting energy.

Ligand design The design of ligands using structural information about the target to which they should bind, often by attempting to maximize the energy of the interaction.

Ligand field Ligand field theory is a modified crystal field theory that assigns certain parameters as variables rather than taking them as equal to the values found for free ions, thereby taking into account the potential covalent character of the metal-ligand bond.

Ligand field theory Ligand field theory is a description of the structure of crystals containing a transition metal ion surrounded by nonmetallic ions (ligands). It is based on construction of molecular orbitals involving the d-orbitals of the central metal ion and combinations of atomic orbitals of the ligands.

Ligase An enzyme of EC class 6, also known as a synthetase, which catalyzes the formation of a bond between two substrate molecules coupled with the hydrolysis of a diphosphate bond of a nucleoside triphosphate or similar co-substrate.

Light year Light year (ly) is a unit of distance used in astronomy, defined as the distance light travels in one year in a vacuum (ly = 9.46052973'"10^{15}).

Lignins Lignins are macromolecular constituents of wood related to lignans, composed of phenolic propylbenzene skeletal units, linked at various sites and apparently randomly.

Lime Lime, also called quicklime or calcium oxide, is a white solid compound formed by heating calcium in oxygen or by the thermal decomposition of calcium carbonate.

Limestone Limestone is a rock composed of calcium carbonate.

Limit of detection (i) The smallest amount, or lowest concentration, of a given substance that a given procedure will detect (WHO, 1980); (ii) for a pesticide residue it is the lowest concentration that can be qualitatively detected in a specified commodity (WHO, 1976).

Limit of determination (Pesticide residue) The limit of determination of a method of analysis is the lowest concentration of a pesticide residue that can be quantitatively measured in the specified commodity with an acceptable degree of certainty (WHO, 1976).

Limit of quantitation (LOD) quantitative detection limit; limit of determination.

The smallest detectable concentration an analytical instrument can determine at a given confidence level. IUPAC defines the quantitative detection limit as C_{ld} = ks/m, where k is 10, s is the standard deviation of instrument readings taken on a "blank" (a solution with zero concentration of analyte), and m is the slope of a plot of instrument response vs. concentration, as calculated by linear regression.

Limiting reactant Limiting reagent.

The reactant that limits the amount of product produced in a chemical reaction. For example, mixing one mole of $H_2(g)$ with one mole of O_2 produces one mole of steam ($H_2O(g)$), with half a mole of $O_2(g)$ remaining. The hydrogen gas limits the amount of steam produced in this case.

Limiting reagent The reactant that will be exhausted first.

Limiting reagents The limiting reagents in a chemical reaction is that reactant which are those that determines the maximum quantity of the product that may be formed.

Line formula A two-dimensional representation of molecular entities in which atoms are shown joined by lines representing single or multiple bonds, without any indication or implication concerning the spatial direction of bonds. For example, methanol is represented as

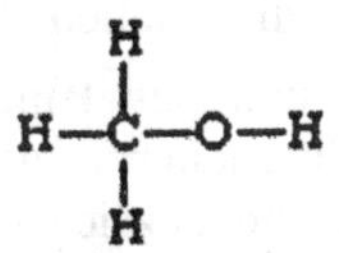

(The term should not be confused with the representation of chemical formulae by the "Wiswesser line notation", a method of string notation. Formulae in this notation are also known as "Wiswesser line formulae".)

Line spectra The line spectra of an element are the specific wavelengths of light which are emitted when the element is excited in an electric arc. The spectra are the discontinuous lines produced by excited atoms and ions as they fall back to a lower energy level.

Line spectrum Line spectra; line emission spectrum. Compare with band spectrum and continuous spectrum.

A emission spectrum that contains very sharp peaks, corresponding to transitions between states in free atoms. For example, the line spectrum of hydrogen contains 4 sharp lines in the visible part of the spectrum.

Linear free-energy relation A linear correlation between the logarithm of a rate constant or equilibrium constant for one series of reactions and the logarithm of the rate constant or equilibrium constant for a related series of reactions. Typical examples of such relations (also known as linear Gibbs energy relations) are the Brønsted relation , and the Hammett equation (see also -value).

The name arises because the logarithm of an equilibrium constant (at constant temperature and pressure) is proportional to a standard free energy (Gibbs energy) change, and the logarithm of a rate constant is a linear function of the free energy (Gibbs energy) of activation. The area of physical organic chemistry which deals with such relations is commonly referred to as "Linear Free-Energy Relationships".

Linear gibbs energy relation See linear free-energy relation .

Linear solvation energy relationships Equations involving the application of solvent parameters in linear or multiple (linear) regression expressing the solvent effect on the rate or equilibrium constant of a reaction.

Line-shape analysis Determination of rate constants for a chemical exchange from the shapes of spectroscopic lines of dynamic processes. The method is most often used in nuclear magnetic resonance spectroscopy.

Linker Bifunctional chemical moiety attaching a compound to a solid support or soluble support which can be cleaved to release compounds from the support. A careful choice of linker allows cleavage to be performed under appropriate conditions compatible with the stability of the compound and assay method.

Lipids Lipids are a loosely defined term for substances of biological origin that are soluble in nonpolar solvents. They consist of saponifiable lipids, such as glycerides (fats and oils) and phospholipids, as well as nonsaponifiable lipids, principally steroids.

Lipinski's rule of 5 Set of criteria for predicting the oral bioavailability of a compound on the basis of simple molecular features (MolWt $<=$ 500, clogp $<=$ 5.0, Hbond donors $<=$ 5, Hbond acceptors $<=$5, Free-rotation bonds $<=$ 10). Often used to profile a library or virtual library with respect to the proportion of drug - like members which it contains. An algorithm, developed by Christopher A. Lipinski (of Pfizer) and colleagues, in which many of the cutoff numbers are five or multiples of five. There are actually four rules, and Pfizer has developed a additional number of criteria for adoption of lead candidates.

Lipophilic Lipophilicity.

Refers to a substance's solubility in fat. Lipophilicity can be measured by shaking the substance with a two-phase mixture of water and 1-octanol and observing the final concentrations of the substance in the two layers. Lipophilic substances will move into the 1-octanol layer, while hydrophilic substances stay in the water.

Lipophilicity The affinity of a molecule or a moiety for a lipophilic environment. It is commonly measured by its distribution behavior in a biphasic system, either liquid-liquid (e.g., partition coefficient in 1-octanol/water) or solid/liquid (retention on reversed-phase high performance liquid chromatography (RP-HPLC) or thin-layer chromatography (TLC) system).

Lipoxygenase A non-heme iron enzyme which catalyzes the insertion of O_2 into polyunsaturated fatty acids to form hydroperoxy derivatives.

Liquefaction of air In this process of the liquefaction of air, air from the atmosphere is drawn through purifiers to remove carbon dioxide and the resulting gas mixture is compressed. The heat of compression is removed by a cooler and the compressed air is passed through a chilled condenser to remove moisture. The dry compressed air is then allowed to expand through a jet and is cooled by adiabatic expansion. This adiabatic process is repeated a number of times, where the temperature drops at each stage. Finally, the air becomes so cold that it liquefies on leaving the jet. The liquid air obtained from the liquefier is richer in oxygen than gaseous air.

Liquid A state of matter that has a high density and is incompressible compared to a gas. Liquids take the shape of their container but do not expand to fill the container as gases do. Liquids diffuse much more slowly than gases.

Liquid phase chemistry Synthetic process employing a macromolecular soluble support.

L-isomer Compare with D-.

Liter Liter (l, L) is a synonym for cubic decimeter ($L = dm^3$).

Lithium. (Li) Li.

Element 3, atomic weight 6.939. The lightest alkali metal, used in special-purpose metal alloys and other industrial applications.

Lithology The branch of mineralogy that deals with the study of rocks and their composition.

Lithosphere Lithosphere is the outer layer of the solid earth, extending from the base of the mantle to the surface of the crust.

Litmus A mixture of pigments extracted from certain lichens that turns blue in basic solution and red in acidic solution.

Litmus paper Litmus test. Paper impregnated with litmus, usually cut in narrow strips. Dipping red litmus paper into a basic solution turns it blue; dipping blue litmus paper into an acidic solution turns it red.

Loading Characteristic property of a solid support which describes the amount of a specific chemical species per unit mass of the support.

Lock and key model Lock-and-key.

A model that explains the role of enzymes in chemical reactions by assuming that the reactants fit into the enzyme like a key fits into a lock.

London forces Attractive forces between apolar molecules, due to their mutual polarizability. They are also components of the forces between polar molecules. Also called "dispersion forces". See also van der Waals forces .

Long range transport This type of transport involves large amounts of air pollutants in the atmosphere. If the pollutants are light enough or are propelled high enough in the atmosphere, they may be picked up and transported long distances. If the air pollutants travel within the moving air mass greater than 100 kilometers, than this is considered long range transport.

Loose ion pair See ion pair.

Low-spin In any coordination entity with a particular d^n ($1 < n < 9$) configuration and a particular geometry, if the n electrons are distributed so that they occupy the lowest possible energy levels, the entity is a low-spin complex. If some of the higher energy d orbitals are occupied before all the lower energy ones are completely filled, then the entity is a high-spin complex.

Low spin complex Low-spin complex. Compare with high spin complex.

A metal-ligand complex with fewer unpaired electrons than the uncomplexed metal ion. When a strong ligand complexes the metal ion, the crystal field splitting is large and some electrons pair rather than occupying the higher energy d orbitals.

Lower explosive limit (LEL) The lower limit of flammability of a gas or vapour at normal ambient temperatures expressed as percentage of the gas or vapour in air by volume. This limit is assumed constant for temperatures up to 130oC.

LPG (liquid petroleum gas) is a mixture of short hydrocarbons with most of the volume being propane and butane. LPG is considered an alternative fuel that burns cleaner than gasoline because of the very short hydrocarbon chains that tend to not produce a high volume of CO_2, NOx radicals. LPG is now being found in the air in the unburned state most likely from leaking pipes and tanks, in this physical state LPG is now being seen a contributor to urban air pollution.

Ludwig gattermann (1860-1920) Born in Goslar, the son of a baker. He studied chemistry briefly at Leipzig and Heidelberg, before coming under the influence of Carl Liebermann (1842-1914) at the polytechnic (Gewerbe Akademie) in Berlin. He took his doctorate at Göttingen where he met Sandmeyer. He became Meyer's assistant and followed him to Heidelberg. He became Professor of Chemistry at the University of Freiburg im Breisgau in 1900, where he was succeeded by Hermann Staudinger (1881-1965). Gattermann's name is associated with the Gattermann reaction (1890), the Gattermann-Koch reaction (1897), the Gattermann aldehyde synthesis (1898) and the Gattermann-Stika pyridine synthesis (1916). He was very interested in aromatic acids, but also prepared various exotic inorganic compounds, notably the preparation of pure (and highly explosive) nitrogen trichloride. His famous textbook of practical organic chemistry was fondly known (or perhaps more accurately derided) as "Gattermanns Kochbuch".

Ludwig knorr (1859-1921) Studied chemistry at the university in his home town, Munich, and at Heidelberg. He became a close colleague of Emil Fischer and followed him to the Universities of Erlangen and Würzburg, before being appointed Professor of Chemistry at Jena in 1888. Knorr gave his name to syntheses of pyrazoles (1883), pyrroles (1884) and quinolines (1886). He was the brother-in-law of Oskar Piloty.

Ludwig wolff (1857-1919) Born in Neustadt in the Palatinate and was educated at the University of Würzburg and Munich Polytechnic. He was one of the ambitious young German chemists who flocked to the recently Germanised University of Strassburg, where Rudolf Fittig (1835-1910) and Adolf von Baeyer were professors and Emil Fischer was a fellow student. In 1891, he became Professor of Analytical Chemistry at Jena, where he worked with Ludwig Knorr. His name is associated with the Wolff-Kishner reaction (1911-2) and the Wolff rearrangement (1912).

Lumen Lumen (lm) is the SI derived unit of luminous flux. The lumen is the luminous flux emitted in a solid angle of one steradian by a point source having a uniform intensity of one candela (1lm =1 cd'"sr).

Luminous flux Luminous flux (Φ) is the intensity of light from a source multiplied by the solid angle. The SI unit is lumen.

LUMO See frontier orbitals .

Lux Lux (lx) is the SI derived unit of illuminance. The lux is the illuminance produced by a luminous flux of one lumen uniformly distributed over a surface of one square metre ($lx = lm \cdot m^{-2}$).

Lyase An enzyme of EC class 4, which catalyzes the separation of a bond in a substrate molecule.

Lyate ion The anion produced by hydron removal from a solvent molecule. For example, the hydroxide ion is the lyate ion of water.

Lyman series Lyman series is the series of lines in the spectrum of the hydrogen atom which corresponds to transitions between the ground state (principal quantum number n = 1) and successive excited states.

Lyonium ion The cation produced by hydronation of a solvent molecule. For example, $CH_3OH_2^+$ is the lyonium ion of methanol.

Lysine Lys.

A naturally occurring amino acid with an amine group on its side.

M

M See helicity.

MacromoleculeMacromolecule is a molecule of high relative molecular mass (molecular weight), the structure of which essentially comprises the multiple repetition of units derived, actually or conceptually, from molecules of low relative molecular mass.

Macrophages Blood cells which are able to ingest a wide variety of particulate materials. They are a type of phagocyte.

Macroporous resin Polymer which contains a permanent network of pores independent of the state of swelling of the resin. This class of resin thus displays much better solvent tolerance than gel-type resins.

Macroscopic diffusion control See mixing control.

Magic acid See superacid.

Magnetic circular dichroism (MCD) A measurement of circular dichroism of a material which is induced by a magnetic field applied parallel to the direction of the measuring light beam.Materials which are achiral still exhibit MCD (the Faraday effect), since the magnetic field leads to the lifting of the degeneracy of electronic orbital and spin states and to the mixing of electronic states. MCD is frequently used in combination with absorption and CD studies to effect electronic assignments. The three contributions to the MCD spectrum are the A-term , due to Zeeman splitting of the ground and/or excited degenerate states, the B-term, due to field-induced mixing of states, and the C-term, due to a change in the population of molecules over the Zeeman sublevels of a paramagnetic ground state. The C-term is observed only for molecules with ground-state paramagnetism, and becomes intense at low temperatures; its variation with field and temperature can be analyzed to provid magnetic parameters of the ground state, such as spin, g-factor, and zero-field splitting. Variable-temperature MCD is particularly effective in identifying and assigning electronic transitions originating from paramagnetic chromophores.

Magnetic equivalence Nuclei having the same resonance frequency in nuclear magnetic resonance spectroscopy and also identical spin-spin interactions with the nuclei of a neighbouring group are magnetically equivalent. The spin-spin interaction between magnetically equivalent nuclei does not appear, and thus has no effect on the multiplicity of the respective NMR signals. Magnetically equivalent nuclei are necessarily also chemically equivalent, but the reverse is not necessarily true.

Magnetic resonance imaging (MRI) The visualization of the distribution of nuclear spins (usually water) in a body by using a magnetic field gradient (NMR imaging). A similar technique, but less widely used, is to visualize the distribution of paramagnetic centres(EPR imaging).

Magnetic susceptibility For paramagnetic materials the magnetic susceptibility may be measured experimentally and used to give information on the molecular magnetic dipole moment, and hence on the electronic structure of the molecules in the material. The paramagnetic contribution to the molar magnetic susceptibility of a material, χ , is related to the molecular magnetic dipole moment m by the Curie relation : χ = constant m^2/T.

Magnetization transfer NMR method for determining kinetics of chemical exchange by perturbing the magnetization of nuclei in a particular site or sites and following the rate at which magnetic equilibrium is restored. The most common perturbations are saturation and inversion, and the corresponding techniques are often called "saturation transfer" and "selective inversion-recovery".

Magnetotactic Able to orient in a magnetic field.

Main group elements The main group elements in the periodic table can be classified as either being metals or non-metals. The vertical columns in the table are called groups. Elements within the same group all have the same number of electrons in their outer shell. Group I Elements are known as alkali metals. Group II Elements are known as the alkaline earth metals and the group of elements between group II and III are known as the Transition Element.

Mainstream smoke (tobacco smoking) The smoke that is inhaled.

Malignant Adjective describing cells in a cancerous growth. See tumour.

Manufacture Iodine is manufactured industrially

from the liquor remaining after the extraction of potassium nitrate from saltpeter found in deposits in Chile,

by extraction from the ash from the combustion of seaweed. In the process, the sea-weeds are burnt in shallow pits and the resulting kelp (i.e. the ash from the process) contains potassium salts and from 0.4 per cent to 1.3 per cent of iodine in the form of iodide salts (e.g. potassium iodide),

in a wet extraction of the seaweed with sodium carbonate solution is prepared and an organic adhesive is precipitated on acidifying of this solution, and

by extraction from salt brine.

Manufacture Lead is manufactured by smelting the gelena in admixture with sand, limestone and iron ore.

3 PbS + 3 O2 $\Rightarrow$ 2 PbO + 2 SO2

PbS + 2 O2 $\Rightarrow$ PbSO4

PbSO4 + SiO2 $\Rightarrow$ PbSiO3 + 2 SO3

The resulting sinter is mixed with coke and discharged into the top of a blast furnace, while air is forced through the bottom.

2 C + O2 $\Rightarrow$ 2 CO

PbO + CO $\Rightarrow$ Pb + CO2

PbSiO3 + CaO + CO $\Rightarrow$ Pb + CaSiO3 + CO2

The products of the smelting, lead and slag are removed at the bottom of the furnace.

Manufacture of ammonia The manufacture of ammonia involves the fixation of atmospheric nitrogen, there by making it available for incorporation into a variety of other compounds. The Haber Process is used industrially for this fixation.

Manufacture of steel The manufacture of steel, which is an alloy of iron and carbon uses the basic-oxygen process, ld process, by blowing oxygen onto molten scrap and pig iron at high pressure. This process removes impurities from the metal in the form of slag and gases. The gases are burned off at high temperature. Various other materials are then added to the iron in order to make different types of steel. Steels containing over 11 per cent of chromium are known as stainless steel.

Marcus equation A general expression which correlates the Gibbs energy of activation (G) with the driving force ($_r$G^{o}') of the reaction:

G = (/4)(1 + $_r$G^{o}'/)^2

where is the reorganization energy and $_rG^{o}$' is the standard free energy of the reaction corrected for the electrostatic work required to bring the reactants together. /4 is the intrinsic barrier of the reaction. Originally developed for outer-sphere electron transfer reactions, the Marcus equation has later been applied also to atom and group transfer reactions.

Martin henry fischer (1879-1962) Born in Gaarden near Kiel, but took his MD at Chicago in 1901. Later became Professor of Physiology at the University of Cincinnati and was closely involved with Wolfgang Ostwald and the promotion of colloid chemistry in the interwar period .

Mask Device which acts as a barrier to the passage of a reagent (often light). A pattern of holes in the mask allows selective passage of reagent and results in a corresponding pattern of reagent deposition or photodeprotection on a surface placed behind the mask. This allows the generation of spatially addressable libraries.

Mass A measurement of the quantity of matter contained in a particle or body regardless of its location in the universe and the gravitational force applied to it. Mass is constant, while weight is affected by the distance of the object from the center of the earth (or other such body). Mass is not affected by gravity.

Mass fraction Mass fraction (w_A) is the ratio of the mass of substance A to the total mass of a mixture. $w_A = \frac{m_A}{\sum m_i}$

Mass mean diameter The diameter of a particle with a mass equal to the mean mass of all the particles in the population.

Mass median diameter The diameter of a particle with the median mass (IAEA, 1978).

Mass number (M,A) Compare with atomic number and atomic weight.The total number of protons and neutrons in an atom or ion. In nuclide symbols the mass number is given as a leading superscript. In isotope names (e. g. carbon-14, sodium-23) the mass number is the number following the element name.

Mass spectrometer The a mass spectrometer is an instrument for the separation and analysis of ions and atoms. In the mass spectrometer a sample (usually gaseous) is ionised and the positive ions produced are accelerated into a high vacuum which is exposed to electric and magnetic fields. Ions of different types can be sorted via a detector into a mass spectrum. The mass spectrum consists of peaks of varying intensity to which mass/charge values can be assigned, and from which molecular structure can be deduced.

Francis William Aston (1877-1945) invented the Mass Spectrometer

Mass spectrometry Mass spectrometry is an analytical technique in which ions are separated according to the mass/charge (m/e) ratio and detected by a suitable detector.

In a mass spectrometer a sample is ionized and the positive ions produced are accelerated into a high-vacuum region containing electric and magnetic fields. These fields deflect and focus the ions onto a detector. A mass spectrum is thus obtained consisting of a series of peaks of variable intensity to which m/e values can be assigned. Different molecules can be identified by their characteristic pattern of lines.

Mass spectrum mass spectra.A plot showing the results of a mass spectrometry experiment, which shows the presence of particles with different masses as a series of sharp, separate peaks. The position of the peaks on the x-axis indicates the mass of the particles; the peak heights indicate the relative abundance of the particles.

Mass-law effect At equilibrium, the product of the activities (or concentrations) of the reacting species is constant. Thus for the equilibrium

A + B C + D

K = (C)(D)/(A)(B) Matrix isolation A term which refers to the isolation of a reactive or unstable species by dilution in an inert matrix (argon, nitrogen, etc.), usually condensed on a window or in an optical cell at low temperature, to preserve its structure for identification by spectroscopic or other means.

Matter Matter is defined as anything that occupies space and that has mass.

These three forms are called the States of Matter :

Solid,

Liquid, and

Gas.

Examples of matter are a book, pen, ruler, the sun, the moon, air, water, milk, etc., since each has mass and occupies space.

Radio waves, heat and light are not matter since they have no mass.

Max georg matthais bamberger (1861-1927) Born in Kirchbichl, near Kufstein in the Tirol, he studied chemistry at the Hochschule für Bodenkultur and the Technische Hochschule in Vienna, before taking his doctorate at Giessen. he then spent the rest of his career at the Technische Hochschule in Vienna. He studied the chemistry of resins, gas-masks, nitrogen-free explosives and the radioactivity of spa waters and minerals. Bamberger was also a member of the International Atomic Weight Commission.

Maximum allowable concentration (MAC) Exposure concentration not to be exceeded under any circumstances.

Maximum residue limit The maximum concentration of a pesticide residue resulting from the use of a pesticide according to good agricultural practice directly or indirectly for the production and/or protection of the commodity for which the limit is recommended. The maximum residue limit should be legally recognized. It is expressed in milligrams of the residue per kilogram of the commodity (WHO, 1976).

Maximum-boiling mixtures A maximum-boiling mixture is an azeotrope having a boiling point higher than those of its components.

MCD See magnetic circular dichroism.

Mean free path Mean free path is the average distance a gas molecule travels between collisions.

Mean life (synonym: mean time, turnover time) The average lifetime of an atomic or nuclear system in a specified state. For an exponentially decaying system, it is the average time for the number of atoms or nuclei in a specified state to decrease by a factor of e (ISO, 1972).

Mechanism A detailed description of the process leading from the reactants to the products of a reaction, including a characterization as complete as possible of the composition, structure, energy and other properties of reaction intermediates, products, and transition states. An acceptable mechanism of a specified reaction (and there may be a number of such alternative mechanisms not excluded by the evidence) must be consistent with the reaction stoichiometry, the rate law, and with all other available experimental data, such as the stereochemical course of the reaction. Inferences concerning the electronic motions which dynamically interconvert successive species along the reaction path (as represented by curved arrows, for example) are often included in the description of a mechanism.

It should be noted that for many reactions all this information is not available and the suggested mechanism is based on incomplete experimental data. It is not appropriate to use the term mechanism to describe a statement of the probable sequence in a set of stepwise reactions. That should be referred to as a reaction sequence, and not a mechanism. See also Gibbs energy diagram.

Mechanism-based inhibition Irreversible inhibition of an enzyme due to its catalysis of the reaction of an artificial substrate. Also called "suicide inhibition".

Median lethal concentration Statistically derived concentration of a chemical in water solution that can be expected to cause death in 50% of given population of organism under defined set of experimental conditions.

Median lethal dose Statistically derived single dose of a chemical that can be expected to cause death in 50% of given population of organism under defined set of experimental conditions (for example oral administration, rat).

Medicinal chemistry A chemistry-based discipline, also involving aspects of biological, medical and pharmaceutical sciences. It is concerned with the invention, discovery, design, identification and preparation of biologically active compounds, the study of their metabolism, the interpretation of their mode of action at the molecular level and the construction of structure-activity relationships.

Medium The phase (and composition of the phase) in which chemical species and their reactions are studied in a particular investigation.

Meisenheimer adduct A cyclohexadienyl derivative formed as Lewis adduct from a nucleophile (Lewis base) and an aromatic or heteroaromatic compound, also called Jackson-Meisenheimer adduct. In earlier usage the term "Meisenheimer complex" was restricted to the typical Meisenheimer alkoxide adducts of nitro-substituted aromatic ethers. Analogous cationic adducts, considered to be reaction intermediates in electrophilic aromatic substitution reactions, are called "Wheland intermediates", and sometimes, inappropriately, -complexes.

Meissner effect Meissner effect is the complete exclusion of magnetic induction from the interior of a superconductor.

Melting Melting is the process by which a solid changes state from a solid to a liquid.

Melting point Melting point (or freezing point) is the temperature at which the solid and liquid phases of a substance are in equilibrium at a specified pressure (normally taken to be atmospheric unless stated otherwise). A pure substance under standard condition of pressure has a single reproducible melting point. The terms melting point and freezing point are often used interchangeably, depending on whether the substance is being heated or cooled.

Member a) Specific compound which is included in a library;b) the uncharacterized physical product of a library synthesis.

MeniscusMeniscus is the concave curve of a liquid surface in a graduate or narrow tube. Caused by surface tension.

Menkes' disease A sex-linked inherited disorder, causing defective gastrointestinal absorption of copper and resulting in copper deficiency early in infancy.

Mercaptans Mercaptans are a traditional term abandoned by IUPAC, synonymous with thiols. This term is still widely used.

Meso A prefix to indicate the presence of a symmetry element of the second kind (see chirality) in a chemical species which is usually one of a set of diastereoisomers that also includes a chiral member. See meso-compound.

Meso compounds Molecules that have chiral centers but are achiral as a result of one or more planes of symmetry in the molecule.

Mesolytic cleavage Cleavage of a bond in a radical ion whereby a radical and an ion are formed. The term reflects the mechanistic duality of the process, which can be viewed as homolytic or heterolytic depending on how the electrons are attributed to the fragments.

Mesomeric effect The effect (on reaction rates, ionization equilibria, etc.) attributed to a substituent due to overlap of its p or pi orbitals with the p or pi orbitals of the rest of the molecular entity. Delocalization is thereby introduced or extended, and electronic charge may flow to or from the substituent. The effect is symbolized by M.

Strictly understood, the mesomeric effect operates in the ground electronic state of the molecule. When the molecule undergoes electronic excitation or its energy is increased on the way to the transition state of a chemical reaction, the mesomeric effect may be enhanced by the electromeric effect, but this term is not much used, and the mesomeric and electromeric effects tend to be subsumed in the term resonance effect of a substituent. See also electronic effect, field effect, inductive effect.

Mesomerism Essentially synonymous with resonance. The term is particularly associated with the picture of pi electrons as less localized in an actual molecule than in a Lewis formula.

The term is intended to imply that the correct representation of a structure is intermediate between two or more Lewis formulae.

Mesopause The upper limit and the coldest portion of the mesosphere. The transition zone between the mesosphere and the thermosphere.

Mesophase The phase of a liquid crystalline compound between the crystalline and the isotropic liquid phase.

Mesosphere In the atmosphere, the region immediately above the stratosphere and innemdiately below the thermosphere. The mesosphere begins about 50 kilometers high at the stratopause and ends about 80 kilometers high at the mesopause. The temperature in the mesosphere decreases sharply with increased altitude.

Mesothelioma A rare cancer known that affects thin membranes surrounding lungs and other internal organs. Mesothelioma has been associated with asbestos exposure. title="Asbestos: A Manufacturing Health Hazard Dating to Prehistoric Times">More information)

Mesozoic Period Period spanning from 225 to 65 million years before present (Myr BP), this period was marked by widespread warmth of the planet's surface, especially in the higher latitudes. During this time the continents were joined in a single supercontinent. This situation most likely facilitated the meridional transport of heat by ocean currents, thus resulting in an even distribution of global heat.

Met A qualifying prefix indicating the oxidized form of the parent protein, e.g. methemoglobin.

Meta Describes the positions of two substituents on a benzene ring that are separated by one carbon.

Metabolic activation Biotransformation of relatively inert chemicals to biologically reactive metabolites (compounds).

Metabolic half-life (Synonym:metabolic half-time) The time required for one half of the quantity of the compound in the body to be metabolically transformed into a modified chemical compound (WHO, 1979).

Metabolic model An analysis and theoretical reconstruction of the way in which the body deals with a specific substance, showing the proportion of the intake that is absorbed, the proportion that is stored and in what tissues, the proportion and rate of breakdown in the body and the subsequent fate of the metabolic products, and the proportion of the substance and the rate at which it is eliminated by different organs (WHO, 1979).

Metabolic transformation (synonym: biotransformation) The chemical transformation of substances that takes place within an organism (WHO, 1979).

Metabolism In general, the sum total of all physical and chemical processes that take place within an organism; in a narrower sense, the physical and chemical changes that take place in a given chemical sub- stance within an organism. It includes the uptake and distribution within the body of chemical compounds, the changes (biotransformations) undergone by such substances, and the elimination of the compounds and of their metabolites (WHO, 1979).

Metabolite A substance resulting from chemical transformation in an organism (WHO, 1979).

Meta directing substituent Any substituent on an aromatic ring that directs in coming electrophiles to the meta position.

Metal A metal is an element which has the physical properties of having 1. a lustrous appearance,

2. being malleable,

3. being tenacity,

4. having high densities, and

5. being good conductors of heat and

6. being good conductors of electricity.

7. Metals react with oxygen to form basic oxides.

Metals react with halogens to form halogen compounds which are stable in presence of water.

Metals are electro-positive elements and form complex salts.

Metallic bond The atoms in a metal are arranged in an orderly manner. Because metals have low ionisation potential, they can easily lose their outermost electrons, which are then free to travel in a cloud between all the atoms of the metal. The electrostatic attractive force between the positively charged ionised atoms of the metal and the negatively charged electronic cloud is responsible for the bonding in the metal. The metallic bond is very strong and accounts for the mechanical strength of metals.

The metallic bond accounts for the physical and chemical properties of metals. The following structure is proposed for the metallic bond. 1. When forming this arrangement each atom of the metal releases its valiancy electrons, which are now free to roam through the intervening spaces between the positive ions. This space of moving electrons is often referred to as the electron gas or electron atmosphere.

2. The positive ions are not free to move, but they are able to vibrate.

3. The mutual sharing of this electron atmosphere by all the positive ions of the metal establishes a kind of diffuse bonding, which is non-directional, and is called the metallic bond.

Metallic chlorides Those metals which are high on the electrochemical series form chlorides which are non-volatile ionic solids and have cubic ionic lattices (e.g. sodium chloride). These ionic crystal lattices, which are composed of metal ions and chloride ions, dissociate in water but the ions themselves are not hydrolysed (they do not react with water).

Many metallic chlorides liberate chlorine when treated with sulphuric acid and manganese dioxide.

Many metallic chlorides liberate hydrogen chloride gas when warmed with concentrated sulphuric acid.

Metalloenzyme An enzyme that, in the active state, contains one or more metal ions which are essential for its biological function.

Metalloid Metalloid (semimetal) is any of a class of chemical elements intermediate in properties between metals and nonmetals. The classification is not clear cut, but typical metalloids are boron (B), silicon (Si), germanium (Ge), arsenic (As), and tellurium (Te). They are electrical semiconductors and their oxides are amphoteric.

Metallo-immunoassay A technique in which antigen-antibody recognition is used, with attachment of a metal ion or metal complex to the antibody. The specific absorption or (radioactive) emission of the metal is then used as a probe for the location of the recognition sites.

Metallothionein A small, cysteine-rich protein that binds heavy metal ions, such as zinc, cadmium and copper in the form of clusters.

Metastable (chemical species) See transient (chemical species).

Metastable state A object that returns to its original position after force is applied.

Metathesis A bimolecular process formally involving the exchange of a bond (or bonds) between similar interacting chemical species so that the bonding affiliations in the products are identical (or closely similar) to those in the reactants. For example:

(The term has its origin in inorganic chemistry with a different meaning, but this older usage is not applicable in physical organic chemistry.)

Meteorology The science of the atmosphere and its direct effects upon the earth's surface. Meteorology is especially concerned with how atmospheric conditions effect the weather.

Methaemoglobin/methaemoglobinaemia In some poisonings, haemoglobin (the substance in the red blood cells to which the oxygen is fixed) may be transformed to methaemoglobin. Methaemoglobin lacks the ability of binding the oxygen, concequently lack of oxygen occures in the organism. When a certain amount of haemoglobi has been transformed to methaemoglobin, mucous memranes and skin become bluish and discolored.

Methane CH_4, a colorless, odorless, flammable, greenhouse gas. It is the simplest of all hydrocarbons with a formula of CH_4. Methane is released naturally into the air from anaerobic environments such as marshes, swamps, and rice fields, and from symbiotic microbes in the guts of ruminant animals (such as cattle, sheep, and camels), and sewage sludge. Methane is released from methane producing bacteria (methanogens) that live in these anaerobic places. Methanogens in termite guts are the source of methane released by termites. The discovery that termites may be a significant source of atmospheric methane is attributed to work by Patrick Zimmerman and his research group members at the National Center for Atmospheric Research in Boulder Colorado

Methane flux The (continuous) flow of methane from one place in the atmosphere to another. It is usually measured as a rate of flow across a given area where methane is produced; for instance, the average amount of methane being released from a rice field of a specific area every hour would be considered the methane flux rate for that field. **Methane mono-oxygenase** A metalloenzyme that converts methane and dioxygen to methanol using NADH as co-substrate. Two types are known, one containing a dinuclear oxo-bridged iron center, the other is a copper protein.

Methanesulfonic acid (MSA) This may be a significant component of atmospheric aerosol particles that form from the atmospheric oxidation of dimethylsulfide (DMS), when this DMS is produced by phytoplankton. These aerosol particles can lead to the formation of cloud condensation nuclei, water droplet formation, and hence clouds and storm formation. This therefore may contribute to the turbulent mixing of oceanic surface waters and subsequent upwelling of nutrients that feed those microorganisms (a positive feedback cycle). It is thought that increased emissions of dimethylsulfide, and subsequent methane-sulfonic acid over the oceans from an increase in sea water temperature may have a negative feedback on global warming by increasing cloud formation; although the net effect of clouds on global warming is very unclear at this point.

Methanogens Strictly anaerobic archaea, able to use avariety of substrates (e.g., dihydrogen, formate, methanol, methylamine, carbon monoxide or acetate) as electron donors for the reduction of carbon dioxide to methane.

Methanol Methanol, CH3OH, (i.e. methyl alcohol) is the simplest aliphatic alcohol, and is the first member of the homologous series.

Methanol is a colourless liquid, completely miscible with water and organic solvents and is very hydroscopic.

Methanol has an agreeable odour, and a burning taste.

Methanol is a potent nerve poison.

Methanol has the physical properties

Melting Point : -97 degC

Boiling Point : 65 deg0C

Relative Density : 0.8

Manufacture of Methanol Methanol is made by a method different from that used in the preparation of the other alcohol. The naphtha fraction from the distillation of crude petroleum, is used as a raw material for the manufacture of methanol. When naphtha is reacted with a high steam ratio, under pressure and at high temperature, synthesis gas of low methane content is obtained. Most of the carbon from the naphtha is converted to carbon monoxide and carbon dioxide which can be removed from the gaseous mixture to leave hydrogen of high purity.

1000 degC

C + H2O $\Rightarrow$ CO + H2

Carbon Hydrogen

Monoxide

When a mixture of hydrogen and carbon monoxide in the ratio 2:1 is passed over a catalyst (e.g. a mixture of zinc oxide and chromium oxide) under high pressure and at high temperature, methanol is formed.

400 degC

CO + 2H2 $\Rightarrow$ CH3OH

Methanol

Chemical Properties of Methanol

Combustion of Methanol

Methanol burns with a pale-blue, non-luminous flame to form carbon dioxide and steam.

$2CH_3OH + 3O_2 \Rightarrow 2CO_2 + 4H_2O$

Oxidation of Methanol

Methanol is oxidised

with acidified Potassium Dichromate, $K_2Cr_2O_7$, or

with acidified Sodium Dichromate, $Na_2Cr_2O_7$, or

with acidified Potassium Permanganate, $KMnO_4$,

to form formaldehyde.

(O)

$CH_3OH \Rightarrow H.CHO + H_2O$

Methanol Formaldehyde

If the oxidising agent is in excess, the formaldehyde is further oxidised to formic acid and then to carbon dioxide and water.

(O) (O)

$H.CHO \Rightarrow H.COOH \Rightarrow CO_2 + H_2O$

Formaldehyde Formic

Acid

Catalytic Oxidation of Methanol

The catalytic oxidation of methanol using platinum wire is of interest as it is used in model aircraft engines to replace the sparking plug arrangement of the conventional petrol engine. The heat of reaction is sufficient to spark the engine.

Dehydrogenation of Methanol

Methanol can also be oxidised to formaldehyde by passing its vapour over copper heated to 300 degC. Two atoms of hydrogen are eliminated from each molecule to form hydrogen gas and hence this process is termed dehydrogenation.

Cu

300 degC

$CH_3OH \Rightarrow H.CHO + H_2$

Methanol Formaldehyde

Dehydration of Methanol

Methanol does not undergo dehydration reactions. Instead, in reaction with sulphuric acid the ester, dimethyl sulphate is formed.

concentrated

H_2SO_4

$2\ CH_3OH \Rightarrow (CH_3)_2SO_4 + H_2O$

Methanol Dimethyl Water

Sulphate

Esterification of Methanol

Methanol reacts with organic acids to form esters.

$H(+)$

$CH_3OH + H.COOH \Rightarrow H.COO.CH_3 + H_2O$

Methanol Formic Methyl Water

Acid Formate

Substitution of Methanol with Sodium

Methanol reacts with sodium at room temperature to liberate hydrogen. This reaction is similar to the reaction of sodium with ethanol.

$2\ CH_3OH + 2\ Na \Rightarrow 2CH_3ONa + H_2$

Methanol Sodium Sodium Hydrogen

Methoxide

Substitution of Methanol with Phosphorus Pentachloride

Methanol reacts with phosphorus pentachloride at room temperature to form hydrogen chloride, methyl chloride, (i.e. chloroethane) and phosphoryl chloride.

$CH_3OH + PCl_5 \Rightarrow HCl + CH_3Cl + POCl_3$

Methanol Phosphorus Hydrogen Methyl Phosphoryl

Pentachloride Chloride Chloride Chloride

Substitution of Methanol with Hydrogen Chloride

Methanol reacts with hydrogen chloride to form methyl chloride (i.e. chloromethane) and water.

A dehydrating agent (e.g. zinc chloride) is used.

ZnCl2

CH3OH + HCl ⇒ CH3Cl + H2O

Ethanol Methyl

Chloride

Uses of Methanol

Methanol is an important industrial material and it is used

in the manufacture of formaldehyde, 1. as a solvent in the paint and varnish industry, and

2. as an anti-Freeze in car radiators.

Nitrogen

Nitrogen, N2, is a colourless odourless gaseous non-metallic main group element which belong to Group Vb of the periodic table.

Atomic Number : 7

Atomic Mass : 14.007

Methods of detecting analytes 1. physical means

mass

color

refractive index

thermal conductivity

2. with electromagnetic radiation (Spectroscopy)

absorption

emission

scattering

3. by an electric charge

electrochemistry

mass spectrometry

As you can see there are a limited number of ways to detect an analyte. However, in each of the above general categories there are a large multitude of specific analytical techniques.

Analytical Standards(h)

Standards are materials containing a known concentration of an analyte. They provide a reference to determine unknown concentrations or to calibrate analytical instruments.

The accuracy of an analytical measurement is how close a result comes to the true value. Determining the accuracy of a measurement usually requires calibration of the analytical method with a known standard. This is often done with standards of several concentrations to make a calibration or working curve.

Methyl bromide CH_3Br, this halocarbon is released, to a degree, naturally from the oceans, but is more commonly released from its anthropogenic use as a soil fumigant or pesticide. Methyl bromide is persistent enough to reach the stratosphere where it photochemically decomposes to yield atomic bromine (radical) and proceeds to destroy stratospheric ozone in the same manner as the atomic chlorine radical. On an atom-for-atom basis, stratospheric bromine is more efficient at destroying ozone than is chlorine because the HBr reservoir species is more photochemically active than HCl; however, there is much less of hydrogen bromide in the stratosphere.

Methyl chloride CH_3Cl, this compound supplies chlorine to the stratosphere by occasional volcanic eruptions and by tropospheric to stratospheric transport. Methyl chloride is also produced by seaweed. The natural chlorine content of the stratosphere as a consequence of these sources is about 0.6 ppbv.

Methyl chloroform CH_3CCl_3, a synthetic organic compound that has been used as a substitute for earlier solvents that contributed to air pollution. Methyl chloroform is now known to cause destruction to stratospheric ozone and is scheduled to be phased out by the year 2005.

Methyl radical- $CH_3^{\cdot}$ is formed, along with water, from the reaction of methane (CH_4) and the hydroxyl radical ($OH^{\cdot}$). This begins a series of reactions that eventually produces ozone(O_3).

Methylene See carbene.

Methylidyne See carbyne.

Metre Metre (m) is the SI base unit of length.

The meter is the length of the path travelled by light in vacuum during a time interval of 1/299 792 458 s.

This definition, adopted by the General Conference on Weights and Measure in October 1983, replaced the 1967 definition based on the krypton lamp.

Mho Mho is an archaic name for the SI unit siemens (reciprocal ohm).

Micellar catalysis The acceleration of a chemical reaction in solution by the addition of a surfactant at a concentration higher than its critical micelle concentration so that the reaction can proceed in the environment of surfactant aggregates (micelles). (Rate enhancements may be due, for example, to higher concentration of the reactants in that environment, more favourable orientation and solvation of the species, or enhanced rate constants in the micellar pseudophase of the surfactant aggregate.) Micelle formation can also lead to a decreased reaction rate. See also catalyst.

Micelle Micelle is an electrically charged colloidal particle, usually organic in nature, composed of aggregates of large molecules, e.g., in soaps and surfactants. For aqueous solutions, the hydrophilic end of the molecule is on the surface of the micelle, while the hydrophobic end (often a hydrocarbon chain) points toward the center.

Michaelis-menten kinetics The dependence of the initial rate of conversion of a substrate (S), of the product (P) by an enzyme or other catalyst (E). The simplest mechanism :

$$E + S \underset{k_{-1}}{\overset{k_1}{\rightleftharpoons}} ES \underset{k_{-2}}{\overset{k_2}{\rightleftharpoons}} E + P$$

yields, under initial steady state conditions, and (P) = 0, the Michaelis-Menten equation,

$\nu = \dfrac{V[S]}{K_m + [S]}$ where ν is the rate of conversion (Ms^{-1}), $V = k_2(E)$ is the maximum rate at (S) = ∞ for a particular enzyme/catalyst concentration, k_2 is the turnover number (s^{-1}), and $K_m = (k_{-1} + k_2)/k_1$ is the Michaelis constant under the conditions used. In the case of an impure enzyme or catalyst, (E) is given as gl^{-1} instead of M. This equation leads to a hyperbolic dependence of ν upon (S), which is frequently observed in practice even when (S) is not in great excess over (E).

Microchannel plate A microchannel plate (MCP) consists of an array of glass capillaries (10-25 m inner diameter) that are coated on the inside with a electron-emissive material. The capillaries are biased at a high voltage and like the channeltron, an ion that strikes the inside wall one of the capillaries creates an avalanche of secondary electrons. This cascading effect creates a gain of 10^3 to 10^4 and produces a current pulse at the output.

Schematic of a microchannel plate

Microchannel plates (MCP) are also used as an intensifier for low-intensity light detection with array detectors.

Diatomic Molecule(h)

Two atoms that can form a bond will do so to create a diatomic molecule when they approach each other closely. We describe the energy minimum with a potential energy curve, called a potential well.

Micron Micron (μ) is an obsolete name for micrometer (μm).

Micronutrient A compound essential for cellular growth, being present in concentrations less than about 1 mM in the growth medium (see also trace elements).

Microscopic chemical event See chemical reaction, molecularity.

Microscopic diffusion control (Encounter control) The observable consequence of the limitation that the rate of a bimolecular chemical reaction in a homogeneous medium cannot exceed the rate of encounter of the reacting molecular entities.

If (hypothetically) a bimolecular reaction in a homogeneous medium occurred instantaneously when two reactant molecular entities made an encounter, the rate of reaction would be an encounter-controlled rate, determined solely by rates of diffusion of reactants. Such a hypothetical "fully diffusion controlled rate" is also said to correspond to "total microscopic diffusion control", and represents the asymptotic limit of the rate of reaction as the rate constant for the chemical conversion of the encounter pair into product (or products) becomes large relative to the rate constant for separation (or dissociation) of the encounter pair.

"Partial microscopic diffusion control" is said to operate in a homogeneous reaction when the rates of chemical conversion and of separation are comparable. (The degree of microscopic diffusion control cannot usually be determined with any precision.) See also mixing control.

Microscopic reversibility, principle of In a reversible reaction, the mechanism in one direction is exactly the reverse of the mechanism in the other direction. This does not apply to reactions that begin with a photochemical excitation. See also chemical reaction, detailed balancing.

Midocean ridge A major, elevated, linear feature of the seafloor, consisting of many small, slightly offset segments, with a total length of 200 to 200,000 km. This type of plate boundary occurs in a divergence zone, which is a site where two plates are being pulled apart and new oceanic lithosphere is being created.

Mie scattering Processes by which particles of similar size and electrical characteristics separate or disperse different wavelengths (read colors) of light. First described by Gustav Mie in 1808. Since the sun's visible spectrum contains a mixture of (traditionally ordered) red, orange, yellow, green, blue, indigo, and violet colors, these wavelengths are differentially scattered by particles as they travel through the atmosphere. Red—longer wavelength—light is scattered not much and blue—shorted wavelength—light is scattered much more. This is why the sky appears blue: the sun's blue light is scattered back towards your eyes from atmospheric particles so when you look up—that is, away from the sun, the light you see is light scattered to your eye from atmospheric particles. (You might ask yourself why the sky—viewed at an off angle from the sun—isn't black instead of blue.) Mie scattering is also the reason why sunsets appear red: the sun's red light is NOT scattered as much as blue light by atmospheric particles and so solar blue light is scattered away from your eyes on its way from the sun and red light is scattered less. The result is that more solar red light hits your eyes than blue and sunsets appear red. Notice also that the redness of sunsets increases at the amount of atmospheric particles between you and the sun increase, that is as the sun "goes down" mie scattering increases as the amount of particles between you and the sun increases.

Migration Migration is the movement of ions through the solution towards the electrodes during electrolysis.

Migratory aptitude The term is applied to characterize the relative tendency of a group to participate in a rearrangement. In nucleophilic rearrangements (migration to an electron-deficient centre), the migratory aptitude of a group is loosely related to its capacity to stabilize a partial positive charge, but exceptions are known, and the position of hydrogen in the series is often unpredictable.

Millimeter of mercury Millimeter of mercury (mmHg) is a non-SI unit of pressure, equal to 133.322 Pa. The name is generally considered interchangeable with torr.

Minimum boiling mixture A minimum boiling mixture is an azeotrope have a boiling point lower than those of its components.

For example, 95% alcohol, boils at a lower temperature than either pure ethanol or water.

Miscible When two liquids mix readily to form a solution, they are said to be miscible.

Mitochondria Cytoplasmic organelles of most eukaryotic cells, they are surrounded by a double membrane and produce adenosine 5'-triphosphate as useful energy for the cell by oxidative phosphorylation. The proteins for the ATP-generating electron transport of the respiration chain are located in the inner mitochondrial membrane. Mitochondria contain many enzymes of the citric acid cycle and for fatty acid β-oxidation. They also contain DNA which encodes some of their proteins, the remainder being encoded by nuclear DNA.

Mixed valency This is one of several names, such as 'mixed oxidation state' or 'non-integral oxidation state' used to describe coordination compounds and clusters in which a metal is present in more than one level of oxidation. The importance in biology is due to the often complete delocalization of the valence electrons over the cluster, allowing efficient electron-transfer processes. See also oxidation number.

Mixing control The experimental limitation of the rate of reaction in solution by the rate of mixing of solutions of the two reactants. It can occur even when the reaction rate constant is several powers of 10 less than that for an encounter-controlled rate. Analogous (and even more important) effects of the limitation of reaction rates by the speed of mixing are encountered in heterogeneous (solid/liquid, solid/gas, liquid/gas) systems.

Mixing height The height in the atmosphere where pollution gets mixed and dispersed through the atmosphere. Factors controlling this phenomenon include solar radiation, wind speed, and local surface roughness. The mixing height is located at the top of a boundary layer and at the base of a temperature inversion.

Mixing ratio The fixed proportions in which two or more substances may become combined, such as the amount of nitrogen in the air compared to the rest of the air. Atmospheric scientists routinely discuss the gas phase concentrations of trace components in mixing ratios expressed in ratios of VOLUMES, as in "the present tropospheric mixing ratio for methane is approximately 1.7 ppmv."

Mixture A mixture is composed of two or more constituent substances which are not bound together by a chemical bond. Mixtures can occur in each of the states of matter.

For example, air is a mixture of gases, consisting of approximately 78% nitrogen, 20% oxygen, and smaller quantities of carbon dioxide, argon and water vapour. The composition of the atmosphere is changing gradually over time, due to the burning of fossil fuels and emissions from industry.

Mixtures can be found everywhere, as well as air. Mixtures can be found in practically all foods, the soil in the ground, oil and petrol and many other everyday substances.

Alloys are a very important kind of mixture in modern life. Alloys are made by mixing two or more molten metals uniformly together, and allowing the metal in metal solution to solidify on cooling.

Mobile source (also call non-point source) consist of emissions from on-road and off-road sources. Cars, light, medium and heavy-duty trucks are considered on-road mobile sources, off-road mobile sources; are tractors, lawn mowers, boats, and airplanes. Although the off road sources usually put out more pollution per hour of work than the on-road, the great number of on-road source put out a larger overall volume into the atmosphere.

Möbius aromaticity A monocyclic array of orbitals in which there is a single out-of-phase overlap (or, more generally, an odd number of out-of-phase overlaps) reveals the opposite pattern of aromatic character to Hückel systems; with 4n electrons it is stabilized (aromatic), whereas with 4n + 2 it is destabilized (antiaromatic). In the excited state 4n + 2 Möbius pi-electron systems are stabilized, and 4n systems are destabilized. No examples of ground-state Möbius pi systems are known, but the concept has been applied to transition states of pericyclic reactions The name is derived from the topological analogy of such an arrangement of orbitals to a Möbius strip.

Model A synthetic coordination entity that closely approaches the properties of a metal ion in a protein and yields useful information concerning biological structure and function. Given the fact that the term is also loosely used to describe various types of molecular structures, constructed, for example, in the computer, the term biomimetic is more appropriate.

Moieties Chemical compounds or functional groups or portions of thosecompounds.

Moiety In physical organic chemistry moiety is generally used to signify part of a molecule, e.g. in an ester R^1COOR^2 the alcohol moiety is R^2O. The term should not be used for a small fragment of a molecule.

Molal concentration (Molarity) Molal concentration is the amount of substance dissolved per unit volume and has units of mol per dm cubed or mol per liter cubed. It has the symbol c.

Molality Is the number of moles of solute dissolved in one kilogram of solvent. Note: be careful not to confuse molality and molarity. Molality is represented by a small "m," whereas molarity is represented by an upper case "M."

Molar Mass is a unit that enables scientists to calculate the weight of any chemical substance, be it an element or a compound. Molar mass is the sum of all of the atomic masses in a formula.

Molar mass Molar mass is the mass of one mole of a substance.

M = m/n

It is normally expressed in units of g mol^{-1}, in which case its numerical value is identical with the relative molecular mass.

Molar quantity Molar quantity is often convenient to express an extensive quantity (e.g., volume, enthalpy, heat capacity, etc.) as the actual value divided by amount of substance (number of moles). The resulting quantity is called molar volume, molar enthalpy, etc.

Molar solution A molar solution, M, is one which contains one gram molecular weight (i.e. one mole) of the solute in a liter of solution.

For example, 0.5M HCl means a hydrochloric acid solution containing 0.5 moles of hydrochloric acid per liter of solution.

Molar volume Molar volume is the volume occupied by substance per unit amount of substance. The volume of the gas at 0 °C and 101 325 Pa is 22.4 dm^3mol^{-1}.

Molarity The number of moles of a solute dissolved in a liter of solvent. Note: be careful not to confuse molality and molarity. Molality is represented by a small "m," whereas molarity is represented by an upper case "M."

Mole (*mol*) Is the basic unit of measurement in chemistry. By definition, in modern chemistry, one mole represents the number of carbon atoms in exactly 12 grams of carbon 12 (C_{12}).

Mole fraction Mole fraction (x_A) is the ratio of the amount of substance (number of moles) of substance A to the total amount of substance in a mixture.

$$\chi_A = \frac{n_A}{\sum n_i}.$$

Molecular entity Any constitutionally or isotopically distinct atom, molecule, ion, ion pair, radical, radical ion, complex, conformer etc., identifiable as a separately distinguishable entity.,The degree of precision necessary to describe a molecular entity depends on the context. For example "hydrogen molecule" is an adequate definition of a certain molecular entity for some purposes, whereas for others it is necessary to distinguish the electronic state and/or vibrational state and/or nuclear spin, etc. of the hydrogen molecule.

Molecular formula formula; chemical formula.A notation that indicates the type and number of atoms in a molecule. The molecular formula of glucose is C6H12O6, which indicates that a molecule of glucose contains 6 atoms of carbon, 12 atoms of hydrogen, and 6 atoms of oxygen.

Molecular geometry Shape of a molecule, based on the relative positions of the atoms.

Molecular ion The fragment in a mass spectrum that corresponds to the cation radical (M+)of the molecule. The molecular ion gives the molecular mass of the molecule.

Molecular mechanics calculation An empirical calculational method intended to give estimates of structures and energies for conformations of molecules. The method is based on the assumption of "natural" bond lengths and angles, deviation from which leads to strain, and the existence of torsional interactions and attractive and/or repulsive van der Waals and dipolar forces between non-bonded atoms. The method is also called "(empirical) force-field calculations"..

Molecular metal A non-metallic material whose properties resemble those of metals, usually following oxidative doping; e.g. polyacetylene following oxidative doping with iodine.

Molecular model stick model; ball and stick model; spacefilling model A representation of a molecule. The model can be purely computational or it can be an actual physical object. Stick models show bonds, ball-and-stick models show bonds and atoms, and spacefilling models show relative atomic sizes.

Molecular modeling A technique for the investigation of molecular structures and properties using computational chemistry and graphical visualization techniques in order to provide a plausible three-dimensional representation under a given set of circumstances.

Molecular orbital A one-electron wavefunction describing an electron moving in the effective field provided by the nuclei and all other electrons of a molecular entity of more than one atom. Such molecular orbitals can be transformed in prescribed ways into component functions to give "localized molecular orbitals". Molecular orbitals can also be described, in terms of the number of nuclei (or "centres") encompassed, as two-centre, multi-centre, etc. molecular orbitals, and are often expressed as a linear combination of atomic orbitals.

An orbital is usually depicted by sketching contours on which the wavefunction has a constant value (contour map) or by indicating schematically the envelope of the region of space in which there is an arbitrarily fixed high (say 96%) probability of finding the electron occupying the orbital, giving also the algebraic sign (+ or -) of the wavefunction in each part of that region.

Molecular orbital theory Model for depict ing the location of electrons that allows electrons to delocalize across the entire molecule. A more accurate but less user-friendly theory than the valence-bond model.

Molecular oxygen A molecule that is composed of two oxygen atoms, O_2, and has no color, odor, or taste. It is present in both the atmosphere and dissolved in the oceans and freshwater sources exposed to the atmosphere. Solar radiation with wavelengths less then 242 nm can break it back into oxygen atoms, $O_2 \longrightarrow O + O$. One of these oxygen radicals in turn can combine with O_2 to form ozone, $O_2 + O \longrightarrow O_3$.

Molecular weight Molecular weight, Mr, (also called relative molecular mass) is the ratio of the average mass per molecule of naturally occurring form of an element or compound to one twelfth (i.e. 1 / 12) of the mass of a Carbon-12 atom.

The molecular weight is equal to the sum of the relative atomic masses of all the atoms that comprise the molecule.

Molecular weight The combined weight (as given on the periodic table) of all the elements in a compound.

Molecularity The number of reactant molecular entities that are involved in the "microscopic chemical event" constituting an elementary reaction. (For reactions in solution this number is always taken to exclude molecular entities that form part of the medium and which are involved solely by virtue of their solvation of solutes.) A reaction with a molecularity of one is called "unimolecular", one with a molecularity of two "bimolecular" and of three "termolecular". See also chemical reaction, order of reaction.

Molecule A complete chemical unit. While a molecule is often thought of as consisting of more than one atom, this is not always true. For instance, helium has only one atom per molecule. The oxygen molecule (O_2) contains two atoms, as do chlorine (Cl_2), hydrogen (H_2) and nitrogen (N_2). Some molecules are huge having a molecular weight in the millions. Macromolecules contain literally thousands of atoms. Some macromolecules must remain in tact in order to retain their chemical entity. These are usually proteins. Polymers are chains or networks of repeating sequences of chemical units known as molecules. Examples are polypropylene or cellulose. Molecules of polymers can be broken without destroying the chemical integrity.

Molluscicide A chemical used to kill molluscs.

Molybdenum cofactor (Moco) The molybdenum complex of the molybdopterin prosthetic group (ligand). In the molybdenum cofactor the minimal coordination of the Mo atom is thought to be provided by the chelating dithiolenato group of the molybdopterin and either two oxo or one oxo and one sulfido ligands.

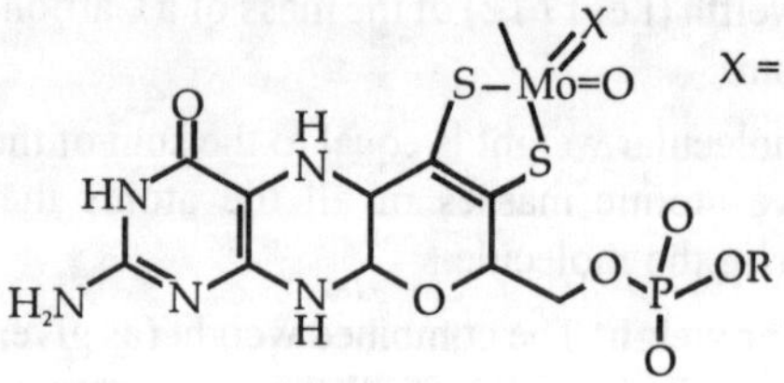

Molybdopterin The prosthetic group associated with the Mo atom of the molybdenum cofactor found in all molybdenum-containing enzymes except nitrogenase. Many of the enzymes catalyze two-electron redox reactions that involve the net exchange of an oxygen atom between substrate and water. The molybdopterin prosthetic group contains a pterin ring bound to a dithiolene functional group on the 6-alkyl side chain. In bacterial enzymes a nucleotide is attached to the phosphate group.

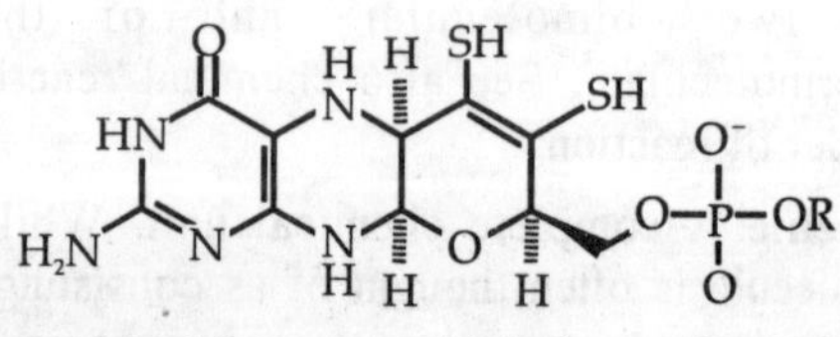

Momentum Momentum (p) is the product of mass and velocity (p = m'''v).

Monatomic Elements that are present in the gaseous state as single atoms. These elements are the noble gasses: He, Ne, Ar, Kr, Xe and Rn.

Monitoring (For health, environmental, and associated technical purposes) The repetitive and continued observation, measurement, and evaluation of health and/or environmental or technical data for defined purposes, according to prearranged schedules in space and time, and using comparable methods for sensing and data collection (WHO, 1980).

Monobasic acids Monobasic Acids are acids that contain one ionisable hydrogen atom in each molecule.

Monohydric alcohol A monohydric alcohol contain one hydroxyl group in the molecule. The general formula of saturated aliphatic alcohols is CnH2n+1OH, where n=1,2,3, etc.

Monomer Member of a building block set which can be repeatedly incorporated into a library to give a set of compounds of repeating structure, e.g. amino acids in a peptide library.

Mono-oxygenase An enzyme that catalyzes the insertion of one atom of oxygen, derived from O_2, into an aromatic or aliphatic compound. The reaction is coupled to the oxidation of a co-substrate such as NAD(P)H or 2-oxoglutarate.

Monosaccharides Monosaccharide is a charbohydrate that cannot be split into smaller units by the action of dilute acids.

Monosaccharide is any of several simple sugars having the formula $C_6H_{12}O_6$; the best-known are glucose, fructose, and galactose. Monosaccharides combine to form more-complex sugars known as oligo- and polysaccharide.

Monsoon Heavy winds characterized by a pronounced seasonal change in direction. Winds usually blow from land to sea in the winter, while in the summer, the flow reverses and precipitation is more common. Monsoons are most typical in India and southern Asia.

Montreal protocol A international environmental agreement (one of the first) to prevent the use of substances that are harmful to ozone, such as chlorofluorocarbons and halons, in order to protect the ozone layer in a global manner. It was agreed upon in 1987 and has been amended repeatedly since that time.

Morbidity Any departure, subjective or objective, from a state of physiological or psychological well-being. In this sense, sickness, illness, and morbid condition are similarly defined and synonymous (Last, 1988).

The WHO Expert Committee on Health Statistics noted in its Sixth report (1959) that morbidity could be measured in terms of three units: (i) persons who were ill; (ii) the illnesses (periods or spells of illnesses) that these persons experienced; and (iii) the duration (days, weeks, etc.) of these illnesses .

Morbidity survey A method for the estimation of the prevalence and/or incidence of disease or diseases in a population. A morbidity survey is usually designed simply to ascertain the facts as to disease distribution, and not to test a hypothesis .

More O'ferrall-jencks diagram Visualization of the potential energy surfaces for a reacting system, as a function of two chosen coordinates. It is particularly useful to discuss structural effects on the transition state geometry for processes occurring either by stepwise or concerted routes. The use of such diagrams, first suggested for elimination reactions (More O'Ferrall (1970)), was later extended to acid-base catalysis and to certain other reactions (JENCKS (1972, 1980)). . More O'Ferrall-Jencks diagrams of energy contours for - elimination reactions as a function of lengths of the two bonds broken. Reaction coordinates follow the dotted lines. Left: E_2 mechanism with "central" character, simultaneous fission of C-Nu and C-H bonds. Right: The result of stabilization of the carbanion, Nu-C-C^-; a continuation of this trend would result in a switch to the E1cb mechanism Structural changes influencing vibrational modes of the transition states cause changes in transition state geometry. Changes in the direction of the reaction coordinate (reactant or product stabilizing- or destabilizing factors) cause changes according to the Hammond principle. Structural changes perpendicular to the reaction coordinate (anti-Hammond effects, perpendicular effects) cause changes opposite to the Hammond behaviour, i.e., the easier the process related to the structural change, the more advanced it will be at the transition state.

Mortar Mortar is a hydraulic cement, the best known of which is known as Portland Cement.

Mössbauer effect Resonance absorption of gamma radiation by specific nuclei arranged in a crystal lattice in such a way that the recoil momentum is shared by many atoms. It is the basis of a form of spectroscopy used for studying coordinated metal ions. The principal application in bioinorganic chemistry is ^{57}Fe. The parameters derived from the Mössbauer spectrum (isomer shift, quadrupole splitting, and the hyperfine coupling) provide information about the oxidation, spin and coordination state of the iron.

Mossbauer spectroscopy(h) The Mossbauer effect is the recoil-free emission of gamma radiation from a solid radioactive material. Since the gamma emission is recoil-free, it can be resonantly absorbed by stationary atoms, i.e., also in a solid. The nuclear transitions are very sensitive to the local environment of the atom and Mossbauer spectroscopy is a sensitive probe of the different environments an atom occupies in a solid material

Approximately 90% of the ^{57}Fe nuclear excited state decays through the intermediate level to produce 14.4 keV gamma radiation. These gamma photons can then be absorbed by ^{57}Fe in a sample.

Motif A pattern of amino acids in a protein sequence which has a specific function, e.g. metal binding.

MRF (Municipal recycle facility) Facilities for recycling municipal waste. There are dirty MRFs, which take all garbage as a collective and sort out the recycables at a central facility, and there are clean MRFs that require recycables to be separated out by the generator and are collected separately.

MRI See magnetic resonance imaging.

MSW (Municipal solid waste) Includes non-hazardous non-liquid waste generated in households, commercial businesses, light industry and commercial establishments. The exact definition varies from one municipality to another.

MTBE The compound methyl tertiary butyl ether (MTBE) was blended in gasoline originally in

1979 to serve as an oxygenate. This would theoretically cause the combustion process in internally combustion engines to be more complete. This effort to improve air quality by producing a fuel that was more readily oxidized and decrease the resulting troposheric ozone production was somewhat effective. A problem arises, however, in the fact that MTBE moves extremely well in water (it may actually move upstream because of its high affinity for water molecules), causes fish kills due to its toxicity to fish, and even at low concentrations, it can render large surface or groundwater sources to be useless. Its use has since been banned in California.

Mu(μ)-symbol See bridging ligand.

Multi-centre bond Representation of some molecular entities solely by localized two-electron two-centre bonds appears to be unsatisfactory. Instead, multi-centre bonds have to be considered in which electron pairs occupy orbitals encompassing three or more atomic centres. Examples include the three-centre bonds in diborane, the delocalized pi bonding of benzene, and bridged carbocations.

Multi-centre reaction A synonym for pericyclic reaction. The number of "centres" is the number of atoms not bonded initially, between which single bonds are breaking or new bonds are formed in the transition state. This number does not necessarily correspond to the ring size of the transition state for the pericyclic reaction. Thus, a Diels-Alder reaction is a "four-centre reaction". This terminology has largely been superseded by the more detailed one developed for the various pericyclic reactions.

Multicopper oxidases A group of enzymes that oxidize organic substrates and reduce dioxygen to water. These contain a combination of copper ions with different spectral features, called type 1 centers, type 2 centers, and type3 centers, where the type 2 and type 3 sites are clustered together as a trinuclear unit. Well-known examples are : laccase, ascorbate oxidase and ceruloplasmin.

Multident See ambident. μ (mu) and μ Notation for a ligand (prefix) that bridges two or more metal centres. The symbol μ is used for dipole moments.

Multienzyme A protein possessing more than one catalytic function contributed by distinct parts of a polypeptide chain (domains), or by distinct subunits, or both.

Multigeneration study Toxicity test in which at least 3 generations of the test organism are exposed to the substance being assessed. Exposure is usually continuous.

Multiheme Refers to a protein containing two or more heme groups.

Multiple (or multiphasic) screening This procedure has evolved by combining single screening tests, and is the logical corollary of mass screening. Where much time and effort has been spent by a population in attending for a single test (e.g., mass radiography), it is natural to consider the economy of offering other tests at the same time. Multiple (or multiphasic) screening implies the administration of a number of tests, in combination, to large groups of people.

Multiple bonds Multiple bonds are formed between atoms in a molecule where more than one pair of electrons are shared between the atoms. Examples of molecules containing multiple bonds include, the alkenes, the alkynes, and the arenes.

Absolute Temperature

Absolute Temperature is temperature measured on the absolute scale of temperature (i.e. the thermodynamic scale of temperature), and may be reported by adding 273 to the Centigrade temperature.

Multistage cluster sampling Cluster sampling with more than two stages, each sampling being made on aggregates (or clusters) in which the clusters already obtained by the preceding sampling have been divided (ISO, 1977).

Multistep synthesis Synthesis of a compound that takes several steps to achieve.

Mutagen An agent that induces mutation (WHO, 1979).

Mutagenesis The introduction of permanent heritable changes, i.e. mutations into the DNA of an organism. In case of site-directed mutagenesis the substitution or modification of a single amino acid at a defined location in a protein is performed by changing one or more base pairs in the DNA using recombinant DNA technology.

Mutagenicity The property of a physical, chemical, or biological agent to induce mutations in living tissue (WHO, 1979).

Mutarotation The change in optical rotation accompanying epimerisation. In sugar chemistry this term usually refers to epimerisation at the hemiacetal carbon atom.

Mutation A heritable change in the nucleotide sequence of genomic DNA (or RNA in RNA viruses), or in the number of genes or chromosomes in a cell, which may occur spontaneously or be brought about by chemical mutagens or by radiation (induced mutation).

MWC (Municipal waste combustor) Are modern waste incinerators that are highly regulated in terms of permissible emissions. They also produce energy either in the form of steam or electricity.

Myocrysin See gold drugs.

Myoglobin A monomeric dioxygen-binding hemeprotein of muscle tissue, structurally similar to a subunit of hemoglobin.

N

N- σ * delocalization (or n-* no bond resonance) Delocalization of a free electron pair (n) into an antibonding -orbital (s*).

See hyperconjugation, resonance.

N+1 rule Rule for predicting the coupling for a proton = in ^{1}H NMR spectroscopy. An NMR signal will split into n+1 peaks, where n is the number of equivalent adjacent protons.

Nacreous clouds Clouds that occur in the stratosphere at altitudes above 20 km and are usually iridescent and luminous in color. They may also be called mother of pearl clouds.

NAD Oxidized form of nicotinamide adenine dinucleotide. Note that despite the plus sign in the symbol, the coenzyme is anionic under normal physiological conditions.

NADH Reduced form of nicotinamide adenine dinucleotide.

NADP$^+$ Oxidized form of nicotinamide adenine dinucleotide phosphate. Note that despite the plus sign in the symbol, the coenzyme is anionic under normal physiological conditions.

NADPH Reduced form of nicotinamide adenine dinucleotide phosphate.

Nano (n) Prefix used in the SI system meaning "multiply by 10^{-9}". For example 1 nm means "0.000000001 m"; 2.8 ng could also be written "2.8×10^{-9} g".

Nanometer (NM) A unit of length, equal to 10^{-9} meters, and equal to 10 Å (Angstroms).

Naphthalene a naturally occurring volatile organic compound, is a component of petroleum and coal and is released by the burning of wood or tobacco. It is also used in moth repellants, dyes and pharmaceuticals. Like other nonmethane hydrocarbons, it reacts with NO in the atmosphere to produce NO_2. The NO_2 is photolzed to produce NO and O. The O then reacts with O_2 to produce ozone.

Narcissistic reaction A chemical reaction that can be described as the conversion of a reactant into its mirror image, without rotation or translation of the product, so that the product enantiomer actually coincides with the mirror image of the reactant molecule. Examples of such reactions are cited under the entries fluxional and degenerate rearrangement..

National center for atmospheric research (NCAR) A non-profit organization dedicated to furthering understanding of the Earth's atmosphere. Located in Boulder, Co., NCAR is operated by the University Corporation for Atmospheric Research (UCAR) and sponsored by the National Science Foundation (NSF).

Native Naturally occuring forms of precious metals, for example, native copper, native gold, and native silver. Native metals are often very impure.

Natural abundance Compare with isotopic abundance.

The average fraction of atoms of a given isotope of an element on Earth.

Natural gas Natural gas is a naturally occurring mixture of gaseous hydrocarbons. The approximate composition of natural gas is 85% methane, 10% ethane, 3% propane, with lesser amounts of butane, and other higher alkanes. Natural gas is used as a fuel and for the manufacture of chemicals.

Natural occurrence The occurrence in nature of a compound, when there are no man-made sources of the compound. The contamination of nature by some compounds may be so widespread that it is virtually impossible at the present time to get access to biota with a natural level and only "normal" levels can be measured, i.e., the levels that are usually prevalent at places where there is no obvious local contamination (WHO, 1979).

Natural product A compound produced by a living organism.

Nature of reactants The nature of reactants determine the course of a chemical reaction.

Nautical mile Nautical mile is a legal international unit of length temporarily maintained with the SI. It is still used in navigation (mercantile marine, aviation). It is equal to the length of an arc of one minute measured at a latitude of N45° (mile = 1852 m). The international nautical mile has been taken equal to the nautical mile.

Necrosis Mass death of areas of tissues surrounded by otherwise healthy tissue.

Needle valve A valve which allows fine control over the rate of gas or liquid flowing through it. The valve contains a thin needle with a point that fits into a conical cup. When the valve is closed, the needle blocks an orifice in the bottom of the cup.

Néel temperature Néel temperature (T_N) is the critical temperature above which an antiferromagnetic substance becomes paramagnetic. The phenomenon was discovered around 1930 by L.E.F. Néel.

Negative Ions Negative ions are atoms or groups of atoms that have acquired one or more electrons.

Neighbouring group participation The direct interaction of the reaction centre (usually, but not necessarily, an incipient carbenium centre) with a lone pair of electrons of an atom or with the electrons of a sigma or pi bond contained within the parent molecule but not conjugated with the reaction centre. A distinction is sometimes made between n, sigma, and pi participation.

A rate increase due to neighbouring group participation is known as "anchimeric assistance". "Synartetic acceleration" is the special case of anchimeric assistance ascribed to participation by electrons binding a substituent to a carbon atom in a -position relative to the leaving group attached to the - carbon atom. According to the underlying model, these electrons then provide a three-centre bond (or "bridge") "fastening together" (as the word "synartetic" is intended to suggest) the - and -carbon atoms between which the charge is divided in the intermediate bridged ion formed (and in the transition state preceding its formation). The term synartetic acceleration is not widely used.

Neil bartlett (b.1932) Born in Newcastle-upon-Tyne, he studied chemistry at King's College, Durham, where he took his doctorate. He went to the University of British Columbia at Vancouver before briefly taking up a chair at Princeton in conjunction with a position at Bell Laboratories. He then joined Lawrence Berkeley National Laboratory, with a Chair of Chemistry at University of California at Berkeley. Bartlett made the first noble-gas compound xenon hexafluoroplatinate. He has also worked on the reaction between halogen compounds and boron nitride or graphite.

Neil kensington adam (1891-1973) Born in Cambridge, England, the son of a Classics don. After studying chemistry at Trinity College, Cambridge, he became a fellow of the college. During the First World War, he served as a chemist at the Royal Naval Airship Service at

Kingsnorth in Kent. As the Royal Society Sorby Research Fellow at the University of Sheffield, Adam extended his studies of surface films. He became a lecturer at University College, London, where he also carried out research on surface-active agents and detergents. He was then called to a Chair at the University of Southampton. Adam was an active Christian Scientist.

Nematicide A chemical used to kill nematodes.

Neoplasm Any formation of tissue assiciated with disease such as tumour. See malignant, tumour.

Neoprene Neoprene is a synthetic ruber made by polimerizing the compound 2-chlorobuta-1,2-diene. Neoprene is often used in place of natural rubber in applications requiring resistance to chemical attack.

Nessler's reagent Nessler's reagent is a solution of mercury(II) iodide (HgI_2) in potassium iodide and potassium hydroxide named after Julius Nessler (1827-1905). It is used in testing for ammonia, with which it forms a brown coloration or precipitate.

Net chemical reaction Net reaction. Compare with elementary reaction.

A reaction that actually occurs as several elementary steps. Equations for net reactions often omit intermediates and catalysts.

Net ionic equation Compare with net ionic equation and molecular equation.

A net ionic equation is an ionic equation with all DEFINE(spectator ions">spectator ions eliminated. For example, $Ag^+(aq) + NO_3^-(aq) + Na^+(aq) + Cl^-(aq) = AgCl(s) + Na^+(aq) + NO_3^-(aq)$ is an ionic equation; the net ionic equation would be $Ag^+(aq) + Cl^-(aq) = AgCl(s)$ because the sodium and nitrate ions are spectators (they appear on both sides of the ionic equation.

Network covalent solid Network covalent substance. A substance which consists of an array of atoms held together by an array of covalent bonds. A crystal of a network covalent solid is actually a single, gigantic molecule. top"Diamond and top"quartz are examples.

Neurotransmitter Neurotransmitters are molecules that are used to carry signals from one neuron to another. One neuron releases the neurotransmitter near another neuron's receptors. The neurotransmitter diffuses across the gap between the neurons and locks into a receptor site on the surface of the downstream neuron. This induces a change in the downstream neuron.

Neutral 1. having no net electrical charge. Atoms are electrically neutral; ions are not.

2. A solution containing equal concentrations of H+ and OH-.

Neutral oxides Neutral oxides are those oxides, which show neither basic nor acidic properties when they react with water. Examples include carbon monoxide and nitrous oxide which are only slightly soluble in water and nitric oxide which is appreciably soluble in cold water.

Neutralization Neutralization is the process in which an acid reacts with a base to form a salt and water.

$$H^+ + OH^- \rightleftarrows H_2O$$

Neutralization reaction Neutralization; acid-base reaction.

A chemical change in which one compound aquires H^+ from another. The compound that receives the hydrogen ion is the base; the compound that surrenders it is an acid.

Neutrino An elementary particle produced by certain nuclear decay processes. Neutrinos have no charge and extremely small masses compared to other subatomic particles.

Neutron (n, ^1_0n) A particle of matter that has a mass of 1.009 amu but no electrical charge. These particles are constituents of the nucleus of all elements except for hydrogen (H). Neutrons can exist in a free state. Because neutrons have no electrical charge, free neutrons have great penetrating capabilities and are highly damaging to living tissue.

The number of neutrons an element contains can vary (see neutron variability), however, the average number of neutrons any particular element contains can be figured by subtracting the atomic number of an element from it's atomic mass average and rounding to the nearest whole number. When doing this, the resulting number should be noted something similar to: "On average this element contains X neutrons."

Neutron activation analysis Activation analysis.

An extremely sensitive technique for analyzing trace amounts of elements in a sample. The sample is bombarded with neutrons in a nuclear reactor, making it radioactive. Different elements produce different 'signatures' in the induced radiation which makes determination of their concentration in the sample possible.

Neutron number Neutron number (N) is a characteristic property of a specific isotope of an element, equal to the number of neutrons in the nucleus.

Newman projection A projection formula representing the spatial arrangement of bonds on two adjacent atoms in a molecular entity. The structure appears as viewed along the bond between these two atoms, and the bonds from them to other groups are drawn as projections in the plane of the paper. The bonds from the atom nearer to the observer are drawn so as to meet at the centre of a circle representing that atom. Those from the further atom are drawn as if projecting from behind the circle.

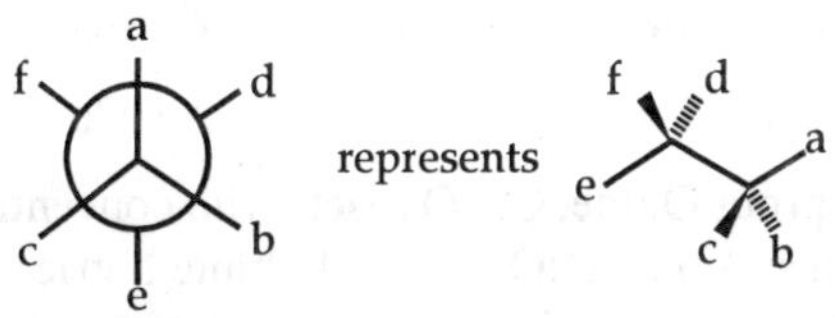

Newton Newton (N) is the SI unit of force, being the force required to give a mass of one kilogram an acceleration of 1 m s^{-2} (N = kg m s^{-2}). It is named after British scientist Sir Isaac Newton (1642-1727).

Newtonian fluid Compare with non-Newtonian fluid.

A fluid whose viscosity doesn't depend on gradients in flow speed. Gases and low-molecular weight liquids are usually Newtonian fluids.

Nif A set of about 20 genes required for the assembly of the nitrogenase enzyme complex.

NIH shift The intramolecular hydrogen migration which can be observed in enzymatic and chemical hydroxylations of aromatic rings. It is evidenced by appropriate deuterium labelling

In enzymatic reactions the NIH shift is generally thought to derive from the rearrangement of arene oxide intermediates, but other pathways have been suggested.

(NIH stands for National Institutes of Health where the shift was discovered.) .

Nikodem caro (1871-1935) Born in Lodz, and studied chemistry in Berlin. Worked with Adolf Frank and Friedrich Rothe on the development of calcium cyanamide as a means of fixing nitrogen, hence the Frank-Caro process. Became an important figure in the nitrogen fixation industry and a bitter rival of Fritz Haber. All the Caros claim descent from a famous sixteenth century rabbi in Prague, but our two Caros were not closely related.

Nikolai dimitrieviè zelinski (1861-1953) Born in Tiraspol near Kishinev in Bessarabia and educated at the University of Odessa, before studying at Leipzig and Göttingen. After taking his doctorate at Odessa, he became Professor of Organic and Analytical Chemistry at the University of Moscow. In 1911, he resigned in protest at the government's education policies, like Borodin he was a strong supporter of women's education, and became director of a government laboratory in St Petersburg. After the revolution in 1917, he returned to Moscow University, where he helped to develop the Soviet petrochemical and synthetic rubber industries. His name is associated with the Hell-Volhard-Zelinski bromination (1881-1887).

Nikolai matveeviè kishner (1867-1935). Born in Moscow and educated at the university there under Valdimir Vasilieviè Markownikov (1838-1904). He became a professor at Tomsk Institute of Technology and in 1914, at the short-lived Municipal University of Moscow. After the Communists took over, he was given the task of developing the Soviet dye industry. His anme is associated with Wolff-Kishner reaction (1911-2).

Nikolaus wolfgang fischer (1782-1850) Born in Grossmeseritz near Mähren. Educated as a physician, he became the first Professor of Chemistry at the University of Breslau. He studied the action of light on silver chloride and the displacement of one metal by another. His name is associated with Fischer's salt (also called Fischer's yellow), potassium cobaltinitrite (potassium hexanitrocobaltate(III)).

Nimbostratus clouds Cloud formation consisting of dark-gray layers of clouds and occurring at altitudes which produce prolonged rain and snow.

Nimbus clouds Nimbus is a term used to describe a cloud that is precipitating.

Nitrate (NO_3^-) 1. The NO_3^- ion, formed by reaction of nitric acid with a base.

2. A compound containing the NO_3^- ion, for example ammonium nitrate, NH_4NO_3.

Nitrate reductase A metalloenzyme, containing molybdenum, that reduces nitrate to nitrite.

Nitrates Nitrates are the salts of nitric acid, and are strong oxidising agents.

Nitrene Generic name for HN: and substitution derivatives thereof, containing an electrically neutral univalent nitrogen atom with four non-bonding electrons. Two of these are paired; the other two may have parallel spins (triplet state) or antiparallel spins (singlet state). The name is the strict analogue of carbene and, as a generic name, it is preferred to a number of alternatives proposed ("imene", "imine radical", "azene", "azylene", "azacarbene", "imin", "imidogen").

Nitrenium ion The cation H_2N^+ and its N-hydrocarbyl derivatives R_2N^+, in which the nitrogen has a positive charge, and two unshared electrons. A synonymous term is aminylium ion.

Nitric acid HNO_3, this is a corrosive, non-volatile, and inorganic acid. It is a strong acid (dissociates completely in aqueous solution) and is also an oxidizer. In the atmosphere it is formed by the conversion of nitrogen monoxide into nitrogen dioxide, and ultimately into nitric acid:

$2NO + O_2 \longrightarrow 2NO_2$

$3NO_2 + H_2O \longrightarrow 2HNO_3 + NO$

Nitrogen monoxide in this process most often come from (fossil fuel-based) combustion processes that use atmospheric air (containing 78.1% N_2) which combines with atmospheric oxygen in those high temperature combustion process (see nitrogen monoxide and nitrogen dioxide entries below). Nitric acid is highly water soluble. This solubility with water allows easy removal of nitric acid from the troposphere by atmospheric precipitation. Commonly, this is referred to as acid rain or snow. Nitric acid has a relatively low concentration in the atmosphere but provides an important role in the production of sulfuric acid. It acts as a catalyst in the conversion of sulfur dioxide to sulfuric acid. Sulfuric acid is usually formed within cloud droplets by oxidation of dissolved sulfur dioxide again most foten released by fossil fuel combustion.

Nitric acid

Properties of nitric acid

Nitric Acid displays the properties of a strong acid and it is a powerful oxidising agent.

Reactions of Nitric Acid

Cuprous Oxide, Cu2O reacts with dilute Nitric Acid, HNO3, in the cold to form a solution of Cupric Nitrate, Cu(NO3)2, and Copper, Cu.

$Cu2O + 2\ HNO3 \Rightarrow Cu(NO3)2 + Cu + H2O$

Cuprous Oxide, Cu2O reacts with concentrated Nitric Acid, HNO3, or with dilute Nitric Acid, HNO3, on heating, when the Cuprous Oxide, Cu2O dissolves with evolution of Nitric Oxide, NO.

3Cu2O + 14HNO3 6Cu(NO3)2 + 2NO + 7H2O

Dinitrogen Pentoxide, N2O5, is best prepared by dehydrating concentrated Nitric Acid, HNO3, by Phosphorus Pentoxide, P2O5.

2 HNO3 + P2O5 N2O5 + 2 HPO3

Nitric Oxide, NO is prepared by the action of Copper, Cu, or Mercury, Hg, on dilute Nitric Acid, HNO3, and was called Nitrous Air.

3 Cu + 8 HNO3 3 Cu(NO3)2 + 2 NO + 4 H2O

Nitrogen Dioxide, NO2, is a mixed acid anhydride and reacts with water to give a mixture of nitrous and nitric acids.

2 NO2 + H2 HNO2 + HNO3

If the solution is heated the nitrous acid decomposes to give nitric acid and nitric oxide.

3 HNO2 HNO3 + 2 NO + H2O

Sulphur Dioxide, SO2, and Nitrogen Oxides, NOx, are toxic acidic gases, which readily react with the Water, H2O in the atmosphere to form a mixture of Sulphuric Acid, H2SO4, Nitric Acid, HNO3, and Nitrous Acid, HNO2, . The dilute solutions of these acids which result give rain water a far greater acidity than normal, and is known as Acid Rain.

Nitrates are the salts of nitric acid, and are strong oxidising agents.

The Oswald Process is the tree stage process by which Nitric Acid, HNO3, is manufactured. Firstly, Ammonia, NH3, is oxidised, at high temperature (900 deg.C.) over a platinum-rhodium catalyst, to form Nitrogen Monoxide, NO.

4 NH3 (g) + 5O2 (g) $\Rightarrow$ 4 NO (g) + 6H2O

The Nitrogen Monoxide, NO, cools and reacts with oxygen, O2, to produce Nitrogen Dioxide, NO2.

2 NO (g) + O2 $\Rightarrow$ 2 NO2 (g)

Finally, the Nitrogen Dioxide, NO2 reacts with Water, , and Oxygen, O2, oxygen to produce Nitric Acid, .

4 NO2 (g) + 2 H2O (l) + O2 $\Rightarrow$ 4 HNO3 (l)

Boron, B, is oxidised by Nitric Acid, HNO3, to Boric Acid, H3BO3.

Detection and Analysis

Boron, B, is detected by converting the material under analysis to Borax, Na2B4O7.10H2O, by heating with concentrated Nitric Acid, HNO3, and then heating with concentrated Sulphuric Acid, H2SO4, and Ethanol, C2H5OH, to form Ethyl Borate, C2H5H2BO3, which burns with a green flame. Copper, Cu, is below Hydrogen, H2, in the electrochemical series and thus, does not react with acids to liberate Hydrogen, H2, unless they are also strong oxidising agents. Thus, Copper metal, Cu, will not react with dilute Sulphuric Acid, H2SO4, or with Hydrochloride Acid, HCl, but it reacts with Nitric Acid, HNO3, to form Oxides of Nitrogen, NOx. Cuprous Oxide, Cu2O reacts with dilute Nitric Acid, HNO3, in the cold to form a solution of Cupric Nitrate, Cu(NO3)2, and Copper, Cu.

Cu2O + 2 HNO3 $\Rightarrow$ Cu(NO3)2 + Cu + H2O

Cuprous Oxide, Cu2O reacts with concentrated Nitric Acid, HNO3, or with dilute Nitric Acid, HNO3, on heating, when the Cuprous Oxide, Cu2O dissolves with evolution of Nitric Oxide, NO.

3 Cu2O + 14 HNO3 $\Rightarrow$ 6 Cu(NO3)2 + 2 NO + 7 H2O

When glycerol is injected carefully, in the form of a spray, into a well-cooled mixture of concentrated Nitric Acid, HNO3, and Sulphuric Acid, H2SO4, Glyceryl Trinitrate, (i.e. the nitric acid ester of glycerol), is obtained. This nitrated compound is wrongly called nitro-glycerin.

CH2OH
CH2ONO2

CHOH + 3HONO2 $\Rightarrow$ CHONO2

CH2OH
CH2ONO2

Glyceryl Trinitrate

(Nitro-glycerin)

Nitric oxide radical This radical is a primary pollutant from many sources, the most common and abundant is automobile exhaust. Its importance in the atmosphere is great because it is a catalyst in the destruction of ozone.

Nitrification The conversion of nitrogen organic compounds to inorganic compounds of nitrogen. This conversion is accomplished, in the main, by bacteria

Nitrile A compound contain a cyano group: a carbon triply-bonded to a nitrogen (CN). Also a functional group

Nitrite (NO_2^-) 1. The NO_2^- ion, formed by reaction of nitrous acid with a base.

2. A compound containing the NO_2^- ion.

Nitrite reductase A metalloenzyme that reduces nitrite. Dissimilatory nitrite reductases contain copper and reduce nitrite to nitrogen monoxide. Assimilatory nitrite reductases contain siroheme and iron-sulfur clusters and reduce nitrite to ammonia.

Nitrogen N_2, a colorless, tasteless, odorless gas which makes up 78.1% of the atmosphere. Atmospheric nitrogen is converted by nitrogen fixation and nitrification into compounds used by plants and animals. In the far upper atmosphere, N_2 is broken down when large numbers of energetic secondary electrons are produced and available to react with the N_2. This leads to the eventual production of NO in that part of the atmosphere and is not—by definition—anthropogenic in nature.

Nitrogen cycle The nitrogen cycle involves the fixation of atmospheric nitrogen and the transport of the ammonia and nitrates so formed to the soil as nutrients for the growth of plants.

Nitrogen dioxide NO_2, is the lesser of the two emitted NO_x gases from high temperature combustion in air. It is an important species in the atmosphere. Since it absorbs in the visible wavelength region—creating the Brown Cloud and can be photolyzed and yield oxygen atoms that can react with molecular oxygen to create ozone, NO_2 and the NO/NO_2 ratio is important in tropospheric chemistry.

Nitrogen fixation The assimilation of dinitrogen by microbial reduction to ammonia and conversion into organonitrogen compounds such as amino acids. Only a limited number of microorganisms are able to fix nitrogen.

Nitrogen monoxide (nitric oxide) NO is the principal emitted NO_x gas from high temperature combustion in air. This gas can acts a catalyst in the reactions that cause the destruction of ozone. Reacting with ozone and tropospheric radicals, NO is inextricably linked with the polluted urban atmospheric production of NO_2, ozone, and other secondary pollutants such as peroxyacetyl nitrate.

Nitrogen oxides NO_x (pronounced "nox") Are produced from high temperature combustion in air. They are nitrogen oxide and nitrogen dioxide.

Nitrogenase An enzyme complex (EC 1.18.6.1) from bacteria that catalyzes the reduction of dinitrogen to ammonia: $N_2 + 8e^- + 10H^+ \longrightarrow 2\ {}^+NH_4 + H_2$ with the simultaneous hydrolysis of at least 16 ATP molecules. The electron donor is reduced ferredoxin or flavodoxin. Dihydrogen is always a co-product of the reaction. Ethyne (acetylene) can also be reduced to ethene (ethylene) and in some cases ethane. All nitrogenases are iron-sulfur proteins. Three different types which differ in the type of cofactor present, have been identified: molybdenum-nitrogenase (the most common, which contains the iron-molybdenum cofactor), vanadium-nitrogenase, and iron-only nitrogenase.

Nitro-glycerine Nitro-glycerine is an oil which crystallises on cooling. The liquid is highly explosive, and detonates violently when subjected to slight shock.

Nitrous oxide N_2O This is a by-product of biological activity of a symbiotic bacteria living in leguminous plant roots. It is a principal greenhouse gas that absorbs in the infrared wavelength region and unfortunately falls in an IR "window" between IR absorbing features of water and carbon dioxide (a characteristic of all the "trace" greenhouse gases with significant radiative forcing). It is also laughing gas used in medicine as a gentle general anesthetic.

Nm An abbreviation for nano-meters. A nanometer is equal to 10^{-9} meters.

NMR Nuclear magnetic resonance sprectroscopy. A technique th at measures radiofrequency light absorption by molecules. A powerful structure-determining method.

NMRSee Nuclear magnetic resonance

No acceptable daily intake allocated This expression is applicable to substances for which the available information is not sufficient to establish their safety or when the specifications for identity and purity are not adequate. The fact that an ADI for an additive was not established should not be interpreted as casting doubt on its safety nor should it be considered for its withdrawal for use .

No observed adverse effect level (NOEL) The greatest concentration or amount of a chemical, found by experiment or observation, that causes no detectable adverse alteration of morphology, functional capacity, growth, development, or life span of the target (WHO, 1979).

Nobel gas The noble gas elements, whose atoms have full outer orbitals tend to be stable, inert and not chemically active, are found in Group 0 Elements of the periodic table. The elements in this group are

Helium

Neon

Argon

Krypton

Xenon

Radon

Helium is an inert gas, because it has two electrons in its outer orbital. Only two electrons may reside in an s-orbital of helium and therefore the outer orbital is full.

The noble gases are characterised by their stable electronic configurations. The small atomic radii and high ionisation potential values suggest the presence of strongly bound electrons. There are no molecules and the inert gases are all monatomic. Low values for the boiling points indicate the presence of weak Van der Waal's Forces. Melting points and boiling points increase with atomic size.

They were called the inert gases, because it was thought that they would not react with other elements. Helium and Neon are not known to combine with other elements. However, compounds of the higher members of the noble gas group have been found, this is due to the presence of d orbitals (e.g. xenon tetrafluoride and krypton difluoride).

Helium is used along with oxygen by divers. Neon is used in neon sign electrical discharge tubes. Argon, krypton and xenon, are used in incandescent lamps. Radon a radioactive noble gas is used in the treatment of malignant growths.

Noble gas core ((X), where X is the symbol of an inert gas element) core configuration. Compare with valence shell.

All completely filled shells underneath the valence shell.

No-bond resonance See hyperconjugation.

Noctulucent clouds Relatively unusual wavy, thin, bluish-white clouds that form at altitudes of about 80 to 90 km.

Node A point, region, or surface where the amplitude of a standing wave is zero. The probability of finding an electron at an orbital node is zero.

Nomenclature A system for naming things. For example, "organic nomenclature" is the system used to name organic compounds.

Nomenclature of alkenes The nomenclature rules for alkenes requires that the names of all alkanes end in "-ene". The nomenclature system agreed by the International Union of Pure and Applied Chemistry, for the naming of the alkenes can be summarised as follows:

1. Select the longest continuous chain containing the carbon to carbon double bond as the parent hydrocarbon and name the compound as a derivative of this structure when hydrogen atoms have been replaced by alkyl groups. The names of alkenes all end in "-ene", (e.g. ethene, propene, butene, pentene, etc.).

2. Give the position of the first carbon atom attached to a C=C the lowest possible number, e.g.

3. HH H HH HH

4. HCCC=CH HCC=CCH

5. HH H H

6. But-1-ene But-2-ene

7. H HCH3

8 HCC=CH HCC=CH

9. HH H H H

10. Propene 2-Methylpropene

Nomenclature of alkynes 1. Select the longest continuous chain containing the carbon to carbon double bond as the parent hydrocarbon and name the compound as a derivative of this structure when hydrogen atoms have been replaced by alkyl groups. The names of alkenes all end in "-yne", (e.g. ethyne, propyne, butyne, pentyne, etc.).

2. Give the position of the first carbon atom attached to a C C triple bond the lowest possible number, e.g.

3. HH H H

4. HCCCCH HCCCCH

5. HH H H

6. But-1-yne But-2-yne

Nonbonded interactions Intramolecular attractions or repulsions between atoms that are not directly linked to each other, affecting the thermodynamic stability of the chemical species concerned.

Nonclassical carbocation A carbocation the ground state of which has delocalized (bridged) bonding pi- or sigma-electrons. (N.B.: Allylic and benzylic carbocations are not considered nonclassical.).

Nonelectrolyte A nonelectrolyte is a substance which does not ionize in solution.

Nonmetal (metal,metalloid) non-metal. A nonmetal is a substance that conducts heat and electricity poorly, is brittle or waxy or gaseous, and cannot be hammered into sheets or drawn into wire. Nonmetals gain electrons easily to form anions. About 20% of the known chemical elements are nonmetals.

Nonmethane hydrocarbons (NMHCs) These are hydrocarbons such as ethylene, butane, hexane, propane and, by definition, exclude the first member of that analogous series, CH_4. Large quantities of NMHCs are emitted from vegetation, the vast majority as isoprene, C_5H_8. The natural emission of isoprene is significant compared to that of anthropogenic NMHC.

Non-newtonian fluid Compare with Newtonian fluid. A fluid whose viscosity changes when the gradient in flow speed changes. Colloidal suspensions and polymer solutions like ketchup and starch/water paste are non-Newtonian fluids.

Nonparticulate Not composed of distinct particles.

Nonpolar Having a relatively even or symmetrical distribution of charge.

Nonpolar molecule A molecule in which the center of positive charge and the center of negative charge coincide. Examples are CCl_4 and CO_2; counterexamples are $CHCl_3$ and H_2O.

Normal kinetic isotope effect See isotope effect.

Normality (N) normal. Compare with molarity and equivalent.

A measure of solution concentration, defined as the number of equivalents of solute per liter of solution.

N-terminal amino-acid residue See amino-acid residue.

Nuclear atom The nuclear atom was proposed by Rutherford to explain the scattering of alpha particles by gold foil, and the theory now forms the basis of our understanding of the chemical properties of the elements.

Nuclear binding energy Energy needed to break an atomic nucleus into separate protons and neutrons.

Nuclear fission Fission. Compare with nuclear fusion. Splitting of a nucleus into two smaller nuclei and neutrons. The smaller nuclei have higher binding energy than the original nucleus, and fission results in the release of energy.

Nuclear magnetic resonance (NMR) NMR spectroscopy makes it possible to discriminate nuclei, typically protons, in different chemical environments. The electron distribution gives rise to a chemical shift δ, of the resonance frequency. The chemical shift of a nucleus is expressed in parts per million (ppm) by its frequency, ν_n, relative to a standard, ν_{ref}, and defined as $\delta = 10^6 (\nu_n - \nu_{ref}) / \nu_o$, where ν_o is the operating frequency of the spectrometer. It is an indication of the chemical state of the group containing the nucleus.

Nuclear magnetic resonance (NMR) spectroscopy NMR spectroscopy makes it possible to discriminate nuclei, typically protons, in different chemical environments. The electron distribution gives rise to a chemical shift of the resonance frequency. The chemical shift, , of a nucleus is expressed in parts per million (ppm) by its frequency, $_{n}$, relative to a standard, $_{ref}$, and defined as = 10^6 ($_{n}$ - $_{ref}$)/$_{o}$, where $_{o}$ is the operating frequency of the spectrometer. It is an indication of the chemical state of the group containing the nucleus. More information is derived from the spin-spin couplings between nuclei, which give rise to multiplet patterns. Greater detail may be derived from two- or three-dimensional techniques. These use pulses of radiation at different nuclear frequencies, after which the response of the spin system is recorded as a free-induction decay (FID). Multi-dimensional techniques, such as COSY and NOESY, make it possible to deduce the structure of a relatively complex molecule such as a small protein (molecular weight up to 25 000). In proteins containing paramagnetic centres, nuclear hyperfine interactions can give rise to relatively large shifts of resonant frequencies known as contact and pseudo-contact (dipolar) shifts, and considerable increases in the nuclear spin relaxation rates. From this type of measurement, structural information can be obtained about the paramagnetic site.

Nuclear reactor Nuclear reactor is an assembly of fissionable material (uranium-235 or plutonium-239) designed to produce a sustained and controllable chain reaction for the generation of electric power. The essential components of a nuclear reactor are: 1. The core, metal rods containing enough fissionable material to maintain a chain reaction at the necessary power level (as much as 50 t of uranium may be required).

2. A source of neutrons to initiate the reaction (such as a mixture of polonium and beryllium)

3. A moderator to reduce the energy of fast neutrons for more efficient fission (material such as graphite, beryllium, heavy water, and light water are used)

4. A coolant to remove the fission-generated heat (water, sodium, helium, and nitrogen may be used)

5. A control system such as rods of boron or cadmium that have high capture cross sections (to absorbs neutrons)

6. Adequate shielding, remote-control equipment, and appropriate instrumentation are essential for personnel safety and efficient operation.

Nuclear winter In the event of a significant nuclear war, researchers (Paul Crutzen and John Birks initially) predicted that a barrier of smoke (from incinerated urban centers and forests) would fill tropospheric skies, and eventually surround the earth, reducing the tropospheric temperature and causing damage to ecosystems and atmospheric components such as stratospheric ozone.

Nuclearity The number of central atoms joined in a single coordination entity by bridging ligands or metal-metal bonds is indicated by dinuclear, trinuclear, tetranuclear, polynuclear, etc.

Nucleation A process in which water condenses on a small atmospheric particle. Nucleation aides in the process of condensation (see methanesulfonic acid).

Nucleation scavenging Term for water vapor that removes particulate matter from the atmosphere.

Nucleation The process by which nuclei are formed, usually in solution. The term "nucleus" as used here is defined as the smallest solid phase aggregate of atoms, molecules or ions which is formed during a precipitation and which is capable of spontaneous growth.

Nucleic acids Macromolecules composed of sequences of nucleotides that perform several functions in living cells, e.g. the storage of genetic information and its transfer from one generation to the next (DNA), and the expression of this information in protein synthesis (mRNA, tRNA), and may act as functional components of subcellular units such as ribosomes (rRNA). RNA contains D-ribose, DNA contains 2-deoxy-D-ribose as the sugar component. Currently, synthetic nucleic acids can be made consisting of hundreds of nucleotides.

Nucleobases See nucleosides.

Nucleofuge A leaving group that carries away the bonding electron pair. For example, in the hydrolysis of an alkyl chloride, Cl^- is the nucleofuge. The tendency of atoms or groups to depart with the bonding electron pair is called nucleofugality.

The adjective is nucleofugal.

Nucleon Nucleons are the constituents of the nucleus in an atom, and include neutrons and protons.

Nucleophile, nucleophilic A nucleophile (or nucleophilic reagent) is a reagent that forms a bond to its reaction partner (the electrophile) by donating both bonding electrons.

A "nucleophilic substitution reaction" is a heterolytic reaction in which the reagent supplying the entering group acts as a nucleophile. For example

MeO^- (nucleophile) + Et-Cl MeOEt + Cl^- (nucleofuge)

The term "nucleophilic" is also used to designate the apparent polar character of certain radicals, as inferred from their higher relative reactivity with reaction sites of lower electron density.

Nucleophilic reagents are Lewis bases.

Nucleophilic aromatic substitution Nucleophilic aromatic substitution occurs when an atom or group which is attached to an aromatic ring is replaced by another atom or group. Benzene, C6H6, is the simplest aromatic compound and is often presented as a six-member ring with alternating single and double bonds. However, these bonds exist as resonance hybrids, leading to significant stabilisation of the ring structure. Because of this stability of the aromatic ring, arenes tend to undergo substitution reactions to preserve the aromatic character of the ring, and the attacking nucleophile displaces a hydrogen atom from the ring. Therefore, the double bonds in an aromatic systems do not undergo the type of addition reactions observed for the Alkenes Alkynes.

Nucleophilic catalysis Catalysis by a Lewis base, involving formation of a Lewis adduct as a reaction intermediate. For example, the hydrolysis of acetic anhydride in aqueous solution catalysed by pyridine:

C_5H_5N + $(CH_3CO)_2O$ $(C_5H_5NCOCH_3)^+$ + $CH_3CO_2^-$

$(C_5H_5NCOCH_3)^+$ + H_2O C_5H_5N + CH_3CO_2H + H^+_{aq}

See also electrophilic, nucleophilicity.

Nucleophilicity 1. The property of being nucleophilic.

2. The relative reactivity of a nucleophilic reagent. (It is also sometimes referred to as "nucleophilic power".) Qualitatively, the concept is related to Lewis basicity. However, whereas Lewis basicity is measured by relative equilibrium constants,

B: + A B^+-A^- (equilibrium constant K)

nucleophilicity of a Lewis base is measured by relative rate constants of different nucleophilic reagents towards a common substrate, most commonly involving formation of a bond to carbon,

B: + A-Z B^+-A^- + $Z:^-$ (rate constant k)

See also electrophilicity,

Nucleoside Compare with nucleotide.

A nucleotide base bound to a five-carbon sugar.

Nucleosides Compounds in which a purine or pyrimidine base is β-N-glycosidically bound to C-1 of either 2-deoxy-D-ribose or of D-ribose, but without any phosphate groups. The common nucleosides in biological systems are adenosine, guanosine, cytidine, and uridine (which contain ribose) and deoxyadenosine, deoxyguanosine, deoxycytidine and thymidine (which contain deoxyribose).

Nucleotide Compare with nucleoside. A molecule which is a basic building block of nucleic acids and which plays a key role in energy transfer in biochemical reactions. Nucleotides consist of a five-carbon sugar, a heterocyclic nitrogen-containing organic base, and a phosphate group.

Nucleotide base A heterocyclic nitrogen-containing base that is a constituent of nucleotides. Examples are adenine, guanine, thymine, uracil, and cytosine.

Nucleus The core of an atom that contains at least one proton and one neutron (except for hydrogen (H), which its most common nuclide doesn't have a neutron). This core is positively charged and contains almost all of an atom's mass.

Nuclide Compare with atomic nucleus and nuclide symbol.

An atom or ion with a specified mass number and atomic number. For example, uranium-235 and carbon-14 are nuclides.

Nuclide symbol Compare with atomic nucleus, nuclide and element symbol.

A symbol for an nuclide that contains the mass number as a leading superscript and the atomic number as a leading subscript. For ions, the ionic charge is given as a trailing superscript. For example, the nuclide symbol for the most common form of the chloride ion is ${}^{35}{}_{17}Cl^{-}$, where 35 is the mass number, 17 is the atomic number, and the charge on the ion is -1. The atomic number is sometimes omitted from nuclide symbols.

Nuclides (*isotopes*) Forms or species of an element that has the same numbers of protons (hence the same atomic number), but different numbers of neutrons, thus different atomic masses. Technically speaking if an element has only one form that is there is no variability in the number of neutrons that exist in the element, it has no isotopes. There are 21 elements that do not very in the number of neutrons in their nuclei.

Null reagent Concept whereby one of a set of pools is subjected to no reaction at a particular stage of a combinatorial synthesis. It is often necessary to record this as a null event to maintain congruence in computational records of the library.

Nutritional table method A method of evaluating the dietary intake of a large number of people. The accuracy of the method largely depends on the accuracy with which records of the food consumption can be established and the accuracy of the nutritional tables specifying the concentration of various nutrients, vitamins, and essential and non-essential substances including pesticide residues. For each record of quantity of food consumed during a certain time period, the daily intake of the substance in question is calculated by multiplying the substance concentration in the food item (as obtained from the nutritional table) by the quantity of food consumed and dividing by the time of observation (WHO, 1979).

O

Objective environment The actual physical, chemical, and social environment as described by objective measurements, such as noise levels in decibels and concentrations of air pollutants (WHO, 1979).

Occupational environment The environment at a work place (WHO, 1979).

Occupational hygiene The applied science concerned with recognition, evaluation and control of chemical, physical or biological factors arising from the workplace and which may affect the well-being of those at work or in the community.

Occurrence In epidemiology, a general term describing the frequency of a disease or other attribute or event in a population without distinguishing between incidence and prevalence .

Occurrence (Iodine) Because of its reactivity, iodine does not exist in the free elemental state in nature, but small quantities are combined in many silver ores found in Mexico.

Compounds of iodine occurs naturally in seaweed.

Iodide salts are found in liquor from the extraction of saltpeter from deposits in Chile.

Occurrence (Cadmium) is often found in association with Zinc ores.

Cadmium is also found in Greenocklite, CdS.

Occurrence (Gadolinium) occurs with the other rare earths and in Gadolinite.

Occurrence (Hydrogen) is the lightest element and the most abundant element in the universe.

Hydrogen occurs naturally as a mixture of the three isotopes :

Protium, H,

Deuterium, D, (which is also called Heavy Hydrogen) and

Tritium, T.

Occurrence (Iron) is widely distributed in the universe, and exists in all the planets, in meteorites and in the sun. Its principal ore is Hematite.

Occurrence (Lead) is known to occur in elemental form in some locations. However, the principal ores are

Gelena, PbS,

Cerussite, PbCO3,

Anglesite, PbSO4,

Crocoisite, PbCrO4,

Wulfenite, PbMoO4,

Pyromorphite, PbCl2.3Pb(PO4)2, and

Matlockite, PbCl2.

Occurrence (Mercury) is found in cinnabar at a wide number of locations around the world.

Occurrence (Nitrogen) is the most abundant gas in the atmosphere, which is composed of approximately 78% nitrogen gas by volume, or 75.5% by weight.

Occurrence (Sulphur) is widely distributed as the free element and combined in compounds.

Sulphur is found near volcanoes, where it is formed by the reaction of sulphur dioxide and hydrogen sulphide which are associated with the volcanoes.

Sulphur also occurs in many metal ores, including

gelena, PbS,

zinc blende, ZnS,

cinnebar, HgS,

stibnite, Sb2S3,

copper pyrites, Cu2S.Fe2S3, and

iron pyrites, Fe2S.

The important sulphate ores include

gypsum, CaSO4, and

heavy spar, BaSO4.

Occurrence (Helium) Traces of Helium occur in the atmosphere.

Occurrence (Uranium) can be found as

Pitchblende or Uraninite, U3O8, and

Carnotite, KUO2(VO4).3H2O.

Ocean The oceans cover approximately 70% of the Earth's surface and are apparently unique in the solar system (although keep an eye on some of the giant planets' moons...). The oceans are important for the regulation of tropospheric heat and the storage and production of oxygen and carbon dioxide.

Octahedron See coordination.

Octane (C_8H_{18}) Compare with alkane and hydrocarbon.

Flammable liquid compounds found in petroleum and natural gas. There are 18 different octanes- they have different structural formulas but share the molecular formula C_8H_{18}. Octane is used as a fuel and as a raw material for building more complex organic molecules. It is the eighth member of the alkane series.

Octane number The octane number of a fuel is an experimentally determined value for the fuel, which describes its combustion properties (i.e. knocking characteristics) relative to octane.

Octet In Lewis structures the goal is to make almost all atoms have an octet. This means that they will have access to 8 electrons regularly, even if they do have to share some of them.

Octet rule The octet rule states that the chemical properties of the elements repeat on a regular basis with increasing atomic mass, and that the chemical properties of each eight element is similar. Since the inert gases, with the exception of helium have eight electrons in their outer shells, this stable electronic configuration is called the octet rule. In chemical reactions atoms of elements tend to react in such a way as to achieve the electronic configuration of the inert gas nearest them in the periodic table. There are a number of exceptions to the octet rule.

The noble gases are known to be very inert chemically. In chemical reactions, the atoms of elements tend to react in such a way as to achieve the electronic configuration of the inert gas nearest them in the periodic table. Since all the inert gases, except helium, have eight electrons in the outer level, this concept is often referred to as the Octet Rule, to which there are exceptions. This rule is a guide to understanding bonding. In order to become stable, atoms with incomplete outer shells will attempt to join chemically with another atom, and to share the electrons of that atom in its own outer shell, so as to fill it. This is called chemical bonding.

Ocular The adjective applying to anything pertaining to the eye.

Odds ratio (Synonymcross product ratio, relative odds) The ratio of two odds. The term "odds" is defined differently according to the situation under discussion. Consider the following notation for the distribution of a binary exposure and a disease in a population or a sample.

Exposed

Unexposed

Disease

ab

No disease

cd

The odds ratio (cross-product ratio) is ad/bc.

The exposure-odds ratio for a set of case control data is the ratio of the odds in favour of exposure among the cases (a/b) to the odds in favour of exposure among non-cases (c/d). This reduces to ad/bc. With incident cases, unbiased subject selection, and a "rare" disease (say, under 2% cumulative incidence rate over the study period), ad/bc is an approximate estimate of the risk ratio. With incident cases, unbiased subject selection, and density sampling of controls, ad/bc is an estimate of the ratio of the person-time incidence rates (forces of morbidity) in the exposed and unexposed (no rarity assumption is required for this).

The disease-odds (rate-odds) ratio for a cohort or cross-section is the ratio of the odds in favour of disease among the exposed (a/c) to the odds in favour of disease among the unexposed (b/d). This reduces to ad/bc and hence is equal to the exposure-odds ratio for the P> abiotic: This

The prevalence-odds ratio refers to an odds ratio derived cross- sectionally, as, for example, an odds ratio derived from studies of prevalent (rather than incident) cases.

The risk-odds ratio is the ratio of the odds in favour of getting disease, if exposed, to the odds in favour of getting disease if not exposed.

Odds The ratio of the probability of occurrence of an event to that of non-occurrence, or the ratio of the probability that something is so, to the probability that it is not so .

ODMR See optically detected magnetic resonance.

OEC See oxygen-evolving complex.

Oersted Oersted (Oe) is a non-SI unit of magnetic field (H), equal to 79.57747157 A/m. The unit is named after the H.C. Oersted (1777-1851).

Ohm Ohm (W) is the SI derived unit of electric resistance. The ohm is the electric resistance between two points of a conductor when a constant difference of potential of one volt, applied between these two points, produces in this conductor a current of one ampere, this conductor not being the source of electromotive force (W = V/A). The unit is named after the German physicist Georg Simon Ohm (1789-1854).

Ohmmeter An instrument for measuring electrical resistance.

Oil pollution Oil pollution is the degradation of the quality of the water, when crude oil, lubricants, fuel oil, etc. contaminate the aquatic environment.

OlefinsOlefins are acyclic and cyclic hydrocarbons having one or more carbon-carbon double bonds, apart from the formal ones in aromatic compounds. The class olefins subsumes alkenes and cycloalkenes and the corresponding polyenes.

Oleic acid when released into the air from cooking, can react with other atmospheric compounds, such as ozone, and create byproducts that are hazardous to our health.Ý Oleic acid is a fatty acid found in animal and vegetable oils. It is called a mono-unsaturated fatty acid because of the single double bond between the carbons. It's physical properties are determined by the number, geometry, and position of this double bond and the degree of unsaturation.

Oligomer Oligomer is a substance consisting of molecules of intermediate relative molecular mass (molecular weight), the structure of which essentially comprises the multiple repetition of units derived, actually or conceptually, from molecules of low relative molecular mass. In contrast to a polymer, the properties of an oligomer can vary significantly with the removal of one or a few of its units.

Oligonucleotide Macromolecules composed of short sequences of nucleotides that are usually synthetically prepared and used e.g. in site-directed mutagenesis.

Oligosaccharide Compare with monsaccharide and polysaccharide. A carbohydrate that consists of only a few linked monosaccharide units.

Omission library Strategy for identifying active library members by the systematic omission of building blocks from mixtures. Observation of reduced activity in a certain pool suggests that the building block which was omitted in that pool contributes to activity.

One dimensional model An atmospheric modelling technique in which a concentration or another unknown can be calculated relative to certain parameters, such as altitude or temperature.

Onium ion 1. A cation (with its counterion) derived by addition of a hydron to a mononuclear parent hydride of the nitrogen, chalcogen and halogen family, e.g. H_4N^+ ammonium ion.

2. Derivatives formed by substitution of the above parent ions by univalent groups, e.g. $(CH_3)_2S^+H$ dimethylsulfonium, $(CH_3CH_2)_4N^+$ tetraethylammonium.

3. Derivatives formed by substitution of the above parent ions by groups having two or three free valencies on the same atom. Such derivatives are, whenever possible, designated by a specific class name. E.g. $R_2C{=}NH_2^+$ iminium ion.

Onsager relations Onsager relations are an important set of equations in the thermodynamics of irreversible processes. They express the symmetry between the transport coefficients describing reciprocal processes in systems with a linear dependence of flux (J_i) on driving forces (X_j).

$$J_i = \sum_{j=1}^{m} L_{ij} X_j$$

In Onsager's theory the coupling coefficients are equal, $L_{ij} = L_{ji}$. This are known as reciprocal relations. The theory was developed by the Norwegian chemist Lars Onsager (1903-1976) in 1931.

Opacity A measure of light blocking by atmospheric particles.

Operon A functional unit consisting of a promoter, an operator and a number of structural genes, found mainly in prokaryotes. An example is the operon nif. The structural genes commonly code for several functionally related enzymes, and although they are transcribed as one (polycistronic) mRNA, each has its separate translation initiation site. In the typical operon, the operator region acts as a controlling element in switching on or off the synthesis of mRNA.

Opposing reactions See composite reactions.

Optical activity A sample of material able to rotate the plane of polarisation of a beam of transmitted plane-polarised light is said to possess optical activity (or to be optically active). This optical rotation is the classical distinguishing characteristic (sufficient but not necessary) of systems containing unequal amounts of corresponding enantiomers. An enantiomer causing rotation in a clockwise direction (when viewed in the direction facing the oncoming light beam) under specified conditions is called dextrorotatory and its chemical name or formula is designated by the prefix (+)-; one causing rotation in the opposite sense is laevorotatory and designated by the prefix (-)-. Materials with optical activity also exhibit other chiroptic phenomena.

Optical isomerism (Usage strongly discouraged) Obsolescent synonym for stereoisomers with different optical properties. It should be described as diastereoisomerism or enantiomerism.

Optical purity The ratio of the observed optical rotation of a sample consisting of a mixture of enantiomers to the optical rotation of one pure enantiomer. See enantiomeric excess.

Optical resolution (Usage strongly discouraged) See resolution. Optical Rotation See optical activity.

Optical yield In a chemical reaction involving chiral reactants and products, the ratio of the optical purity of the product to that of the precursor, reactant or catalyst. This should not be confused with "enantiomeric excess". The optical yield is in no way related to the chemical yield of the reaction.. See stereoselectivity.

Optically active Rotates plane-polarized light.

Optically detected magnetic resonance (ODMR) A double resonance technique in which transitions between spin sublevels are detected by optical means. Usually these are sublevels of a triplet, and the transitions are induced by microwaves.

Optically labile (Usage strongly discouraged) A term describing a system in which stereoisomerisation results in a change of optical rotation with time.

Optimization The process of synthesizing chemical variations, or analogs, of a lead compound, with the goal of creating those compounds with improved pharmacological properties.

Orbital Orbital is the area in space about an atom or molecule in which the probability of finding an electron is greatest.

The possible atomic orbitals correspond to subshells of the atom. Thus there is one s-orbital for each shell (orbital quantum number l = 0). There are three p-orbitals (corresponding to the three values of l) and five d-orbitals. The shapes of orbitals depend on the value of l.

Orbital steering A concept expressing that the stereochemistry of approach of two reacting species is governed by the most favourable overlap of their appropriate orbitals.

Orbital symmetry The behaviour of an atomic or localized molecular orbital under molecular symmetry operations characterizes its orbital symmetry. For example, under a reflection in an appropriate symmetry plane, the phase of the orbital may be unchanged (symmetric), or it may change sign (antisymmetric), i.e. the positive and negative lobes are interchanged.

A principal context for the use of orbital symmetry is the discussion of chemical changes that involve "conservation of orbital symmetry". If a certain symmetry element (e.g. the reflection plane) is retained along a reaction pathway, that pathway is "allowed" by orbital symmetry conservation if each of the occupied orbitals of the reactant(s) is of the same symmetry type as a similarly (e.g. singly or doubly) occupied orbital of the product(s). This principle permits the qualitative construction of correlation diagrams to show how molecular orbitals transform (and how their energies change) during idealized chemical changes (e.g. cycloadditions).

An idealized single bond is a sigma bond, i.e., it has cylindrical symmetry, whereas a p-orbital or pi-bond orbital has pi symmetry, i.e. it is antisymmetric with respect to reflection in a plane passing through the atomic centres with which it is associated. In ethene, the pi-bonding orbital is symmetric with respect to reflection in a plane perpendicular to and bisecting the C-C bond, whereas the pi*-antibonding orbital is antisymmetric with respect to this operation.

Considerations of orbital symmetry are frequently grossly simplified in that, for example, the pi orbitals of a carbonyl group would be treated as having the same symmetry as those of ethene, and the fact that the carbonyl group in, for example, camphor, unlike that in formaldehyde, has no mirror planes would be ignored. These simplified considerations nevertheless afford the basis of one approach to the understanding of the rules which indicate whether pericyclic reactions are likely to occur under thermal or photochemical conditions. . See also sigma, pi.

Orbital The area around an atom where according to orbital theory the probability of finding an electron is the greatest.

Orbital theory The quantum theory matter that combines Schrodinger's wave mechanics and Heisenburg's uncertainty principle and applies this to the behavior and nature of electrons. Orbital theory was formulated in 1926 and has yielded a better understanding of electrons and their critical role in chemical bonding than is possible with Newtonian mechanics.

Order Order of reaction; reaction order.

The order of a reaction is the sum of concentration exponents in the rate law for the reaction. For example, a reaction with rate law $d(C)/dt = k(A)^2(B)$ would be a third order reaction. Noninteger orders are possible.

Order of filling the atomic Orbitals The sequence in which the atomic orbitals in an atom are filled is as follows

1s 2s 2p 3s 3p 4s 3d 4p 5s 4d 5p 6s 4f 5d 6p 7s 5f 6d 7p

Organic chemistry Organic compounds form the chemical basis for life and are more abundant than inorganic compounds. Carbon compounds are made up of carbon atoms which form covalent bonds with each of its neighboring carbon atoms in either a chain or a ring and also form bonds with other atoms mainly hydrogen, oxygen, nitrogen or sulphur.

Compounds containing only carbon and hydrogen are known as Hydrocarbons.

Organic compound. Compounds that contain carbon chemically bound to hydrogen. They often contain other elements (particularly O, N, halogens, or S). Organic compounds were once thought to be produced only by living things. We now know that any organic compound can be synthesized in the laboratory (although this can be extremely difficult in practice!)

Organochromic indicators Colored organic compounds that change color when they chelate different metals. Organochromic indicators are used to determine the endpoint in complexometric titrations. Examples of organochromic indicators are are Eriochrom Black T, calmagite, and Eriochrom Cyanine R.

Ortho Describes the positions of two substituents on a benzene ring that are on adjacent carbons.

Orthogonality a) Property of protecting groups or linkers allowing removal, modification or cleavage of one such without affecting others; b) pooling strategy whereby library members are incorporated in more than one pool, and are mixed with a different set of other members in each pool. Thus a hit results in two or more active pools with only one member in common.

Ortho-para director An aromatic substituent that directs incoming electrophiles to the ortho or para positions.

Oskar piloty (1866-1915) Was the son of an artist and was educated at the university in his home town of Munich, under Adolf von Baeyer and later married Baeyer's daughter. Piloty was also the brother-in-law of Ludwig Knorr (1859-1921). He took his doctorate under Emil Fischer in Würzburg and followed Fischer to Berlin. In 1899, he joined his father-in-law in Munich as Professor of Inorganic Chemistry. Piloty initially focussed on sugar chemistry, but later turned his attention to the structure of the natural pyrroles. After one of his sons was killed in the early days of the First World War, Piloty volunteered to serve at the western front, and was killed by a bullet while on duty at the front near the Somme.

Osmometry Compare with osmosis. Determination of the average molecular weight of a dissolved substance from measurements of osmotic pressure.

Osmosis Osmosis is the passage of a solvent through a semipermeable membrane which acts as a barrier to the passage of the solute dissolved in the solution. The passage of water through the walls of living cells, which consist of very thin membranes that act as a barrier to substances dissolved in the water, is an essential process in living cells and is an example of osmosis. In general, the molecules of the solute are too large to pass through such semi-permeable membranes.

Osmotic pressure The osmotic pressure of a solution is the pressure which must be applied to the solution to stop osmosis. Osmotic pressure is similar to gas pressure and is caused by the dissolved substances (i.e. the solutes) behaving as if they were gases occupying the space volume of the solvent. Under osmosis, a solvent always diffuses from the less concentrated solution to the more concentrated solution.

Ostwald's dilution law Ostwald's dilution law is a relation for the concentration dependence of the molar conductivity Λ of an electrolyte solution, viz.

$$\frac{1}{\Lambda} = \frac{1}{K_c \Lambda_0^2} c\Lambda + \frac{1}{\Lambda}$$

where c is the solute concentration, K_c is the equilibrium constant for dissociation of the solute, and L_0 is the conductivity at $c\Lambda = 0$. The law was first put forward by the German chemist Wilhelm Ostwald (1853-1932).

Oswald process The Oswald Process is the three stage process by which nitric acid is manufactured. Firstly, ammonia is oxidised, at high temperature (900 deg.C.) over a platinum-rhodium catalyst, to form nitrogen monoxide.

$4\,NH_3\,(g) + 5O_2\,(g) \Rightarrow 4\,NO\,(g) + 6H_2O$

The nitrogen monoxide cools and reacts with oxygen to produce nitrogen dioxide.

$2\,NO\,(g) + O_2 \Rightarrow 2\,NO_2\,(g)$

Finally, the nitrogen dioxide reacts with water and oxygen to produce nitric acid.

$NO_2\,(g) + 2\,H_2O\,(l) + O_2 \Rightarrow 4\,HNO_3\,(l)$

Other measures of precision The quantitative measures of precision described above are the most common for reporting analytical results. You might encounter other measures of precisions, and several other quantities are listed here for completeness.

standard error:

$$s_m = \frac{s}{\sqrt{N}}$$

variance:

$$\text{variance} = s^2 = \frac{\sum_{i=1}^{N}(x_i - \bar{x})^2}{N-1}$$

The advantage of working with variance is that variances from independent sources of variation may be summed to obtain a total variance for a measurement.

All of the equations above are for quantitating the precision of a relatively small numbers of replicate measurements. For 20 or more measurements, the true or population standard deviation, σ , is:

$$\sigma = \sqrt{\frac{\sum_{i=1}^{N}(x_i - \mu)^2}{N}}$$

Continuous-Wave Nuclear Magnetic Resonance (NMR) Spectroscopy(h)

Continuous-wave NMR spectrometers have largely been replaced with pulsed FT-NMR instruments. However due to the lower maintenance and operating cost of cw instruments, they are still commonly used for routine ^{1}H NMR spectroscopy at 60 MHz. (Low-resolution cw instruments require only water-cooled electromagnets instead of the liquid-He-cooled superconducting magnets found in higher-field FT-NMR spectrometers.)

Otto bayer (1902-1982) Born in Frankfurt-on-the-Main, he studied chemistry at Frankfurt University, and after taking his doctorate, he was an assistant to von Braun. He then joined I.G. Farben's Mainkur factory, and he quickly became the leader of the main research laboratory at Leverkusen. He eventually joined the board of Bayer AG. His major achievement was the development of polyurethanes. Despite working for Bayer, He was not related to Friedrich Bayer (compare the similar situation of Alastair Pilkington and Pilkington Brothers).

Outer-sphere (Electron transfer) An outer-sphere electron transfer is a reaction in which the electron transfer takes place with no or very weak (4 -16 kJ mol^{-1}) electronic interaction between the reactants in the transition state. If instead the donor and the acceptor exhibit a strong electronic coupling, the reaction is described as inner-sphere electron transfer. The two terms derive from studies concerning metal complexes and it has been suggested that for organic reactions the term "nonbonded" and "bonded" electron transfer should be used.

Outgassing- The release of "young" gases into the earth's atmosphere by volcanic activity.

Out-isomer See in-out isomerism.

Overpotential Overpotential (η) is a potential that must be applied in an electrolytic cell in addition to the theoretical potential required to liberate a given substance at an electrode. The value depends on the electrode material and on the current density.

Ovotransferrin An iron-binding protein from eggs, structurally similar to the transferrins.

Oxidants Chemical species that readily accept electrons. The electron accepting species is reduced in a chemical reaction.

Oxidase An enzyme which catalyzes the oxidation of substrates by O_2.

Oxidatio The term oxidation originally meant a reaction in which oxygen combines chemically with another substance. More generally, oxidation is a part of a chemical reaction in which a reactant loses electrons (increase oxidation number). Simultaneous reduction of a different reactant must occur (redox reaction).

Oxidation 1. The complete, net removal of one or more electrons from a molecular entity (also called "de-electronation").

2. an increase in the oxidation number of any atom within any substrate (see Hendrickson, Cram and Hammond (1970)).

3. Gain of oxygen and/or loss of hydrogen of an organic substrate.

All oxidations meet criteria 1. and 2., and many meet criterion 3., but this is not always easy to demonstrate.Alternatively, an oxidation can be described as a transformation of an organic substrate that can be rationally dissected into steps or primitive changes. The latter consist in removal of one or several electrons from the substrate followed or preceded by gain or loss of water and/or hydrons or hydroxide ions, or by nucleophilic substitution by water or its reverse and/or by an intramolecular molecular rearrangement.

This formal definition allows the original idea of oxidation (combination with oxygen), together with its extension to removal of hydrogen, as well as processes closely akin to this type of transformation (and generally regarded in current usage of the term in organic chemistry to be oxidations and to be effected by "oxidizing agents") to be descriptively related to definition 1.. For example the oxidation of methane to chloromethane may be considered as follows:

$CH_4 - 2e^- - H^+ + OH^- = CH_3OH$ (oxidation)
CH_3Cl (reversal of hydrolysis)

Oxidation half reaction Oxidation half-reaction. Compare with reduction half reaction.

That part of a redox reaction that involves loss of electrons. In the oxidation half reaction, the oxidation number of one or more atoms within the reactants is increased.

Oxidation number The oxidation number of an element in any chemical entity is the number of charges which would remain on a given atom if the pairs of electrons in each bond to that atom were assigned to the more electronegative member of the bond pair. The oxidation (Stock) number of an element is indicated by a roman numeral placed in parentheses immediately following the name (modified if necessary by an appropriate ending) of the element to which it refers. The oxidation number may be positive, negative or zero. Zero, not a roman numeral, is represented by the usual cipher, 0. The positive sign is never used. An oxidation number is always positive unless the minus sign is explicitly used. Note that it cannot be non-integral (see also mixed valency). Non-integral numbers may seem appropriate in some cases where a charge is spread over more than one atom, but such a use is not encouraged. In such ambiguous cases, the charge number, which designates ionic charge, can be used. A charge (Ewens-Bassett) number is a number in parentheses written without a space immediately after the name of an ion, and whose magnitude is the ionic charge. Thus the number may refer to cations or anions, but never to neutral species. The charge is written in arabic numerals and followed by the sign of the charge. In a coordination entity, the oxidation number of the central atom is defined as the charge it would bear if all the ligands were removed along with the electron pairs that were shared with the central atom. Neutral ligands are formally removed in their closed-shell configurations. Where it is not feasible or reasonable to define an oxidation state for each individual member of a group or cluster, it is again recommended that the overall oxidation level of the group be defined by a formal ionic charge, the net charge on the coordination entity.

Oxidation reaction A reaction where a substance loses electrons

Oxidation reduction reaction A reaction involving the transfer of electrons.

Oxidative addition The insertion of a metal of a metal complex into a covalent bond involving formally an overall two-electron loss on one metal or a one-electron loss on each of two metals, i.e.,

$L_nM^m + XY\ L_nM^{m+2}(X)(Y)$, or

$2\ L_nM^m + XY\ L_nM^{m+1}(X) + L_nM^{m+1}(Y)$

In free-radical chemistry, the term is used to indicate a free radical addition to a carbon-carbon double bond, under oxidative conditions.

Oxidative coupling The coupling of two molecular entities through an oxidative process, usually catalysed by a transition metal compound and involving dioxygen as the oxidant; e.g.,

$2\ CO + 2\ MeOH + 1/2\ O_2$ (+ catalyst) MeOOCCOOMe + H_2O

Oxide A binary compound that contains oxygen in the -2 oxidation state.

Oxidizing This adjective applied to a chemical is a substance which gives off oxygen to another substance. Oxidizing chamicals may increase ans sustain fires. For example, chemicals belonging to following groups may act as oxidizers: bromates, chlorates, chromates, dichromates, iodates, nitrates, oxides, perborate, perbromates, perchlorates, periodates, permanganates and peroxodes.

Oxidizing agent Oxidant; oxidizer. Compare with reducing agent.

A reactant that removing electrons from other reactants in a chemical reaction. Oxidizing agents cause other substances to be oxidized in chemical reactions while they themselves are reduced. For example, nitrate ion is an oxidizing agent in the following reaction:

$Cu(s) + 4\ H^+(aq) + 2\ NO_3^-(aq)\ Cu^{2+}(aq) + 2\ H_2O + 2\ NO_2(g)$

Copper gets oxidized (its oxidation number goes from 0 to +2) while the nitrogen gets reduced (from +5 in nitrate to +4 in nitrogen dioxide).

Oxidoreductase An enzyme of EC class 1, which catalyzes an oxidation-reduction reaction.

Oximes Oximes are organic compounds of structure $R_2C{=}NOH$ derived from condensation of aldehydes or ketones with hydroxylamine. Oximes from aldehydes may be called aldoximes; those from ketones may be called ketoximes.

Oxo compounds Oxo compounds are organic compounds that contain the karbonyl group, C=O. The term thus embraces aldehydes, carboxylic acids, ketones, amides, and esters.

Oxyacid When one or more hydroxide (OH) groups are bonded to a central atom.

Oxygen (O) Element 8, atomic weight 15.9994, a colorless, odorless gas that makes up about 1/5 of the earth's atmosphere and (in combined form) 8/9ths of earth's oceans and almost half of the earth's crust. The name is derived from the French oxygène, which means "acid generating".

Oxygen radical· is the result of the oxygen molecule (O_2) being photodissociated at a wavelength greater than 242 nm. These two oxygen atoms react with O_2 and create ozone (O_3).

Oxygen-evolving complex (OEC) The enzyme which catalyzes the formation of O_2 in photosynthesis. Contains a cluster of probably four manganese ions.

Ozone Ozone, O3, is the triatomic isomer of oxygen, O2, and it is a powerful oxidising agent. The presence of ozone in the stratosphere is responsible for the absorption of ultra-violet wavelengths from the sunlight.

Ozone depletion potential (ODP) The is the ratio of a particular ozone depleting compound environmental effect compared to the depleting ability of a standard compound, CFC-11, which arbitrarily defined as 1.0. An example would be the ODP of of halon-1301 which is about 13.

Ozone destruction In addition to photolytic destruction by infrared light (wavelengths of 1140 nanometers or less), stratospheric ozone is destroyed by interaction with a variety of chemical substances in the stratosphere. Some of these chemicals, such as the chlorine free radical—liberated by the photolysis of chlorofluorocarbons, are not normally produced there in nature at such high concentrations. This leads to the conclusion that human activities, and not natural chemical cycles, are responsible for the current rate of stratospheric ozone destruction. Ozone destruction is most noticeable (alarmingly!) over the South Pole, where seasonal plummeting in stratospheric ozone concentrations have been measured since the 1970s and are apparently still increasing (1999). That said, estimates based on declining CFC use (because of the Montreal Protocol) suggest that the damage to the global ozone layer will max out soon in the 21st century and that that ozone damage will decrease within the next 75 to 100 years.

Ozone distribution- Nearly all of the planet's natural ozone in the atmosphere is found in the stratosphere. The thickness and distribution of this stratospheric ozone vary seasonally and by location. Ozone also occurs in the troposphere as a pollutant, often in photochemical smog.

Ozone formation Molecular oxygen in the stratosphere is broken into a pair of oxygen radicals by light with a wavelength of 240 nanometers or less. If one of these O radicals encounters an oxygen molecule, it can bond to produce ozone. This reaction is only stable if another molecule is present to absorb the excess energy released as the oxygen radical and molecule bond. This is a called a three body reaction, and the third body exhibits its removal of the excess energy by whizzing off at a higher energy and thereby increasing the temperature of the atmosphere where this reaction occurs.

Ozone hole The ozone hole refers to the total amount of ozone in the atmosphere over Antarctica that was discovered to be decreasing at an alarming rate over an observation site in Antarctica. The hole was discovered by J.C. Farman, B.G. Gardiner and J.D. Shanklin.

P

P, M See helicity.

Pangea A huge protocontinent thought to have existed 200 million years ago and from which all present continents are thought to have formed.

Para Describes the poitions of two substituents on a benzene ring that are separated by two carbons.

Paraffin paraffin wax. 1. a waxy substance that is a mixture of alkanes with chains containing 18 to 36 carbon atoms. 2. An alkane.

Parallel reaction See composite reaction.

Parallel synthesis Strategy whereby sets of discrete compounds are prepared simultaneously in arrays of physically separate reaction vessels or microcompartments without interchange of intermediates during the assembly process. Contrast Pool/Split.

Paramagnetic Substances having a magnetic susceptibility greater than 0 are paramagnetic. They are drawn into a magnetic field.

Paramagnetism Paramagnetism is a type of magnetism characterized by a positive magnetic susceptibility, so that the material becomes weakly magnetized in the direction of an external field. The magnetization disappears when the field in removed.

Parent elements The element that underwent radioactive decay to produce a radioactive isotope and eventually produce a more stable product. Can be by the emission of an alpha particle, two protons and two neutrons. or by the emission of a beta particle, emission of an electron which changes a neutron to a proton. Is used in the dating of objects because the decay happens in a predictable known pattern.

Parent isotope An element that undergoes nuclear decay.

Partial library Partly assembled library, or portion thereof, which is reserved to be completed once initial property relationships have been identified. For instance, part of an intermediate pool may not be treated with the final building block until the optimal residue at the final position is known, thus avoiding the need to prepare that pool from the starting materials.

Partial miscibility Partially miscible. Compare with miscible and immiscible.

Two liquids are considered partially miscible if shaking equal volumes of the liquids together results in a meniscus visible between two layers of liquid, but the volumes of the layers are not identical to the volumes of the liquids originally added.

Partial pressures The pressure exerted by a certain gas in a mixture.

Partial rate factor The rate of substitution at one specific site in an aromatic compound relative to the rate of substitution at one position in benzene. For example, the partial rate factor

fp^Z for para-substitution in a monosubstituted benzene C_6H_5Z is related to the rate constants $k(C_6H_5Z)$ and $k(C_6H_6)$ for the total reaction (i.e. at all positions) of C_6H_5Z and benzene, respectively, and% para (the percentage para-substitution in the total product formed from C_6H_5Z) by the relation

$$f_p^z = \frac{6k(C_6H_5Z)}{k(C_6H_6)} \frac{\%para}{100}$$

Similarly for meta-substitution:

$$f_m^z = \frac{6k(C_6H_5Z)}{2k(C_6H_6)} \frac{\%meta}{100}$$

(The symbols pf^Z, mf^Z, of^Z are also in use.) The term applies equally to the ipso position, and it can be extended to other substituted substrates undergoing parallel reactions at different sites with the same reagent according to the same rate law. See also selectivity.

Partial release Cleavage process designed to release a compound from a solid support in discrete portions, e.g. by using orthogonal linkers or by controlled application of cleavage reagent or condition.

Partial vaccuum Compare with absolute vaccuum.

A volume that contains traces of gas at very low pressure.

Particle Small portion of matter

Particulate Composed of distinct particles. Smoke is particulate; pure gases are not.

Particulate matter ($PM_{2.5}$) Aerosol particles that are smaller than or equal to 2.5 micrometers or have an aerodynamic diameter of 2.5 micrometers.Ý Considered fine particle matter, $PM_{2.5}$ has a greater effect of visibility and health than do larger aerosol particles, such as decreased lung function and premature death.

Partition coefficient The constant ratio that is found when a heterogenous system of two phases is in equilibrium; the ratio of the concentrations of the same molecular species (substance) in two phases (usually water and octanol) is constant at given temperature and pressure.

Parts per million (ppm) Concentration expressed as parts of solute per million parts of solution. Usually refers to parts per million by mass. For example, a 10 ppm NaCl solution can be written as: 10 mg NaCl/kg solution, 10 µg NaCl/g solution, 10 ng NaCl/mg solution. In very dilute aqueous solutions, ppm is approximately equal to mg solute per liter of solution.

Parts per million (PPM) and parts per billion (PPB) These terms give scientists a way to describe how much of a substance is contained in a sample: parts of analyte per million parts of sample, for instance. In atmospheric chemistry these become volume parts of analyte per volume parts of atmosphere: ppmv, ppbv, etc. At low analyte gas phase concentrations the analyte is assumed to act as an ideal gas. For instance, a 1 ppmv concentration of formaldehyde would represent 1 liter of formaldehyde per every 1,000,000 liters of air; also equivalent to 1 microliter of H_2CO per 1 L air. In gas phase concentrations these units are also called gas phase mixing ratios. The reason is because they are just that, ratios of analyte volume to sample volume; the volume of the sample doesn't matter.

Pascal (PA) The SI unit of pressure, equal to a force of one newton per square meter. 101325 pascals = 1 atmosphere; 10^5 pascals = 1 bar.

Pascal Paramagnetism is a type of magnetism characterized by a positive magnetic susceptibility, so that the material becomes weakly magnetized in the direction of an external field. The magnetization disappears when the field in removed.

Paschen series Paschen series are the series of lines in the spectrum of the hydrogen atom which corresponds to transitions between the state with principal quantum number n = 3 and successive higher states.

Path length (B) In absorption spectroscopy, the length of a path taken by radiation through a sample.

Patina A thin layer of corrosion products with a distinctive coloration that forms on a metal surface exposed to air and water. Patina usually refers to the greenish coating that forms on copper alloys over time.

Pattern recognition A computational technique used to find patterns and develop classification schemes for data in very large data sets.

Paul A bartlett (b.1948) Born in Trenton New Jersey, but grew up in Boston. He studied chemistry at Harvard, working with David Dolphin, before taking his Ph.D. at Stanford University under William S. Johnson. After working as a NIH Postdoctoral Fellow at University of California at San Diego, Bartlett was appointed to the faculty of the University of California at Berkeley. He is a synthetic organic chemist, working initially on stereospecific synthesis of natural products, before switching to the synthesis of enzyme inhibitors and metabolic intermediates. As a result of his work of his work on enzyme inhibitors, Bartlett has also developed the computer modelling of 3-D chemical structures.

Paul doughty bartlett (1907-1997) Born in Ann Arbor, Michigan, he studied chemistry at Amherst, before taking his Ph.D. at Harvard under Conant. After brief spells at the Rockefeller Institute, Columbia University and the University of Minnesota, Bartlett returned to Harvard. After his retirement from Harvard, he started afresh at Texas Christian University at Fort Worth. He has studied a wide range of organic reaction mechanisms, notably the Wagner-Meerwein rearrangements and carbonium ions. An obituary on a Yale website also cited substitution and rearrangement mechanisms, strained lactones, highly branched molecules, alkyl lithium compounds, and polymerisation and chain reactions, among many many other topics.

Paul gabriel adam (1856-1916) Born in Avesnes (Nord), near the Belgian border. After taking his doctorate at the Sorbonne, he became a principal sanitary inspector in Paris. He was then made Professor of Chemistry at a veterinary college at Alfort (Seine). Adam was interested in the bromination of organic compounds and the chemistry of biphenyl. He should not be confused with Paul Emilé Adam , a French civil engineer and professor of mechanics.

Pauli principle Exclusion principle; Pauli exclusion; Pauli exclusion principle.

No two electrons in an atom can have the same set of 4 quantum numbers. Because the n, , and m quantum numbers address a particular orbital, and because the m_s quantum number has only two possible values, the Pauli principle says that a maximum of two electrons can occupy an atomic orbital- and these electrons must have opposite spins.

PCR See polymerase chain reaction.

PCV systems A PCV system is an air pollution control device in automobile engines. PCV stands for positive crankcase ventilation. Automobile makers began installing PCV systems in all new cars in 1963. It helps to recycle hydrocarbon contaminated gases inside the engine (blowby) back into the engine's combustion process.

Pedosphere This is the area of the Earth comprised of the solid plates of the continental crust, loose rocks and soil.

Peltier effect Peltier effect is the absorption or generation of heat (depending on the current direction) which occurs when an electric current is passed through a junction between two materials.

Penetration Compare with shielding. Electrons in penetrating orbitals can reach the nucleus. The n and quantum numbers determine how well an orbital penetrates. Lower n and lower values mean better penetration. A low n value means the orbital is small. A low value means the orbital has fewer nuclear nodes (planes that pass through the nucleus where the probability of locating the electron is zero).

In order of decreasing penetration, the subshells are s > p > d > f. A 1s orbital penetrates better than a 2s orbital.

Peptide bond peptide A planar, amide linkage between the amino group of one amino acid and the carboxyl group of another, with the elimination of a molecule of water.

PeptidesPeptides are amides derived from two or more amino acids (the same or different) linked by peptide bonds. These bonds are formed by the reaction between adjancent carboxyl (-COOH) i amino (-NH_2) groups with the elimination of water.

Peptoid Oligomer consisting of repeating N-substituted glycine units.

Percent composition Expresses the weight ratio between different elements in a compound.

Percent diastereoisomer excess See diastereoisomer excess.

Percent enantiomer excess See enantiomer excess.

Percent error Percentage error. The relative error times 100%.

Percent yield Percentage yield. Compare with theoretical yield and actual yield.

Percent yield equals experimental yield divided by theoretical yield times 100%.

Perfect crystal A crystal with no defects or impurities, made of completely identical repeating subunits. Further, a perfect crystal has only one possible arrangement of subunits, with every subunit making exactly the same contribution to the total energy of the crystal.

Perfluoroethane (C_2F_6) See hexafluoroethane

Pericyclic reaction A chemical reaction in which concerted reorganization of bonding takes place throughout a cyclic array of continuously bonded atoms. It may be viewed as a reaction proceeding through a fully conjugated cyclic transition state. The number of atoms in the cyclic array is usually six, but other numbers are also possible. The term embraces a variety of processes, including cycloadditions, cheletropic reactions, electrocyclic reactions and sigmatropic rearrangements, etc. (provided they are concerted).

Periods While groups are characterized by the number of electrons present in the outer shell, periods are characterized by the number of energy levels (shells) of electrons surrounding the nucleus. Elements in Period 1 has only one shell. As you probably recall, the elements in the first period have a 2 electrons maximum (hydrogen has 1 electron and helium has 2 electrons. As we move to the first group of the second period, we find that lithium, which has the two electrons in the first shell and one in the second. Neon is in Group 18 of Period 2 and therefore has the two electrons in the first shell and eight electrons in the second shell. Sodium starts Period 3 with 11 electrons, two in the first shell, eight in the second shell and one in the third shell. In other words, the element in Group 1 always has one more electron (in a new shell) than the Group 18 element in the previous period.

Periodic law The electron configurations of the atoms of the elements vary periodically with their atomic number. Consequently, all properties of the elements that depend on their electron configuration tend also to change with the increasing atomic number in a periodic manner.

Periodic table An arrangement of elements in a geometric pattern designed to represent the periodic law by aligning elements into periods and groups. Elements with the same number of orbitals are in the same period and elements with the same number of electrons in the filling orbital and have other similar properties are in the same group.

Periodic trend A regular variation in element properties with increasing atomic number that is ultimately due to regular variations in atomic structure.

Periplanar See torsion angle.

Periplasm The fluid occupying the space between the inside and outside cellular membranes of bacteria.

Periselectivity The differentiation between two symmetry-allowed processes, for example the (2+4) vs. (4+6) cycloaddition of cyclopentadiene to tropone.

Peritoneum Lining of the abdominal organs and cavity.

Permafrost Soil, silt and rock located in perpetually cold areas which remains frozen year-round. Though a thin layer may thaw during summer months, the majority of the permafrost in a given location will remain frozen until the local climate dramatically changes, as from continental drift toward the equator or global warming.

Permanent hardness Permanent water hardness. Compare with temporary hardness and water hardness.

Water hardness that remains after boiling the water, mainly due to dissolved calcium sulfate. Chlorides also contribute to permanent hardness.

Permanganate (MnO_4^-) permanganate ion.

Permanganate ion (MnO_4^-) is a powerful oxidizing agent used in chemical analysis and water treatment. The ion has an intense purple color.

Peroxidase A heme protein (donor:hydrogen-peroxide oxidoreductase, EC class 1.11.1) which catalyzes the one-electron oxidation of a substrate by dihydrogen peroxide. Substrates for different peroxidases include various organic compounds, cytochrome c, halides, and Mn^{2+}.

Peroxides Peroxides, R-O-O-R, are organic molecules which contain the peroxide linkage, -O-O-, within the molecule. Peroxides readily give rise to free radicals, because the oxygen to oxygen bond is easily broken in these organic molecules.

R-O-O-R $\Rightarrow$ 2 R-O*

Peroxide Peroxide

Free Radical

The most important organic peroxide is benzoyl peroxide, (C6H5COO)2, which is the initiator used in the polymerisation of many industrially important polymers (e.g. fillers, etc) and adhesives (e.g. superglue, etc.).

Persistence When applied to a chemical this has a meaning of ability to remain unchanged in the environment.

Persistent The term persistent is used to characterize radicals which have lifetimes of several minutes or greater in dilute solution in inert solvents. Persistence is a kinetic or reactivity property. In contrast, radical stability, which is a thermodynamic property, is expressed in terms of the C-H bond strength of the appropriate hydrocarbon. The lifetime of a radical is profoundly influenced by steric shielding of the radical centre by bulky substituents.. See also transient.

Perspective formula A geometric representation of stereochemical features of a molecule or model which appears as a view from an appropriate direction.

Perspex Perspex is a trade name for plastics made from polymethylmethacrylate.

Pesticide residue A pesticide residue is any substance or mixture of substances in food for man or animals resulting from the use of a pesticide and includes any specified derivatives, such as degradation and conversion products, metabolites, reaction products, and impurities that are considered to be of toxicological significance (WHO, 1976).

Pesticides This is a descriptor applied to chemicals used to kill pests and minimize their impact in agriculture, health and other human interests. Pesticides are often classified according to the organisms which they are used to control, for example as fungicides, herbicides, insecticides, molluscicides, nematicides, rodenticides, etc.

Petrochemicals Petrochemicals are the industrially important organic chemicals which are derived from petroleum or natural gas.

Petrol Petrol (i.e. Gasoline) is a complex mixture of hydrocarbons containing 5 to 10 carbon atoms in each molecule.

Petroleum ether Petroleum ether is the petroleum fraction consisting of C_5 and C_6 hydrocarbons and boiling in the range 35 °C to 60 °C; commonly used as a laboratory solvent.

PH A mean to express and to compare the acidity and alkalinity of a solution. It is expressed in a scale from 0 to 14. The solution of pH 7 is neutral; if the pH is lower than 7 the solution is acidic; if the pH is higher than 7 the solution is alkaline (basic).

PHpH is a convenient measure of the acid-base character of a solution, usually defined by

$pH = -\log c(H^+)$

where $c(H^+)$ is the concentration of hydrogen ions in moles per liter. The more precise definition is in terms af activity rather than concentration.

A solution of pH 0 to 7 is acid, pH of 7 is neutral, pH over 7 to 14 is alkaline.

Phage display Use of genetically engineered phage to present peptides as segments of their native surface proteins. Peptide libraries may be produced by populations of phage with different gene sequences.

Phagocyte A cell which is able to ingest, and often to digest, large particles such as bacteria and dead tissue cells.

Pharmacognosy Identification, isolation, and characterization of biologically active substances in living things.

Pharmacokinetics The study of absorption, distribution, matabolism and excretion (ADME) of bioactive compounds in a higher organism.

Pharmacology The study of drugs, which includes determination of biological activity, biological effects, breakdown and synthesis, and delivery.

Pharmacophore A pharmacophore is the ensemble of steric and electronic features that is necessary to ensure the optimal supramolecular interactions with a specific biological target structure and to trigger (or to block) its biological response. A pharmacophore does not represent a real molecule or a real association of functional groups, but a purely abstract concept that accounts for the common molecular interaction capacities of a group of compounds towards their target structure.

Phase In phase; out of phase; wave phase. 1.A phase is a part of a sample of matter that is in contact with other parts but is separate from them. Properties within a phase are homogeneous (uniform). For example, oil and vinegar salad dressing contains two phases: an oil-rich liquid, and a vinegar-rich liquid. Shaking the bottle breaks the phases up into tiny droplets, but there are still two distinct phases.

2. In wave motion, phase is the fraction of a complete cycle that has passed a fixed point since the current cycle began. The phase is often expressed as an angle, since a full cycle is 360^deg; (2). Two waves are "in phase" if the peaks of one wave align with the peaks of the other; they are "out of phase" if the peaks of one wave align with the troughs of the other.

Phase boundary A phase boundary is a surface where two samples of matter with different properties are in contact. The surface of a gas bubble in water or the surface of a crystal are examples of phase boundaries.

Phase change Phase transition. A change in the state of a sample of matter; for example, solid to liquid or liquid to gas. Phase changes are considered physical rather than chemical changes.

Phase diagram Phase diagram is a graphical representation of the equilibrium relationships between phases (such as vapor-liquid, liquid-solid) of a chemical compound, mixture of compounds, or solution.

Phase switch Strategy for compound isolation, whereby the desired material is rendered sufficiently different from reagents, side-products and other impurities that it may be separated from them by simple physical processes such as filtration or extraction. May be achieved by the attachment of a tag, such as a highly fluorinated component, or other sequestration enabling reagent.

Phase-transfer catalysis The phenomenon of rate enhancement of a reaction between chemical species located in different phases (immiscible liquids or solid and liquid) by addition of a small quantity of an agent (called the "phase-transfer catalyst") that extracts one of the reactants, most commonly an anion, across the interface into the other phase so that reaction can proceed. These catalysts are salts of "onium ions" (e.g. tetraalkylammonium salts) or agents that complex inorganic cations (e.g. crown ethers). The catalyst cation is not consumed in the reaction although an anion exchange does occur.

Phenol 1. A corrosive poisonous crystalline acidic compound present in coal tar and wood tar that in dilute solutions is used as a disinfectant and

2. any of various acidic compounds analogous to phenols and regarded as hydroxyl derivatives of aromatic hydrocarbons. When coal and wood are burned for energy, phenols are released into the atmosphere.

Phenolphthalein An organic compound used as an acid-base indicator. The compound is colorless in acidic solution and pink in basic solution (with the transition occuring around pH 8.3). Phenolpthalein was used for many years as a laxative in very low concentrations- high concentrations are toxic!

PhenolsPhenols are compounds having one or more hydroxy groups attached to benzene or other arene ring.

Phenonium ion See bridged carbocation.

Phenyl A molecular group or fragment formed by abstracting or substituting one of the hydrogen atoms attached to a benzene ring.

-amino-ß%3b- phenylpropionic%20acid)”>

Phenyl ring A benzene ring as a substituent. Abbreviated Ph.

Phenylalanine Phe; -amino-ß-phenylpropionic acid).

A naturally occuring amino acid with a nonpolar side chain terminated by a phenyl group.

Philip schaffner baker (b.1916) Born in Minneapolis, Minnesota, and studied at Occidental College (Los Angeles), Antioch College (Ohio) and DePauw University in Greencastle, Indiana. Like many bright students from small Midwestern colleges, he took his Ph.D. at the University of Illinois. After working for an oil company in Texas and at the Institute for Paper Chemistry in Appleton, Wisconsin, he taught at Bradley University in Peoria, Illinois. He then joined Oak Ridge National Laboratory, where he worked on the applications of isotopes and the production of lithium metal.

Philip sigmon bailey (b.1916) Born in Chickasha, Oklahoma, he studied chemistry at Oklahoma Baptist University in Shawnee and the University of Oklahoma in Norman. He took his Ph.D. under Lutz at the University of Virginia at Charlottesville. After working briefly for an oil company, Bailey joined the Office of Scientific Research and Development. After the Second World War, he went to the University of Texas in Austin. He now lives in Lago Vista, near Austin. Bailey has carried out extensive research on the mechanism of ozonolysis and has published Ozonation in Organic Chemistry.

Phillipp wilhelm fischer (1877-1946) Born in Schweinfurt, he took his Ph.D. at Würzburg under Krauss. Became the official pharmacist in Nürnberg, and published papers on pharmaceutical analysis.

Phosphatase An enzyme that catalyzes the hydrolysis of orthophosphoric monoesters. Alkaline phosphatases (EC 3.1.3.1) have an optimum pH above 7 and are zinc-containing proteins. Acid phosphatases (EC 3.1.3.2) have an optimum pH below 7 and some of these contain a dinuclear center of iron, or iron and zinc.

Phosphate (PO_4^{-3}) 1. The PO_4^{-3} ion.

2. A compound containing the PO_4^{-3} ion.

Phospholipid An ester of glycerol with two fatty acids and phosphoric acid (H_3PO_4) or a derivative of phosphoric acid group (like $H_2PO_4CH_2CH_2N(CH_3)_3$). Phospholipids have a hydrophilic head (the phosphate group) and a lipophilic tail (the fatty acids).

Phospholipid bilayer Bilayer; phospholipid membrane. In an aqueous environment phospholipids can form a two-layered “sandwich” with the hydrophobic lipid tails on the inside and the hydrophilic phosphate heads facing outward. These bilayers are the essential structure for building cell membranes.

Phosphorylation A process involving the transfer of a phosphate group (catalyzed by enzymes) from a donor to a suitable acceptor; in general an ester linkage is formed, for example:

$$\text{ATP} + \text{alcohol} \longrightarrow \text{ADP} + \text{phosphate ester}$$

Photic zone In the ocean (or in lakes) this is the area near the surface with enough light to support photosynthesis. In this zone, both the chemical and biological activity can be enhanced by penetrating solar radiation. This zone is about 20 meters in depth in the ocean. In the upper surface of this zone, the marine photochemistry is most vigorous. Changes in the ultraviolet light content penetrating this region because of changes in the UV-blocking ability of the ozone layer may create unexpected biological results there.

Photochemical smog Natural and artificially emitted hydrocarbons in the presence of oxides of nitrogen undergo photochemical reactions which produce a cloud of toxic chemicals including ozone and a variety of caustic agents. This process is powered by sunlight and some of the products, such as ozone, reach a peak soon after photon flux from the sun reaches a maximum, around midday. The thermal inversions often associated with some cities can lead to a dangerous buildup of smog in urban areas. Human deaths have been attributed to photochemical smog since the Industrial Revolution in cities such as London and New York.

Photochemistry The study of chemical changes caused by light. For example, many of the key reactions that generate smog are photochemical.

Photodissociation This is a specific type of photolysis reaction, in which the destruction of a molecule is initiated strictly by solar energy.

Photoelectric effect Ejection of electrons from an atom or molecule that has absorbed a photon of sufficient energy. The photoelectric effect is the operating principle behind "electric eyes"; it is experimental evidence for particle-like behavior of electromagnetic radiation.

Photoelectron An electron ejected from an atom or molecule that has absorbed a photon.

Photolithography Process by which selective masking generates light patterns which direct chemical transformations to certain areas of a photosensitive surface. Coupling of different building blocks to discrete sites may give rise to spatially addressable arrays of compounds.

Photolysis The cleavage of one or more covalent bonds in a molecular entity resulting from absorption of light, or a photochemical process in which such cleavage is an essential part. For example:

Cl_2 2 Cl·

The term is used incorrectly to describe irradiation of a sample, although in the combination flash photolysis this usage is accepted.

Photon A particle that has no rest mass, no electrical charge and whose energy is determined by its electromagnetic wavelength. A photon is considered to be an amount of electromagnetic radiation energy that is proportional to the frequency of that radiation. It may also be considered as a particle of light with energy, hv, where h is Planck's constant and v is the frequency of light.

Photostationary state This describes a condition in which there is equilibrium in a photochemical reaction. For this condition to take place the rate of dissociation of the reactants has to be equal to the rate of their recombination.

Photosynthesis A metabolic process in plants and certain bacteria, using light energy absorbed by chlorophyll and other photosynthetic pigments for the reduction of CO_2, followed by the formation of organic compounds.

Photosystem A membrane-bound protein complex in plants and photosynthetic bacteria, responsible for light harvesting and primary electron transfer. Comprises light-harvesting pigments such as chlorophyll; a primary electron-transfer center, and secondary electron carriers. In green plant photosynthesis, Photosystem I transfers electrons from plastocyanin to a (2Fe-2S) ferredoxin, and contains iron-sulfur proteins. Photosystem II transfers electrons from the oxygen-evolving complex to plastoquinone, and contains an iron center.

PH-rate profile A plot of observed rate coefficient, or more usually its decadic logarithm, against pH of solution, other variables being kept constant.

Physical change Compare with chemical change. A change which does not transform one substance into another. For example, freezing water is a physical change because both water and ice are H_2O. However, electrolysis of water would not be a physical change because passing a strong electric current through water can decompose it into H_2 and O_2.

Physical chemistry Chemical physics. A branch of chemistry that studies chemical phenomena from a physical and mathematical perspective. Physical chemistry includes chemical thermodynamics, kinetics, spectroscopy, quantum chemistry, and statistical mechanics.

Physical property Physical properties. Compare with chemical property. Measurement of a physical property may change the arrangement but not the structure of the molecules of a material. Examples of physical properties are density, color, boiling point, volume, temperature, and mass.

Phytochelatin A peptide of higher plants, consisting of polymers of 2-11 glutathione (γ-glutamyl-cysteinyl-glycine) groups, which binds heavy metals.

Phytochemistry Phytochemical. The study of substances found in plants. "Phytochemicals" are materials extracted from plant tissue.

Pi bond (bond) Compare with sigma bond. In the valence bond theory, a pi bond is a valence bond formed by side-by-side overlap of p orbitals on two bonded atoms. In most multiple bonds, the first bond is a sigma bond and all of the others are pi bonds.

Picket-fence porphyrin A porphyrin with a protective enclosure for binding oxygen at one side of the ring that is used to mimic the dioxygen-carrying properties of the heme group.

Pico- (P) Prefix used in the SI system meaning "multiply by 10^{-12}". For example, 3 pm means 3×10^{-12} meters.

Picogram PG One picogram is 10^{-12} grams.

Picoliter PL. One picoliter is 10^{-12} liters.

Picometer PM. One picometer is 10^{-12} meters.

Pin An elongated device in which the tip acts as a solid support. An array of pins may typically be held such that sets of pins may be simultaneously inserted or retracted from solvents or reagents allowing library preparation by parallel synthesis.

PKa (pK_a) Compare with acid dissociation constant.

The pK_a of an acid is minus the base-10 log of its acid dissociation constant, $pK_a = - \log K_a$. For example, a pK_a of 5 is equivalent to an acid dissociation constant of 10^{-5}.

PKb (pK_b) Compare with base hydrolysis constant.

The pK_b of an base is minus the base-10 log of its base hydrolysis constant, $pK_b = - \log K_b$. For example, a pK_b of 5 is equivalent to an base hydrolysis constant of 10^{-5}.

Placard A warning sign measuring 10.8 inches on each side and standing on one point (like a diamond), that is placed on the four vertical sides of the truck, rail car, intermodal container or two sides of an intermediate bulk container containing hazardous materials. Placards are required by the USDOT, Transport Canada, Secretariat of Transport (Mexico) and other regulatory agencies around the world when shipping hazardous materials by air, highway, rail or water.

Planar chirality Term used by some authorities to refer to stereoisomerism resulting from the arrangement of out-of-plane groups with respect to a plane (chirality plane). It is exemplified by the atropisomerism of (E)-cyclooctene (chiral plane = double bond and attached atoms) or monosubstituted paracyclophane (chiral plane = substituted ring). The configuration of molecular entities possessing planar chirality is specified by the stereodescriptors R_p and S_p (or P and M).

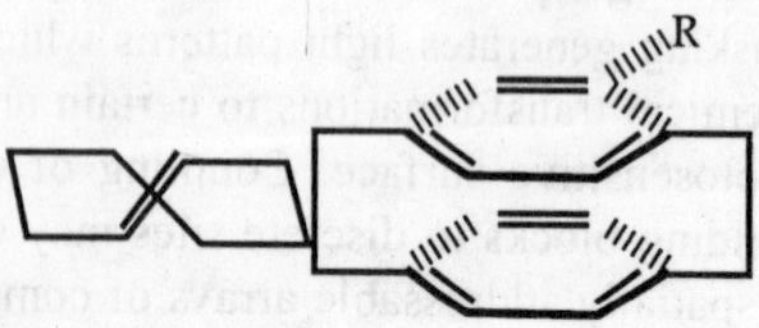

Planck Planck contributed to the understanding of the electromagnetic spectrum by realizing that the relationship between the change in energy and frequency is quantized according to the equation delta E=hv where h is Planck's constant.

Planck constant Planck constant (h) is a constant that when multipled by the frequency of radiation gives the quantity of energy contained in one quantum.

E = hν

Equal to 6.6260755(40)'"10^{-34} J s. It is named after Max Planck (1858-1947).

Planck's law This is a derived formula, from the German physicist Max Planck, that portrays the amount of radiation emitted by a blackbody as theoretically determined by its temperature. It is an equation that produces a curve, termed Planck's blackbody radiation curve, which illustrates that the warmer a body is, the greater is its blackbody emission at each wavelength and the shorter is the wavelength at which emissions peak.

Plane of symmetry A plane cutting through a molecule in which both halves are mirror images of each other.

Plane-polarized light Light that oscillates in a single plane.

Plankton Plankton are forms of marine, organic life that gather vital energy for life through photosynthesis. Since they are very low on the food chain, they are vital to all marine life. They are also vital to the atmosphere in that they consume large amounts of carbon dioxide, which contribute to enhanced global warming when excess amounts are present. Plankton are considered to be one of the more important contributors of oxygen to the atmosphere.

Plasma In biology this term has the following three meanings : 1. Fluid component of blood in which the blood cells and platelets are suspended ('blood plasma').

N.B. Note the distinction between plasma, which describes a part of the blood (the fluid part of blood, outside the blood cells) and serum, which describes a fraction derived from blood by a manipulation (the fluid which separates when blood coagulates).

2. Fluid component of semen produced by the accessory glands, the seminal vesicles, the prostate, and the bulbo-urethal glands.

3. Cell substance outside the nucleus (cytoplasm).

Plasmid An extrachromosomal genetic element consisting generally of circular double-stranded DNA, which can replicate independently of chromosomal DNA. R-plasmids are responsible for the mutual transfer of antibiotic resistance among microbes. Plasmids are used as vectors for cloning DNA in bacteria or yeast host cells.

Plastocyanin An electron transferprotein, containing a type 1 copper site, involved in plant and cyanobacterial photosynthesis, which transfers electrons to Photosystem I.

Plate tectonics The process of planetary surface plate formation, movement, interaction, and destruction. The consequence of this process is that many regions of the planet are situated near interacting—and slowly moving—plates. The process causes earthquakes such as those along the West Coast of the United States.

Pleistocene An epoch in Earth history from about 2-5 million years to 10,000 years ago. Also refers to the rocks and sediment deposited in that epoch.

Pleura The thin membrane enveloping the lungs and lining the thoracic cavity.

Pleural effusion Presence of fluid in the pleural cavity resulting from excessive transudation or exudation from the pleural surfaces. It is a sign of disease and not a diagnosis in itself.

Pleural plaques Areas of fibrosis present on the inner surface of the ribcage and the diaphragm. They may be partly calcified .

Plume A plume is a visible smoke-like structure, which may contain pollutants emitted from an exhaust or smoke stack and released into the atmosphere. This elongated band of smoke has changing characteristics that vary with its local environmental conditions. These conditions may include the physical and chemical nature of the pollutant, weather conditions and downwind topography.

Pneumoconiosis A condition characterized by deposition of large amounts of particulate matter or fibers in the lungs, usually of occupational or environmental origin, and by the tissue reaction to its presence.

POH Measures the basicity of a solution. It is the negative log of the concentration of the hydroxide ions.

Point group The classification of the symmetry elements of an object. It is denoted in the Schoenflies notation by an italic symbol, such as C_3, D_2, T_d, etc.

Point of diversity Portion of a molecule, or step in a synthetic scheme, where different building blocks may be introduced.

Point source Pollutants which are put into the atmosphere by stationary objects, such as refineries, power plants, mills, and the like. In Texas any source that emits more than one ton in a calendar year is considered a point source.

Poise (P) A cgs unit of resistance to fluid flow (viscosity). If a force of 1 dyne is needed to force two fluid layers with 1 cm^2 area that are 1 cm apart past each other at a speed of 1 cm/s, the liquid has a viscosity of 1 poise.

Polar aprotic solvent See dipolar aprotic solvent.

Polar bond Compare with covalent bond and ionic bond.

A bond involving electrons that are unequally shared. Polar bonds can be thought of as intermediate between the extremes represented by covalent bonds and ionic bonds.

Polar effect For a reactant molecule RY, the polar effect of the group R comprises all the processes whereby a substituent may modify the electrostatic forces operating at the reaction centre Y, relative to the standard R^oY. These forces may be governed by charge separations arising from differences in the electronegativity of atoms (leading to the presence of dipoles), the presence of unipoles, or electron delocalization. It is synonymous with electronic effect or "electrical effect" of a substituent as distinguished from other substituent effects, e.g. steric effects.

Sometimes, however, the term "polar effect" is taken to refer to the influence, other than steric, that non-conjugated substituents exert on reaction rates, i.e. effects connected with electron delocalization between a substituent and the molecular framework to which it is attached are excluded. Polar effect is then not synonymous with electronic effect. See also field effect, inductive effect, mesomeric effect.

Polar molecule Polar. Compare with covalent compound, ionic compound and polar bond.

An asymmetric molecule containing polar bonds. H_2O, NH_3, and HCl are examples of polar molecules. Non-examples are CO_2, CCl_4, and BCl_3 which contain polar bonds but are nonpolar because they have symmetric shapes. Alkanes are usually asymmetric but are nonpolar because they contain no polar bonds. Polar molecules are electric dipoles and they attract each other via dipole-dipole forces.

Polar solvent See polarity.

Polar stratospheric clouds PSCs See stratospheric clouds.

Polar vortex This is a phenomenon that occurs during the polar winter in which stratospheric air moves in a circular motion, with an area of relatively still air in its center. The temperature in the vortex is approximately -130 degrees F (-80 degrees C), which assists in the formation of polar stratospheric clouds. Though usually more prolonged and colder over Antarctica, the Arctic polar vortex does form to a degree, and when the temperatures there are coldest and the vortex persists, Arctic stratospheric ozone destruction on Arctic polar stratospheric clouds has also been observed.

Polarity When applied to solvents, this rather ill-defined term covers their overall solvation capability (solvation power) for solutes (i.e. in chemical equilibria: reactants and products; in reaction rates: reactants and activated complex; in light absorptions: ions or molecules in the ground and excited state), which in turn depends on the action of all possible, nonspecific and specific, intermolecular interactions between solute ions or molecules and solvent molecules, excluding such interactions leading to definite chemical alterations of the ions or molecules of the solute. Occasionally, the term solvent polarity is restricted to nonspecific solute/solvent interactions only (i.e. to van der Waals forces).

Polarizability The ease of distortion of the electron cloud of a molecular entity by an electric field (such as that due to the proximity of a charged reagent). It is experimentally measured as the ratio of induced dipole moment (uind) to the field E which induces it:

= ind/E

The units of are C^2 m^2 V^{-1}. In ordinary usage the term refers to the "mean polarizability", i.e., the average over three rectilinear axes of the molecule. Polarizabilities in different directions (e.g. along the bond in Cl_2, called "longitudinal polarizability", and in the direction perpendicular to the bond, called "transverse polarizability") can be distinguished, at least in principle. Polarizability along the bond joining a substituent to the rest of the molecule is seen in certain modern theoretical approaches as a factor influencing chemical reactivity, etc., and parametrization thereof has been proposed.

Pollutant Any undesirable solid, liquid, or gaseous matter in a gaseous, liquid, or solid medium (ISO, 1977). For the meaning of "undesirable" in air pollution contexts, see pollution. A primary pollutant is a pollutant emitted into the atmosphere from an identifiable source. A secondary pollutant is a pollutant formed by chemical reaction in the atmosphere (WHO, 1980).

Pollutants A pollutant may be considered as any substance, usually an unwanted by-product or waste, that is released into the environment as a result of (human) activities that alter the chemical, physical and biological characteristics of the environment. These substances may be found in any of the solid, liquid or gas phases. Their long-term effects are difficult to predict and depend upon future human or cultural judgments.

Poly(ethylene glycol) (PEG) Polymer which has been applied both as a soluble support and (as a graft copolymer with a polystyrene matrix) as a linker for combinatorial synthesis. The soluble support may have hydroxyls at both termini, or one or both may be capped or modified with additional functionality.

Polyatomic ion Compare with molecule, ion and polyatomic molecule.

A polyatomic ion is a charged particle that contains more than two covalently bound atoms.

Polyatomic molecule Compare with polyatomic ion and diatomic molecule.

A polyatomic molecule is an uncharged particle that contains more than two atoms.

Polydent See ambident.

Polydentate Polydentate ligand. A ligand that has more than one atom that coordinates directly to the central atom in a complex. Polydentate ligands are called chelating agents when two or more coordinating atoms are attached to the same metal ion in a complex. For example, EDTA or ethylenediaminotetracetic acid is a hexadentate ligand of calcium ion.

Polyhedral symbol The polyhedral symbol indicates the geometrical arrangements of the coordinating atoms about the central atom. It consists of one or more capital italic letters derived from common geometric terms (tetrahedron, square plane, octahedron, etc.) which denote the idealized geometry of the ligands around the coordination center, and an arabic numeral that is the coordination number of the central atom. The polyhedral symbol is used as an affix, enclosed in parentheses, and separated from the name by a hyphen. Examples are T-4, SP-4, TBPY-5, SPY-5, OC-6, and CU-8.

Polymer Polymer is a substance composed of molecules of high relative molecular mass (molecular weight), the structure of which essentially comprises the multiple repetition of units derived, actually or conceptually, from molecules of low relative molecular mass (monomers). In most cases the number of monomers is quite large and often is not precisely known. A single molecule of a polymer is called a macromolecule.

Polymerase chain reaction (PCR) A laboratory technique, to rapidly amplify pre-determined regions of double stranded DNA. Generally involves the use of a heat-stable DNA polymerase.

Polymerization A process that links smaller molecules together to form a larger molecule.

Polymerize To link smaller molecules together to form a larger molecule.

Polymorph Polymorphism; polymorphic. Compare with isotope and allotrope. Solid substances that occur in several distinct forms. Polymorphs have different chemical and physical properties. allotropes are polymorphs of elements.

Polynuclear aromatic hydrocarbons (PAHs) These compounds are large ring aromatic hydrocarbons that are molecules containing carbon and hydrogen. It could also be any large ring that consists of five or more cyclic compounds of carbons and any elements that may be bonded to the ring. They usually exist in nature in low amounts. Many times these hydrocarbons exist within the atmosphere in the aerosol phase from combustion processes.

Polysaccharides Polysaccharides are compounds consisting of a large number of monosaccharides linked together by glycosidic bonds. Some important examples are starch, glycogen, and cellulose.

Polytopal rearrangement Stereoisomerisation interconverting different or equivalent spatial arrangements of ligands about a central atom or of a cage of atoms, where the ligand or cage defines the vertices of a polyhedron. For example pyramidal inversion of amines, Berry pseudorotation of PF_5, rearrangements of polyhedral boranes.

Pool/split Strategy for assembly of a combinatorial library. The solid support is divided into portions, each of which is subjected to reaction with a single building block. Pooling of these portions results in a single batch of solid support bearing a mixture of components. Repetition of the divide, couple, recombine processes results in a library where each discrete particle of solid support carries a single library member, and the number of members is equal to the product of the number of building blocks incorporated at each step (i.e. fully combinatorial).

Population (general usage) The total number of persons inhabiting a country, town, or other area. A population may also be defined by some other characteristic (such as biological, legal, social, or economic) than living in a particular area, e.g., the male population, the gainfully occupied population.

Population (Statistics) The totality of items under consideration. Every clearly defined part of a population is called a "subpopulation". In the case of a random variable, the probability distribution is considered as defining the population of that variable (ISO, 1977). The term Population Segment is sometimes used as a synonym for subpopulation.

Population at risk The number of people who can develop the adverse health effect under study and who are potentially exposed to the risk factor of interest. For example, all people in a population who have not developed immunity to an infectious disease are at risk of developing the disease, if they are exposed. Similarly, people already having chronic disease are excluded from the population at risk in studies of the incidence of the disease (WHO, 1979).

Population critical concentration (PCC) The concentration of a chemical in the critical organ (toxicology) at which a specified percentage of the exposed population has reached their individual critical organ concentrations. The percentage indicated by PCC-10 for 10%, PCC-50 for 50% etc.

Porphyrin A macrocyclic molecule that contains four pyrrole rings linked together by single carbon atom bridges between the alpha positions of the pyrrole rings. Porphyrins usually occur in their dianionic form coordinated to a metal ion.

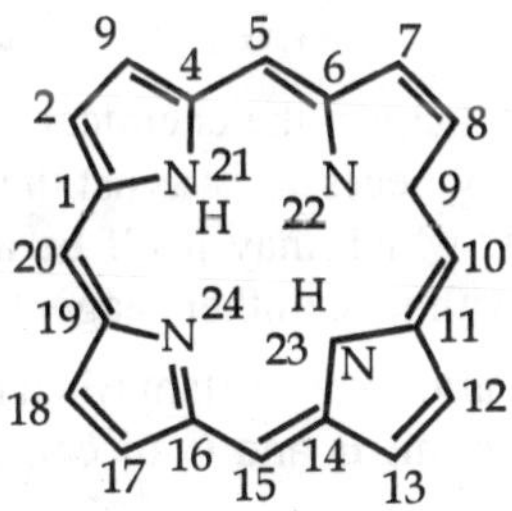

Position of equilibrium Compare with dynamic equilibrium.

When a reaction's equilibrium "lies to the right", the concentrations of products will be greater than the concentrations of reactants when equilibrium is established. Conversely, an equilibrium that lies to the left will have a relatively small fraction of products.

Positional scan Strategy for identifying individual compounds of interest from a library, whereby a collection of sub-libraries is prepared, equal in number to the total number of building blocks used in the entire library. In each pool, one point of diversity is held constant by incorporating a single building block, while the other positions use all possible building blocks.

Positron A positively charged particle of antimatter that has the same mass and spin of an electron.

Postulates of the kinetic theory of gases The Kinetic Theory of Gases is a model that accounts for many of the physical properties of gases. The main concepts of the theory are : 1. that a gas consists of small identical particles which exert no forces on each other except at the instant of collision.

2. that these particles move about at random and travel in straight lines between collisions.
3. that the collisions of particles with each other and with the walls of the containing vessel are perfectly elastic.

4. that the pressure of a gas is created by the particles striking the walls of the container.

5. that the size of a particle is negligible compared with the volume of the container.

6. that the particles are elastic, no energy is gained or lost on collision. In an elastic collision the only change in energy is the kinetic energy of the colliding particle will gain kinetic energy and the other will lose an equal amount.

The kinetic theory of gases is largely the result of the work of Count Rumford, James Joule, and James Clark Maxwell.

Potency Refers to the concentration of an agent (drug) at which it inhibits an enzyme to a defined extent, i.e. IC50 is the concentration at which an inhibitor blocks the activity of an enzyme 50 per cent.

Potential daily intake The potential daily intake of a pesticide is the theoretical intake calculated on the basis of the maximum residue limits and/ or extraneous residue limits and the per caput consumption of the relevant food commodities per day. The same concept applies to food additive intakes.

Potential difference Electrical potential difference.

Work that must be done to move an electric charge between specified points. Electric potential differences are measured in volts.

Potential energy Compare with kinetic energy. Energy an object possesses by virtue of its position. For example, lifting a mass mby h meters increases its potential energy by mgh, where g is the acceleration due to gravity.

Potential-energy (Reaction) surface A geometric hypersurface on which the potential energy of a set of reactants is plotted as a function of the coordinates representing the molecular geometries of the system.

For simple systems two such coordinates (characterizing two variables that change during the progress from reactants to products) can be selected, and the potential energy plotted as a contour map.

For simple elementary reactions, e.g. A-B + C A + B-C, the surface can show the potential energy for all values of the A, B, C geometry, providing that the ABC angle is fixed.

For more complicated reactions a different choice of two coordinates is sometimes preferred, In many qualitative applications it is convenient (although not strictly equivalent) for the third coordinate to represent the standard Gibbs energy rather than potential energy.

Using bond orders is, however, an oversimplification, since these are not well-defined, even for the transition state.

Potential-energy profile A curve describing the variation of the potential energy of the system of atoms that make up the reactants and products of a reaction as a function of one geometric coordinate, and corresponding to the "energetically easiest passage" from reactants to products (i.e. along the line produced by joining the paths of steepest descent from the transition state to the reactants and to the products). For an elementary reaction the relevant geometric coordinate is the reaction coordinate; for a stepwise reaction it is the succession of reaction coordinates for the successive individual reaction steps. (The reaction coordinate is sometimes approximated by a quasi-chemical index of reaction progress, such as "degree of atom transfer" or bond order of some specified bond.)

Potentiation The joint action of two or more chemicals on an organism is more than additive (WHO, 1978a).

Power The rate at which energy is supplied. Power has define(SI) units of J/s, sometimes called "Watts" (W).

Power broadening Power broadening occurs by shortening the lifetime of the excited state due to stimulated emission.

Power saturation A phenomenon used in electron paramagnetic resonance spectroscopy to estimate the electron-spin relaxation times, providing information about distances between paramagnetic centers.

Pre-association A step on the reaction path of some stepwise reactions in which the molecular entity C is already present in an encounter pair or encounter complex with A during the formation of B from A, e.g.

$$A + C \xrightarrow{\text{Pre-association}} \underset{\text{encounter complex}}{(A\cdots C)} \longrightarrow \underset{\text{encounter complex}}{(B\cdots C)} \xrightarrow{\text{rapid}} \text{Product}$$

In this mechanism the chemical species C may but does not necessarily assist the formation of B from A, which may itself be a bimolecular reaction with some other reagent.

Pre-association is important when B is too short-lived to permit B and C to come together by diffusion.

Precambrian All geologic time from the beginning of Earth history to 570 million years ago. Also refers to the rocks that formed in that epoch.

Precipitate PPT. An insoluble substance that has been formed from substances dissolved in a solution. For example, mixing silver nitrate and sodium chloride solutions produces a precipitate, insoluble silver chloride (along with soluble sodium nitrate.

Precipitation What often occurs when clouds form upon reaching 100% relative humidity. If condensation nuclei are present, liquid or solid particles of water will form. When this material becomes heavy enough to fall towards the ground, it is known as the common phenomenon snow, rain, sleet, or hail.

Pressure Pressure (p) is the force acting normally on unit area of a surface. It is measured in pascals in Si units (Pa)

$p = F/A$

Prevalence The number of instances of a given disease or other condition in a given population at a designated time; sometimes used to mean prevalence rate. When used without qualification, the term usually refers to the situation at a specified point in time (point prevalence).

prevalence, annual(an occasionally used index): The total number of persons with the disease or attribute at any time during a year. It includes cases of the disease arising before but extending into or through the year as well as those having their inception during the year.

prevalence, lifetime: The total number of persons known to have had the disease or attribute for at least part of their life.

prevalence, period: The total number of persons known to have had the disease or attribute at any time during a specified period.

prevalence, point: The number of persons with a disease or an attribute at a specified point in time.

prevalence rate (ratio): The total number of individuals who have an attribute or disease at a particular time (or during a particular period) divided by the population at risk of having the attribute or disease at this point in time or midway through the period. A problem may arise with calculating period prevalence rates because of the difficulty of defining the most appropriate denominator.

Primary kinetic electrolyte effect See kinetic electrolyte effect.

Primary kinetic isotope effect See isotope effect.

Primary pollutant This refers to the types of pollutants that are either found in the atmosphere or on the ground and are emitted directly from some sources such as automobile tailpipes or smoke stacks. This is contrasted with secondary pollutants which in and of themselves are not directly emitted from a source. Instead they are created from chemical and photochemical reactions of primary pollutants. Nitrogen monoxide is an example of a primary pollutant and ozone is an example of a secondary pollutant.

Primary protection standard An accepted maximum level of a pollutant (or its indicator) in the target, or some part thereof, or an accepted maximum intake of a pollutant or nuisance into the target under specified circumstances .

Primary standards A primary standard is a reagent that is extremely pure, stable, has no waters of hydration, and has a high molecular weight.

Some primary standards for titration of acids:

sodium carbonate: Na_2CO_3, mol wt. = 105.99 g/mol

tris-(hydroxymethyl)aminomethane (TRIS or THAM): $(CH_2OH)_3CNH_2$, mol wt. = 121.14 g/mol

Some primary standards for titration of bases:

potassium hydrogen phthalate (KHP): $KHC_8H_4O_4$, mol wt. = 204.23 g/mol

potassium hydrogen iodate: $KH(IO_3)_2$, mol wt. = 389.92 g/mol

Some primary standards for redox titrations:

potassium dichromate: $K_2Cr_2O_7$, mol wt. = 294.19 g/mol

Primary structure In the context of macromolecules such as proteins, the constitutional formula, usually abbreviated to a statement of the sequence and if appropriate cross-linking of chains. See also secondary structure, tertiary structure, quaternary structure.

Primitive change One of the conceptually simpler molecular changes into which an elementary reaction can be notionally dissected. Such changes include bond rupture, bond formation, internal rotation, change of bond length or bond angle, bond migration, redistribution of charge, etc.

The concept of primitive changes is helpful in the detailed verbal description of elementary reactions, but a primitive change does not represent a process that is by itself necessarily observable as a component of an elementary reaction.

Principal compounds Hydrogen is widely distributed in industrially important compounds and is present 1 in a wide range of inorganic compounds, including

Ammonia

Hydrogen Sulphide

Water

2. in the strong acids, including

Sulphuric Acid

Nitric Acid

Hydrochloric Acid

Hydrobromic Acid

Hydrofluoric Acid and

3. in almost all organic compounds.

Alkanes

Alkenes

Alkynes

Alcohols etc.

Helium, He, is a colourless gaseous non-metallic element which belong to Group 0 of the periodic table.

Atomic Number : 2

Relative Atomic Mass : 4.0026

Melting Point : -272.2 degC

Boiling Point : -268.93 degC

Relative Density : 0.178

Principal quantum number (N) The quantum number that determines the size and (in hydrogen atoms) the energy of an orbital. n is used to label electron shells. n may take on integer values from 1 to infinity.

Prior equilibrium See pre-equilibrium.

Privileged structure Substructural feature which confers desirable (often drug-like) properties on compounds containing that feature. Often consists of a semi-rigid scaffold which is able to present multiple hydrophobic residues without undergoing hydrophobic collapse.

Prochirality centre An atom of a molecule which becomes a chirality centre by replacing one of the two stereoheterotopic ligands attached to it by a different ligand is said to be a prochirality centre e.g. C-1 of ethanol; C-3 of butan-2-ol.

Prodrug A chemical structure that undergoes conversion to an active drug within a biological system, such conversion usually involving metabolism.

Product Compare with reactant.

A substance that is produced during a chemical change.

Product development control The term is used for reactions under kinetic control where the selectivity parallels the relative (thermodynamic) stabilities of the products. Product development control is usually associated with a transition state occurring late on the reaction coordinate.

Product-determining step The step of a stepwise reaction, in which the product distribution is determined. The product-determining step may be identical to, or occur later than, the rate-controlling step on the reaction coordinate.

Projection formula A formal two-dimensional representation of a three-dimensional molecular structure obtained by projection of bonds (symbolised as lines) onto a plane with or without the designation of the positions of relevant atoms by their chemical symbols.

A projection formula which indicates the spatial arrangement of bonds is called a stereochemical formula or stereoformula. Examples of stereoformulae are Fischer projection, Newman projection, sawhorse projection, wedge projection and zig-zag projection.

Prokaryote A unicellular organism characterized by the absence of a membrane-bound nucleus.

Proline Pro. A naturally occurring amino acid with a heterocyclic ring that is classified as nonessential in the diet.

Promoter The DNA region, usually upstream to the coding sequence of a gene or operon, which binds and directs RNA polymerase to the correct transcriptional start site and thus permits transcription at a specific initiation site.

N.B.In catalysis a promoter is used differently: a co-catalyst usually present in much smaller amounts than the catalyst.

Promotion See pseudo-catalysis.

Propagation In chemical terms, a propagation step is any step but the first, initial step, in a chain reaction (a series of chemical reactions in which the first reaction causes the second which causes the third, which causes the fourth, etc).

Propane (C_3H_8) Compare with alkane and hydrocarbon. A colorless, odorless, flammable gas, found in petroleum and natural gas. It is used as a fuel and as a raw material for building more complex organic molecules. Propane is the third member of the alkane series.

Propellant 1. A mixture of fuel and oxidizing agent that reacts to produce a high-energy stream of product gases that can produce thrust.

2. A compressed gas used to push a material through a nozzle, forming an aerosol or a foam. For example, nitrogen or propane are used as propellants for shaving cream; nitrous oxide is used as a propellant for whipped cream.

Properties of Cadmium is a soft white ductile metal which melts at 320 degC.

Cadmium oxidises slowly in moist air, but burns when heated.

Cadmium is similar to zinc in many of its chemical properties.

Properties of Gadolinium is silvery-white, lustrous metal.

Gadolinium has seventeen isotopes.

Gadolinium reacts slowly with water.

Gadolinium is attacked by dilute acids.

Properties of Helium is a colourless, tasteless odourless inert gas.

Helium can only be liquefied with extreme difficulty, as it does not undergo adiabatic cooling on expansion until very low temperatures are reached.

Helium is very inert chemically due to its electronic configuration.

Properties of Hydrogen is a colourless odourless gaseous element,

sparingly soluble in water and the solubility is not much affected by change of temperature, does not support respiration although it is not poisonous. When hydrogen is breathed mixed with some air for a short time, it weakens the voice and raises its pitch, a better conductor of heat than other gases, its conductivity being about five times that of air, and forms compounds with a large number of elements. In many cases, these compounds are formed by the direct combination of the elements.

Chemically, hydrogen reacts with most elements.

Combustion of Hydrogen

Hydrogen burns in oxygen or air to form water.

$2 H_2 + O_2 \Rightarrow 2 H_2O$

Oxygen will also burn in hydrogen.

Hydrogen does not itself support combustion, as may be shown by passing a lighted taper into an inverted jar of hydrogen, when the taper is extinguished.

A mixture of hydrogen with oxygen or air explodes violently when kindled, provided either gas is not present in too large excess.

Properties of Iodine is a violet dense solid, which is insoluble in water, but it is soluble in non-polar solvents. When heated it gives a violet vapour that sublimes. Iodine is soluble in a solution of potassium iodide due to the formation of the soluble potassium tri-iodide.

$KI + I_2 \Rightarrow KI_3$

Properties of Iron In its pure state is highly reactive, and in its pure state oxidises rapidly in air.

Properties of Lead A soft bluish-gray lustrous metal, which can be easily shaped under pressure into rods and pipes and rolled into sheets.

Lead is stable in dry air, but quickly tarnished in moist air, due to the formation of a film of oxide.

Lead is rendered much harder when it contains trace amounts of different impurities, including antimony, arsenic, zinc, and copper.

Properties of Mercury is a silvery liquid metal of high density, does not tarnish in air at room temperature, but forms its oxide when heated to its boiling point, reacts readily with chlorine, and reacts readily with nitric acid, yielding mercuric nitrate and Oxides of Nitrogen.

Properties of transition metals The Properties of Transition Metals are largely dependent on the electronic configuration of the electrons in the outer shell and in the penultimate outer shell.

The transition elements readily form alloys with themselves and with other elements (e.g. a copper-tin alloy is used for mirrors, brass is a copper-zinc alloy). Tungsten, is used to make tools and filaments in light bulbs.

The atomic size is fairly constant since the electrons in the outer most shells have similar environments.

The low ionisation potentials mean that the elements show variable valency states by loss of electrons from the s and 3d orbitals.

The elements in this group can have different oxidation states which makes them useful as catalysts.

Compounds of the transition elements can be paramagnetic (i.e. attracted by a magnetic field) or diamagnetic (i.e. not attracted by a magnetic field). Paramagnetism in the transition elements is caused by the presence of unpaired electrons in the d sub-orbital. Diamagnetism is characteristic of compounds where all the electrons are paired in the d sub-orbitals.

Properties of Uranium is a hard white solid, but normally exists in the form of a highly divided black powder, is primarily of interest because of its radioactive properties.

Properties of Zinc is a bright bluish-white metal, which is brittle at room temperature.

Is slowly oxidised in moist air.

Is an good conductor of heat and electricity.

Property space Multidimensional representation of a set of compounds in which the axes represent quantifiable properties, such as molecular weight, log P, molar refractivity, etc., and individual compounds are represented by a vector or set of coordinates.

Propionaldehyde (C_3H_6O) A colorless liquid that enters the atmosphere by emissions from the manufacture of plastics, synthesis of rubber, and as a disinfectant and a preservative.

Proportion An equality between two ratios.

Proportionate mortality rate, ratio (PMR) Number of deaths from a given cause in a specified time period, per 100 or 1000 total deaths in the same time period. Can give rise to misleading conclusions if used to compare mortality experience of populations with different distributions of causes of death .

Proprochirality See prochirality.

Prospective cohort study See cohort study.

Prosthetic group A tightly bound, specific nonpolypeptide unit in a protein determining and involved in its biological activity.

Protease An enzyme that hydrolyzes (breaks down a bond and adds water) proteins, especially to peptides.

Protein A molecule composed of a long chain of amino acids. Proteins are the principal constituents of cellular material and serve as enzymes, hormones, structural elements, and antibodies. The molar mass is usually above 100,000.

Protein kinases Enzymes that phosphorylate certain amino acid residues (most often Ser, Thr, or Tyr) in specific proteins.

Protic See protogenic.

Protic solvent A solvent that contains O-H or N-H bonds.

Protogenic (solvent) Capable of acting as a proton (hydron) donor strongly or weakly acidic (as a Brønsted acid). The term is preferred to the synonym "protic" or the more ambiguous expression "acidic" by itself. Also called HBD (hydrogen bond donor) solvent. See protophilic solvent.

Protolysis This term has been used synonymously with proton (hydron)-transfer reaction. Because of its misleading similarity to hydrolysis, photolysis, etc., its use is discouraged.

Proton(p+) A particle of matter with a positive electrical charge and a mass of 1 amu or 1.67E $^{-24}$g. Although a proton has a mass 1837 times greater than an electron, it is nearly identical in mass to a neutron. Protons are constituents of the nuclei of all elements and an elements atomic number is dependent upon how many protons an element has.

Proton affinity The negative of the enthalpy change in the gas phase reaction (real or hypothetical) between a proton (more appropriately hydron) and the chemical species concerned, usually an electrically neutral species to give the conjugate acid of that species. Proton affinity is often, but unofficially, abbreviated as PA.

Proton donor acid.Because a free H+ ion is technically a bare proton, acids are sometimes referred to as "proton donors" because they release hydrogen ions in solution. The term "proton donor" is misleading, since in aqueous solution, the hydrogen ion is never a bare proton-it's covalently bound to a water molecule as an H3O+ ion. Further, acids don't "donate" protons; they yield them to bases with a stronger affinity for them.

Proton transfer reaction A chemical reaction, the main feature of which is the intermolecular or intramolecular transfer of a proton (hydron) from one binding site to another. For example,

$CH_3CO_2H + (CH_3)_2C{=}O\ CH_3CO_2^- + (CH_3)_2C{=}O^+H$

In the detailed description of proton transfer reactions, especially of rapid proton transfers between electronegative atoms, it should always be specified whether the term is used to refer to the overall process (including the more-or-less encounter-controlled formation of a hydrogen bonded complex and the separation of the products;or just to the proton transfer event (including solvent rearrangement) by itself.

Protophilic (solvent) Capable of acting as proton acceptor, strongly or weakly basic (as a Brønsted base). Also called HBA (hydrogen bond acceptor) solvent. See also protogenic solvent.

Protoporphyrin IX The porphyrin ligand of heme b. Heme b is a Fe(II) porphyrin complex readily isolated from the hemoglobin of beef blood, but also found in other proteins including other hemoglobins, myoglobins, cytochromes P-450, catalases, peroxidases as well as b type cytochromes. Protoporphyrin IX contains four methyl groups in positions 2, 7, 12 and 18, two vinyl groups in positions 3 and 8 and two propionic acid groups in positions 13 and 17.

Prototropic rearrangement (or prototropy) See tautomerism.

Pseudo-asymmetric carbon atom The traditional name for a tetrahedrally coordinated carbon atom bonded to four different entities, two and only two of which have the same constitution but opposite chirality sense. The r/s descriptors of pseudo-asymmetric carbon atoms are invariant on reflection in a mirror (i.e. r remains r, and s remains s), but are reversed by the exchange of any two entities (i.e. r becomes s, and s becomes r). An example is C-3 of ribaric (C-3 is r), xylaric acid (C-3 is s) or hyoscyamine (C-3 is r).

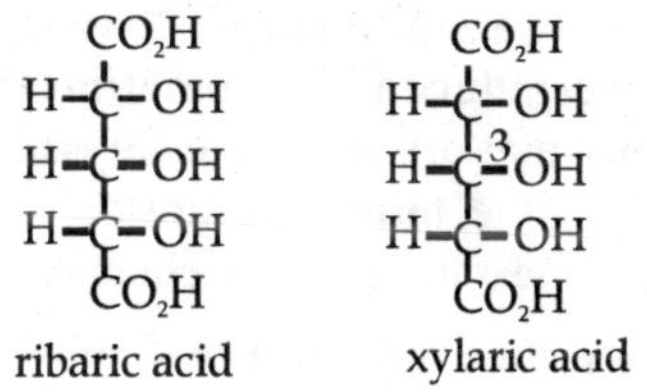

ribaric and xylaric acid

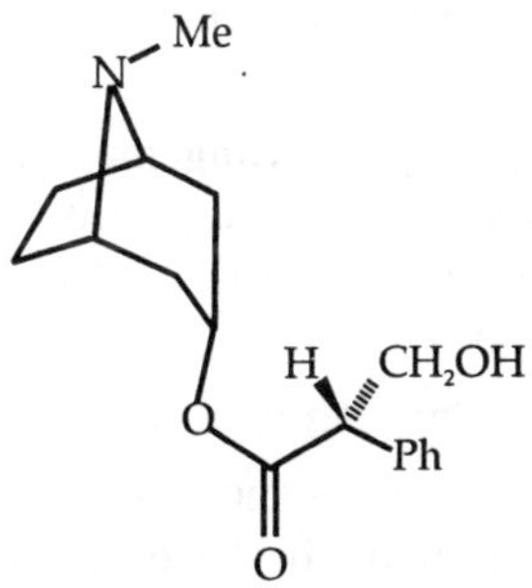

hyoscyamine

Pseudo-catalysis If an acid or base is present in nearly constant concentration throughout a reaction in solution (owing to buffering or the use of a large excess), it may be found to increase the rate of that reaction and also to be consumed during the process. The acid or base is then not a catalyst and the phenomenon cannot be called catalysis according to the well-established meaning of these terms in chemical kinetics, although the mechanism of such a process is often intimately related to that of a catalysed reaction. It is recommended that the term pseudo-catalysis be used in these and analogous cases (not necessarily involving acids or bases). For example, if a Brønsted acid accelerates the hydrolysis of an ester to a carboxylic acid and an alcohol, this is properly called acid catalysis, whereas the acceleration, by the same acid, of hydrolysis of an amide should be described as pseudo-catalysis by the acid: the "acid pseudo-catalyst" is consumed during the reaction through formation of an ammonium ion. The terms "general acid pseudo-catalysis" and "general base pseudo-catalysis" may be used as the analogues of general acid catalysis and general base catalysis.

The term "base-promoted", "base-accelerated, or "base-induced" is sometimes used for reactions that are pseudo-catalysed by bases. However, the term "promotion" also has a different meaning in other chemical contexts.

Pseudocore Electrons in d or f subshells which are outside the noble gas core.

Pseudo-equatorial See axial, equatorial.

Pseudo-first order rate coefficient See order of reaction.

Pseudomolecular rearrangement The use of this awkwardly formed term is discouraged. It is synonymous with "intermolecular rearrangement".

See molecular rearrangement.

Pseudopericyclic A concerted transformation is pseudopericyclic if the primary changes in bonding occur within a cyclic array of atoms at one (or more) of which nonbonding and bonding atomic orbitals interchange roles. Because the pi and sigma atomic orbitals that interchange roles are orthogonal, such a reaction does not proceed through a fully conjugated transition state and is thus not a pericyclic reaction and therefore not governed by the rules that express orbital symmetry restrictions applicable to pericyclic reactions.

Pseudorotation Stereoisomerisation resulting in a structure that appears to have been produced by rotation of the entire initial molecule and is superposable on the initial one, unless different positions are distinguished by substitution, including isotopic substitution.

One example of pseudorotation is a facile interconversion between the many envelope and twist conformers of a cyclopentane due to the out of plane motion of carbon atoms. Another example of pseudorotation (Berry pseudorotation) is a polytopal rearrangemnt that provides an intramolecular mechanism for the isomerisation of trigonal bipyramidal compounds (e.g. ? ^{5}-phosphanes), the five bonds to the central atom E being represented as e^1, e^2, e^3, a^1 and a^2. Two equatorial bonds move apart and become apical bonds at the same time as the apical bonds move together to become equatorial.

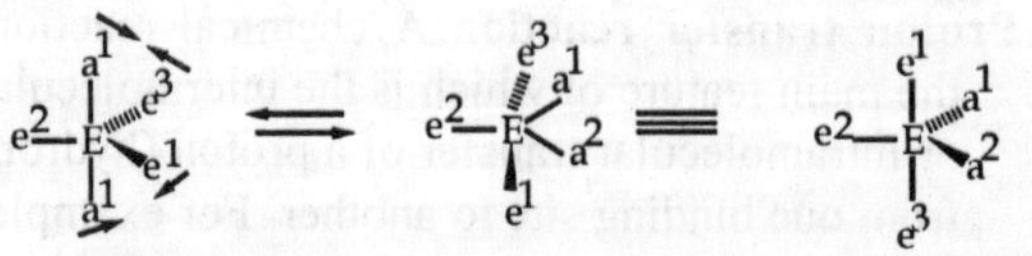

A related conformational change of a trigonal bipyramidal structure is described as turnstile rotation. The process may be visualised as follows. An apical and an equatorial bond rotate as a pair ca. 120° relative to the other three bonds. (Doubts have been expressed about the distinct physical reality of this mechanism.).

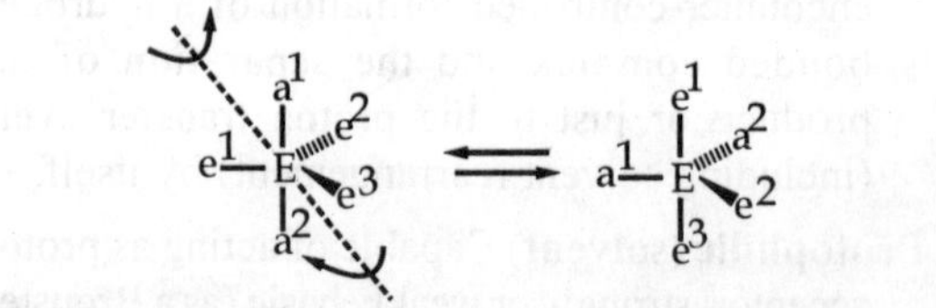

Pseudo-unimolecular A term sometimes used as synonymous with pseudo-first order, but is inherently meaningless. See molecularity, order of reaction.

Pterin 2-Amino-4-hydroxypteridine.

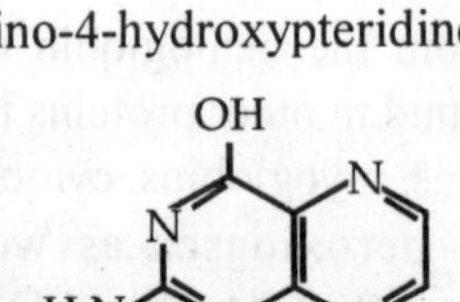

Public health impact assessment Application of risk assessment procedures to a specific target population. The size of the populations needs to be known. The end product is a quantitative statement about the number of people affected in the specific target populations (WHO, 1988).

Pulmonary fibrosis Chronic inflammation and progressive fibrosis of the pulmonary alveolar walls, usually results in death from oxygen lack or right heart failure.

Pure substance A sample of matter that cannot be separated into simpler components without chemical change. Physical changes can alter the state of matter but not the chemical identity of a pure substance. Pure substances have fixed, characteristic elemental compositions and properties.

PVC (Polyvinyl chloride) The polymerized form of chloroethene or vinyl chloride is a matrix of double the bonded carbon molecule $CH_2=CHCl$. The double bonds break and become single bonds, and the vinyl chloride molecules become bonded to each other. The result is PVC or polyvinyl chloride plastic. The burning of PVC results in the formation of acid rain.

Pyramidal inversion A polytopal rearrangement in which the change in bond directions to a three-coordinate central atom having a pyramidal arrangement of bonds (tripodal arrangement) causes the central atom (apex of the pyramid) to appear to move to an equivalent position on the other side of the base of the pyramid. If the three ligands to the central atom are different pyramidal inversion interconverts enantiomers.

Pyrodenitrification This is best defined as way that nitrogen is lost or emitted by the use of heat or radiation, mainly, from the sun. This can occur in the air or on the ground around vegetative regions. It is less likely to occur in arid regions, but it is possible.

Pyrolysis Thermolysis, usually associated with exposure to a high temperature.

Pyrophoric Pyrophoric solid. Catches fire spontaneously when exposed to air at normal room temperature. For example, powdered potassium metal is pyrophoric.

Q

Qualitative analysis Qualitative analysis involves determining the nature of a pure unknown compound or the compounds present in a mixture.

Quantal effect An effect that can be expressed only as "occurring" or "not occurring" . Typical examples of quantal effects are death or occurrence of a tumour.

Quantitative analysis Quantitative analysis is the determination of the amount of substances present in a sample.

Quantitative structure-activity relationships (QSAR) The building of structure-biological activity models by using regression analysis with physicochemical constants, indicator variables or theoretical calculations. The term has been extended by some authors to include chemical reactivity, i.e. activity is regarded as synonymous with reactivity. This extension is, however, discouraged.

Quantum mechanics Quantum theory. A branch of physics that describes the behavior of objects of atomic and subatomic size.

Quantum numbers Quantum numbers are used to describe the different aspect of the electronic shell, they are required to describe the properties of each atomic orbital. There are four quantum numbers :

Principal Quantum Number, n

This quantum number describes the general overall energy level and size of the orbital (i.e. shell).

Azimuthal (or Subsidiary) Quantum Number, l

This quantum number describes the shape of the sub-orbital. Each principal level is divided into sub-shells, which are designated by this number.

Magnetic Quantum Number, ml

This quantum number is used to indicate the direction of the sub-orbital in space.

Spin Quantum Number, s

This quantum number indicates the spin of an electron in the sub-orbital.

Thus, for any atom, each electron in the orbitals about the nucleus in the atom has a unique set of quantum numbers, and no two electrons in the atom can have an identical set of quantum numbers.

Quantum Quantum is the smallest quantity of energy that can be emitted (or absorbed) in the form of electromagnetic radiation. Energy of a quantum (E) is equal to

E = h’”ν

where h is Planck constant and ν is the frequency of the radiation.

Quantum yield The number of defined events which occur per photon absorbed by the system.

The integral quantum yield is

$$\phi = \frac{\text{number of events}}{\text{number of photons absorbed}}$$

For a photochemical reaction,

$$\phi = \frac{\text{moles of reactant consumed or product formed}}{\text{moles of photons absorbed}}$$

The differential quantum yield is

$$\phi = \frac{d[X]/dt}{n}$$

where d(X)/dt is the rate of change of the amount of (substance) concentration of a measurable quantity, and n the amount of photons (mol or its equivalent einsteins) absorbed per unit time. can be used for photophysical processes or photochemical reactions.

Quark A Quark is a fundamental particle.

Quasi-axial See axial, equatorial.

Quasi-biennial oscillation (use buttons at top in this link) Periodic variation in the direction, either easterly or westerly, of tropical lower stratospheric winds. The direction changes every 26 months.

Quasicrystal Quasicrystal is a solid having conventional crystalline properties but whose lattice does not display translational periodicity.

Quasi-enantiomers Constitutionally different yet closely related chemical species MX and MY having the opposite chirality sense of the large common chiral moiety M are called quasi-enantiomers. For example (R)-2-bromobutane is a quasi-enantiomer of (S)-2-chlorobutane. See also quasi-racemic compound.

Quasi-equatorial See axial, equatorial.

Quasi-racemic compound The crystalline product of a 1:1 association between quasi-enantiomers.

Quaternary The period of Earth's history from about 2 million years ago to the present; also, the rocks and deposits of that age.

Quaternary structure The defined organisation of two or more macromolecules with tertiary structure such as a protein that are held together by hydrogen bonds and van der Waals and coulombic forces. See also primary structure, secondary structure, tertiary structure.

Quicklime Quicklime, CaO, (i.e. calcium oxide) is manufactured by burning limestone (i.e. calcium carbonate, CaCO3) in a furnace. It reacts vigorously with water to produce slaked lime (i.e. calcium hydroxide).

R

R group Abbreviation given to an unimportant part of a molecule. Indicates Rest of molecule.

R, S The approved designations (devised by Cahn, Ingold and Prelog) of absolute configuration at four-coordinate (quadriligant) and six-coordinate (sexiligant) stereogenic centres.

Racemate An equimolar mixture of a pair of enantiomers. It does not exhibit optical activity. The chemical name or formula of a racemate is distinguished from those of the enantiomers by the prefix (±)- or rac- (or racem-) or by the symbols RS and SR.

Racemic compound A crystalline racemate in which the two enantiomers are present in equal amounts in a well defined arrangement within the lattice of a homogeneous crystalline addition compound.

Racemie conglomerate An equimolar mechanical mixture of crystals each one of which contains only one of the two enantiomers present in a racemate. The process of its formation on crystallisation of a racemate is called spontaneous resolution, since pure or nearly pure enantiomers can often be obtained from the conglomerate by sorting.

Racemic mixture(Racemic) A 50:50 mixture of two enantiomers. Racemic mixtures are optically inactive (don't rotate plane-polarized light).

Racemisation The production of a racemate from a chiral starting material in which one enantiomer is present in excess.

Radian (rad) Compare with steradian.

An angle with vertex at the center of a circle of radius r that encompasses an arc of length r.

Radiant intensity (P,I) radiant power; intensity; radiation intensity.

Energy of radiation striking a unit area per unit time. The SI unit of radiant power is J m^{-2} s^{-1}.

Radiation (*radiant energy*) Energy in the form of electromagnetic waves. Different forms of radiation comprise the electromagnetic spectrum and are characterized by their different wave frequency and wavelength. Radiation is emitted from matter in the form of photons. The following types of radiation make up the electromagnetic spectrum (in order of decreasing energy and increasing wavelength): cosmic rays, gamma rays, x-rays, UV (ultraviolet) rays, visible light rays, infrared, microwave, radio (Hertzian) waves and electric waves. These electromagnetic waves have no mass and are not electrically charged. The shortest wavelength forms of radiation are most penetrating and are the most damaging to living tissue.

Radiation Energy in the form of photons.

Radiation flux This is defined as the way heat from the sun is absorbed, scattered, returned, or deposited around the earth and can be expressed as some radiation arrival rate per the unit of area of the surface. The radiation flux can be the total amount of radiation that the earth collects, emits, and disperses.

Radiative forcing Reduction of radiation reflecting from earth per units of increase of a chemical component that causes this change in the atmosphere. The result is often used to compare the relative effect of greenhouse gases which can alter climate on earth. This means that components that are present in the atmosphere at low concentrations (e.g. methane at 1.7 ppmv) but with high radiative forcing have a greater effect on global warming than higher concentration components with a lower radiative forcing.

Radical (or free radical) A molecular entity such as $\cdot CH_3$, $\cdot SnH_3$, $Cl\cdot$ possessing an unpaired electron. (In these formulae the dot, symbolizing the unpaired electron, should be placed so as to indicate the atom of highest spin density, if this is possible.) Paramagnetic metal ions are not normally regarded as radicals. However, in the "isolobal analogy" , the similarity between certain paramagnetic metal ions and radicals becomes apparent.

At least in the context of physical organic chemistry, it seems desirable to cease using the adjective "free" in the general name of this type of chemical species and molecular entity, so that the term "free radical" may in future be restricted to those radicals which do not form parts of radical pairs.

Depending upon the core atom that possesses the unpaired electron, the radicals can be described as carbon-, oxygen-, nitrogen-, metal-centred radicals. If the unpaired electron occupies an orbital having considerable s or more or less pure p character, the respective radicals are termed sigma or pi radicals.

In the past, the term "radical" was used to designate a substituent group bound to a molecular entity, as opposed to "free radical", which nowadays is simply called radical. The bound entities may be called groups or substituents, but should no longer be called radicals.

Radical centre(s) The atom (or group of atoms) in a polyatomic radical on which an unpaired electron is largely localized. Attachment of a monovalent atom to a radical centre gives a molecule for which it is possible to write a Lewis formula in which the normal stable valencies are assigned to all atoms.

Radical chain reaction Free radicals are molecules that, through photolysis or chemical reaction, have an unpaired electron in their outer valance shell. These radicals are very reactive and thus have a short life. When a free radical reacts with a more stable molecule, the radical often pulls an atom from it and becomes a stable molecule itself. The original molecule then becomes a free radical and will react with other species of atoms and molecules in a long series (or chain) of reactions until the process reaches the termination phase. In this phase two free radicals combine, sharing the pair of electrons and breaking the chain.

Radical combination See colligation.

Radical ion A radical that carries an electric charge. A positively charged radical is called a "radical cation" (e.g., the benzene radical cation $C_6H_6{\cdot}^+$); a negatively charged radical is called a "radical anion" (e.g., the benzene radical anion $C_6H_6{\cdot}^-$ or the benzophenone radical anion $Ph_2C\text{-}O{\cdot}^-$). Commonly, but not necessarily, the odd electron and the charge are associated with the same atom.

Unless the positions of unpaired spin and charge can be associated with specific atoms, superscript dot and charge designations should be placed in the order ${\cdot}^+$ or ${\cdot}^-$ suggested by the name "radical ion", (e.g $C_3H_6{\cdot}^+$). In mass spectroscopic usage the symbol for the charge precedes the dot representing the unpaired electron.

Radical pair (or geminate pair) The term is used to identify two radicals in close proximity in solution, within a solvent cage. They may be formed simultaneously by some unimolecular process, e.g., peroxide decomposition, or they may have come together by diffusion. While the radicals are together, correlation of the unpaired electron spins of the two species cannot be ignored: this correlation is responsible for the CIDNP phenomenon.

Radical propagation Free radicals react with molecules, atoms, other radicals in long series of chain reactions. A radical is formed in the atmosphere through photolysis or chemical reaction. Radicals are very reactive and will pull atoms or electrons from other molecules to fill their outer valance shell. New radicals may formed from the original molecule which then react with another molecule, etc. This reaction propagates from one species to another until a termination reaction occurs. The termination reaction usually involves the reaction of two free radicals to form a stable product.

Radioactive Substance containing an element which decays.

Radioactive decay The spontaneous breakdown of certain kinds of atomic nuclei into one or more nuclei of different elements, involving the release of energy and subatomic particles.

Radioactive half-life (i) For a single radioactive decay process, the time required for the activity to decrease to half its value by that process (ISO, 1972); (ii) the time taken for the activity of an amount of radioactive nuclide to fall to half its initial value.

Radioactivity Radioactivity is the spontaneous disintegration of an unstable nucleus, by the emission of a small particle (i.e. an alpha-particle or a beta-particle) or by the change of the energy level of the nucleus by the emission of a gamma-ray.

Radioactivity The spontaneous emission of radiation and atomic particles from unstable elements. Radioactivity is measured in terms of half-lives and is not affected by physical state or chemical combination.

Radiocarbon dating Establishing the relative age of various materials with the use of carbon-14. This involves measuring the amount of ^{14}C and of ^{12}C and comparing the measured ratio to the one established by the production of ^{14}C in the upper atmosphere by cosmic rays. When an organism is alive the $^{14}C/^{12}C$ ratio in its biomass is constant (because of constant atmospheric ^{14}C production, difussion to the lower atmsophere, absorption by organisms, and because of constant ^{14}C radioactive decay) but when the organism dies the ratio begins to change—at a very predictable rate. Therefore knowing the carbon-14/carbon-12 ratio now in some artifact gives a very good measure of how long the carbon has been "dead." The object however must obviously have organic material either in it or on it for this method to work. Also the lenth of time one can "look back" is limited because the amount of carbon-14 must be detected with some certainty for the age to be known with confidence, and the longer the sample has been dead the less ^{14}C is present because it is continually decaying.

Radiofrequency encoding Strategy for identifying library members by physically associating them with a set of electronic devices which emit characteristic radiofrequency signals upon stimulation with a radiofrequency energy source. These signals can be used to track the reaction history of each sample in a synthesis.

Radioisotope Compare with isotope.

A radioactive isotope. For example, tritium is a radioisotope of hydrogen.

Radiolysis The cleavage of one or several bonds resulting from exposure to high-energy radiation. The term is also often used loosely to specify the method of irradiation ("pulse radiolysis") used in any radiochemical reaction, not necessarily one involving bond cleavage.

Radionuclide A radioactive nuclide. The term nuclide implies an atom of specified atomic number and mass number. In the study of biochemical processes, radioactive isotopes are used for labelling compounds that subsequently are used to investigate various aspects of the reactivity or metabolism of proteins, carbohydrates and lipids or as sources of radiation in imaging. The fate of the radionuclide in reactive products or metabolites is determined by following (counting) the emitted radiation. Prominent among the radionuclides used in biochemical research are : ^{3}H, ^{14}C, ^{32}P, ^{35}Ca, ^{99m}Tc, ^{125}I and ^{131}I. .

Rainer ludwig claisen (1851-1930) Born in Cologne and studied chemistry at Bonn, and briefly at Göttingen. He took his doctorate at Bonn under August Kekulé (1829-1896). Unusually for a German chemist, he then spent four years at Owens College, Manchester (now Manchester University) working with Henry Enfield Roscoe (1833-1915) and Carl Schorlemmer (1834-1892), and then joined Adolf von Baeyer (1835-1917) in Munich. He held Chairs in Organic Chemistry at Aachen Polytechnic and the University of Kiel, before he went to Berlin in 1904 to work with Emil Fischer in Berlin. In addition to the Claisen (1887) and Claisen-Schmidt (1881) condensations, the Claisen rearrangement (1912) and the Claisen isatin synthesis (1879), his name is associated with the Claisen flask and Claisen's rule which relates acidity to enolisation. The Claisen rearrangement has since been "hyphenated" to yield the Eschenmoser-Claisen rearrangement (1964), the Johnson-Claisen rearrangement (1970), and the Ireland-Claisen rearrangement (1972).

Rainout One item necessary for a rain drop to form is called a condensation nucleus. Pollutants in the air such as nitric acid and sulfuric acid form nucleation sites (particles) where water vapor often condenses and forms a droplet. As these droplets grow they collect more moisture and may ultimately fall out as precipitation. This accounts for some of the acid rain deposition on the planet's surface.

Raman effect Raman effect is a type of scattering of electromagnetic radiation in which light suffers a change in frequency and a change in phase as it passes through a material medium. Is named according to the Indian physicist C. V. Raman (1889-1970). The intensity of Raman scattering is about one-thousandth of that in Rayleigh scattering in liquids.

Random error Indeterminate error. Compare with systematic error, gross error and mistake.

Random errors are errors that affect the precision of a set of measurements. Random error scatters measurements above and below the mean, with small random errors being more likely than large ones.

Rankine cycle Rankine cycle is a thermodynamic cycle which can be used to calculate the ideal performance of a heat engine that uses a condensable vapor as the working fluid.

Raoult's law Raoult's Law states that the vapour pressure of a solvent is reduced in the presence of dissolved solute. The law applies only to dilute solutions.

The reduction of the vapour pressure of a solvent containing dissolved solute is proportional to the numbers of molecules of solute and is independent of their nature.

Rare earth An oxide of a rare earth element.

Rare earth elements Rare earth elements are the elements Sc, Y, and the lanthanides (La, Ce, Pr, Nd, Pm, Sm, Eu, Gd, Tb, Dy, Ho, Er, Tm, Yb, Lu). These elements got their name from the fact that chemists first isolated them in their oxide forms. These oxides somewhat resemble calcium, magnesium and aluminum oxides, sometimes called common earths.

Rare earth metals See Lanthanides

Rare gasses See Noble Gasses

Rate A measure of the frequency of a phenomenon. An expression of the frequency with which an event occurs in a defined population .

Rate coefficient See order of reaction, kinetic equivalence.

Rate constant (k) A rate constant is a proportionality constant that appears in a rate law. For example, k is the rate constant in the rate law d(A)/dt = k(A). Rate constants are independent of concentration but depend on other factors, most notably temperature.

Rate difference (RD) The absolute difference between two rates, for example, the difference in incidence rate between a population group exposed to a causal factor and a population group not exposed to the factor:

$RD = I_e - I_u$

where I_e= incidence rate among exposed, and I_u = incidence rate among unexposed. In comparisons of exposed and unexposed groups, the term excess rate may be used as a synonym for rate difference

Rate law A rate law or rate equation relates reaction rate with the concentrations of reactants, catalysts, and inhibitors. For example, the rate law for the one-step reaction A + B C is d(C)/dt = k(A)(B).

Rate of appearance See rate of reaction.

Rate of migration The rate of migration is the speed at which the ions travel through a solution during electrolysis.

Rate of reaction For the general chemical reaction

aA + bB pP + qQ ...

occurring under constant-volume conditions, without an appreciable build-up of reaction intermediates, the rate of reaction v is defined as

$$v = -\frac{1}{a}\frac{d[A]}{dt} = -\frac{1}{b}\frac{d[B]}{dt} = +\frac{1}{p}\frac{d[P]}{dt} = +\frac{1}{q}\frac{d[Q]}{dt}$$

where symbols placed inside square brackets denote amount (or amount of substance) concentrations (conventionally expressed in units of mol dm^{-3}). The symbols R and r are also commonly used in place of v. It is recommended that the unit of time should always be the second.

In such a case the rate of reaction differs from the rate of increase of concentration of a product P by a constant factor (the reciprocal of its coefficient in the stoichiometric equation, p) and from the rate of decrease of concentration of the reactant A by 1/a. (It should be noted that all coefficients in the stoichiometric equation are positive; those for products (p, q...) will therefore differ in sign from the stoichiometric numbers defined in IUPAC MANUAL (1979).)

The quantity

$$\dot{\xi} = \frac{d\xi}{dt}$$

defined by the equation

$$\dot{\xi} = -\frac{1}{a}\frac{dn_A}{dt} = -\frac{1}{b}\frac{dn_B}{dt} = +\frac{1}{p}\frac{dn_P}{dt} = +\frac{1}{q}\frac{dn_Q}{dt}$$

(where nA designates the amount of substance A, conventionally expressed in units of mole) may be called the "rate of conversion" and is appropriate when the use of concentrations is inconvenient, e.g. under conditions of varying volume. In a system of constant volume, the rate of reaction is equal to the rate of conversion per unit volume throughout the reaction.

For a stepwise reaction this definition of "rate of reaction" (and "extent of reaction",) will apply only if there is no accumulation of intermediate or formation of side products. It is therefore recommended that the term "rate of reaction" be used only in cases where it is experimentally established that these conditions apply. More generally, it is recommended that, instead, the terms "rate of disappearance" or "rate of consumption" of A (i.e. -d(A)/dt, the rate of decrease of concentration of A) or "rate of appearance" of P (i.e. d(P)/dt, the rate of increase of concentration of product P) be used, depending on the concentration change of the particular chemical species that is actually observed. In some cases reference to the chemical flux observed may be more appropriate.

The symbol v (without lettered subscript) should be used only for rate of reaction; v with a lettered subscript (e.g. v_A) refers to a rate of appearance or rate of disappearance (e.g. of the chemical species A). See also chemical relaxation, lifetime, order of reaction.

Rate-determining step, rate-limiting step These terms are best regarded as synonymous with rate-controlling step. However, other meanings that have been given to them should be mentioned, as it is necessary to be aware of them in order to avoid confusion:

Sometimes the term rate-determining is used as a special case of rate-controlling, being assigned only to an initial slow step which is followed by rapid steps. Such a step imposes an upper limit on the rate, and has also been called rate-limiting.

In view of the considerable danger of confusion when special meanings are applied to rate-determining and rate-limiting, it is recommended that they be regarded as synonymous, with the meaning explained under the entry rate-controlling step.

Rates of reactions Rates of reactions are the velocities with which chemical reactions proceed.

Ratio The relative size of two quantities expressed as the quotient of one divided by the other; the ratio of a to b is written as a:b or a/b.

Ratio coding Encoding strategy in which the relative quantities of tags convey information about compound identity. In comparison to binary coding, more information may be obtained from a given tag set, but tag interpretation is more complex.

Rayleigh scattering The scattering of light by a body with a particle diameter (D_p) less than 0.03 micrometers is termed Rayleigh Scattering. The wavelength of light scattered is dependent on the D_p, and the amount of light scattered is dependent on the number of particles present per unit volume. Shorter visible wavelengths, such as blue, are scattered by smaller particles than are the longer wavelengths like red.

RDF (Refuse derived fuel) A product of a mixed waste processing system from which some recycleable wastes and som non-combustible materials have been removed. The remaining material is used as a fuel to create energy.

Re, Si A stereoheterotopic face of a trigonal atom is designated Re if the ligands of the trigonal atom appear in a clockwise sense in order of CIP priority when viewed from that side of the face. The opposite arrangement is termed Si.

Reactants Reactants are the starting materials in a chemical reaction, and during the course of the reaction, these are replaced by a new set of materials called products.

Reacting bond rules 1. For an internal motion of a molecular entity corresponding to progress over a transition state (energy maximum), any change that makes the motion more difficult will lead to a new molecular geometry at the energy maximum, in which the motion has proceeded further. Changes that make the motion less difficult will have the opposite effect. (This rule corresponds to the Hammond principle.

2. For an internal motion of a molecular entity that corresponds to a vibration, any change that tends to modify the equilibrium point of the vibration in a particular direction will actually shift the equilibrium in that direction.

3. Effects on reacting bonds (bonds made or broken in the reaction) are the most significant. The bonds nearest the site of structural change are those most strongly affected.

Reaction coordinate A geometric parameter that changes during the conversion of one (or more) reactant molecular entities into one (or more) product molecular entities and whose value can be taken for a measure of the progress of an elementary reaction (for example, a bond length or bond angle or a combination of bond lengths and/or bond angles; it is sometimes approximated by a non-geometric parameter, such as the bond order of some specified bond).

In the formalism of "transition-state theory", the reaction coordinate is that coordinate in a set of curvilinear coordinates obtained from the conventional ones for the reactants which, for each reaction step, leads smoothly from the configuration of the reactants through that of the transition state to the configuration of the products. The reaction coordinate is typically chosen to follow the path along the gradient (path of shallowest ascent/deepest descent) of potential energy from reactants to products.

The term has also been used interchangeably with the term transition coordinate, applicable to the coordinate in the immediate vicinity of the potential energy maximum.

Reaction mechanisms Reaction mechanisms are the detailed step-by-step descriptions of the progress of a chemical reactions. The exact sequence for the breaking and making of bonds, and the electron transfers which give rise to this process are described.

Reaction path 1. A synonym for mechanism.

2. A trajectory on the potential-energy surface.

3. A sequence of synthetic steps.

Reaction quotient (Q) Compare with equilibrium constant.

The product of the concentrations of the products, divided by the product of the concentrations of the reactants, for a chemical reaction which is not necessarily at equilibrium. For example, the reaction quotient for A + B = C + D is equal to (C)(D) / ((A)(B)), where the parentheses indicate concentrations. Each concentration is raised to a power equal to its stoichiometric coefficient in the expression. The reaction quotient for A + 2B = 3C is equal to $(C)^3/((A)(B)^2)$. For gas phase reactions, partial pressures can be used in the reaction quotient in place of concentrations.

Reaction rate A reaction rate is the speed at which reactants are converted into products in a chemical reaction. The reaction rate is given as the instantaneous rate of change for any reactant or product, and is usually written as a derivative (e. g. d(A)/dt) with units of concentration per unit time (e. g. mol L^{-1} s^{-1}).

Reaction See chemical reaction.

Reaction stage A set of one or more (possibly experimentally inseparable) reaction steps leading to and/or from a detectable or presumed reaction intermediate.

Reaction step An elementary reaction, constituting one of the stages of a stepwise reaction in which a reaction intermediate (or, for the first step, the reactants) is converted into the next reaction intermediate (or, for the last step, the products) in the sequence of intermediates between reactants and products. See also rate-limiting step, reaction stage.

Reaction with phosphorus pentachloride Reaction with phosphorus pentachloride is a characteristic of organic compounds containing a hydroxyl group and this reaction is used to identify these compounds in organic analysis.

Reactions Cadmium is closely resembles zinc in its chemical properties.

Reactions with helium Chemically, helium is totally inert.

Reactions with Lead Lead is not attacked by Sulphuric Acid, H_2SO_4.

Lead is dissolved by concentrated nitric acid and by warm solutions of the dilute acid.

Lead is dissolved by organic acids, including acetic acid.

Reactions with Uranium Uranium is a reactive metal and it resembles Molybdenum and Tungsten in chemical properties.

Uranium is attacked slowly by dilute hydrochloric acid.

Reactions with Zinc Zinc burns in air at 1000 degC to form a bulky mass of zinc oxide which is known as Philosopher's Wool.

Zinc reacts with dilute acids, with the liberation of hydrogen.

Zn + H2SO4 ⇒ ZnSO4 + H2

Zinc in the presence of dilute acid is a powerful reducing agent, probably due to Nascent Hydrogen, H*, liberated in the reaction being a better reducing agent than hydrogen gas.

Zincis amphoteric and forms salts with alkalis. For example, Zinc dissolves in hot concentrated caustic soda solution with the liberation of hydrogen gas and the formation of sodium zincate.

Zn + 2 NaOH ⇒ Na2ZnO2 + H2

Reactive, reactivity As applied to a chemical species, the term expresses a kinetic property. A species is said to be more reactive or to have a higher reactivity in some given context than some other (reference) species if it has a larger rate constant for a specified elementary reaction. The term has meaning only by reference to some explicitly stated or implicitly assumed set of conditions. It is not to be used for reactions or reaction patterns of compounds in general.

The term is also more loosely used as a phenomenological description not restricted to elementary reactions. When applied in this sense the property under consideration may reflect not only rate, but also equilibrium, constants.

Reactivity index Any numerical index derived from quantum mechanical model calculations that permits the prediction of relative reactivities of different molecular sites. Many indices are in use, based on a variety of theories and relating to various types of reaction. The more successful applications have been to the substitution reactions of conjugated systems where relative reactivities are determined largely by changes of pi-electron energy.

Reactivity of a metal The reactivity of a metal is determined to a large extent by its reduction potential.

Reactivity-selectivity principle (RSP) This idea may be expressed loosely as: the more reactive a reagent is, the less selective it is.

Consider two substrates S^1 and S^2 undergoing the same type of reaction with two reagents R^1 and R^2, S^2 being more reactive than S^1, and R^2 more reactive than R^1 in the given type of reaction. The relative reactivities (in log units, see selectivity) for the four possible reactions may notionally be represented as follows: With the positions of ($S^1 + R^1$), ($S^2 + R^1$), and ($S^1 + R^2$) fixed, there are three types of positions for ($S^2 + R^2$):

In position (i) the selectivity of R^2 for the two substrates, measured by a is the same as the selectivity of R^1 for the two substrates, also a.

In position (ii) the selectivity of R^2 for the two substrates, measured by b, is less than the selectivity of R^1 for the two substrates, i.e. $b < a$. It is this situation which is in accord with the RSP.

In position (iii) the selectivity of R^2 for the two substrates, measured by c, is greater than the selectivity of R^1 for the two substrates, i.e. $c > a$. This situation may be described as anti-RSP.

There are many examples in which the RSP is followed, but there are also many examples corresponding to situations (i) and (iii). The RSP is in accord with intuitive feeling and certainly holds in the limiting case when reactivity is controlled by diffusion. However, the validity of the RSP is a matter of great controversy "...and diverse opinions have been expressed, from declaring the reactivity-selectivity principle as a universal law up to 'virtually useless in practice as a general rule'."

Reagent A substance or mixture that is useful in chemical analysis or synthesis.

Reagent efficiency The ratio of the number of library members prepared compared to the number which would have been prepared in a fully combinatorial library using the same building blocks. Lower reagent efficiency may be desirable in order to reduce the number of compounds to be synthesized or tested, for example by maximizing the number of members expected to have high activity in a library prepared by parallel synthesis.

Reagent partitioning Phenomenon whereby the concentration of a compound within, for instance, a particle of solid support is higher or lower than of the bulk solution due to the physicochemical properties of the solid support.

Rearrangement See degenerate rearrangement, molecular rearrangement, sigmatropic rearrangement.

Rearrangement reaction Isomerization; isomerize.

A reaction in which a reactant and product are isomers of each other. Chemical bonds within the reactant are broken and reformed to produce the product.

Rearrangement stage The elementary reaction or reaction stage (of a molecular rearrangement) in which there is both making and breaking of bonds between atoms common to a reactant and a reaction product or a reaction intermediate. If the rearrangement stage consists of a single elementary reaction, this is a "rearrangement step".

Receptor A molecule within a cell or on a cell surface to which a substance (such as a hormone or a drug) selectively binds, causing a change in the activity of the cell.

Recombination In chemical terms, a recombination is one of the ways in which free radicals are lost (the other being disproportionation). In this case, two free radicals join to create one product. This could be from a molecule that broke apart (ionized) and reconstituted, or two free radicals from different sources that form a new molecule.

Recommended exposure limit (*REL*) The maximum recommended exposure limit. With the occasional exception in some jurisdictions, REL's are only suggested industry practices and are not enforced by legal regulations .

Recursive partitioning Process for identifying complex structure-activity relationships in large sets by dividing compounds into a hierarchy of smaller and more homogeneous sub-groups on the basis of the statistically most significant descriptors.

Redox indicator Oxidation-reduction indicator.

An organic molecule that has reduced and oxidized forms with different colors; interconversion of the reduced and oxidized forms of the indicator must be reversible. Ferroin is an example.

Redox potential Any oxidation-reduction (redox) reaction can be divided into two half reactions: one in which a chemical species undergoes oxidation and one in which another chemical species undergoes reduction. If a half-reaction is written as a reduction, the driving force is the reduction potential. If the half-reaction is written as oxidation, the driving force is the oxidation potential related to the reduction potential by a sign change. So the redox potential is the reduction/oxidation potential of a compound measured under standard conditions against a standard reference half-cell. In biological systems the standard redox potential is defined at pH = 7. 0versus the hydrogen electrode and partial pressure of hydrogen = 1 bar. See also electrode potential.

Redox reaction Redox reaction occur when Oxidation and Reduction takes place in the one reaction. If one substance loses electrons, or atoms of oxygen or hydrogen, then another substance must gain them. Redox reactions always involve the swapping of electrons.

Redox titration Oxidation-reduction titration.

A titration based on a redox reaction. For example, iron in water can be determined by converting dissolved iron to Fe^{2+} and titrating the solution with potassium permanganate ($KMnO_4$), a powerful oxidizing agent.

Reduction Reduction is the process in which an atom or ion gains electrons. This definition applies only to reactions in which electron transfer occurs. Originally, reduction was simply regarded as a chemical reaction with the loss of oxygen.

Reduction by carbon Reduction by carbon is an important method for the extraction of metals from their ores.

Reduction half reaction Reduction half-reaction. Compare with oxidation half reaction.

That part of a redox reaction that involves gain of electrons. In the oxidation half reaction, the oxidation number of one or more atoms within the reactants is reduced.

Reduction reaction A reaction in which a substance gains at least one electron.

Reductive elimination The reverse of oxidative addition.

Reference population The standard against which a population that is being studied can be compared.

Reforming Reforming is the conversion of straight chain alkanes into branched chain alkanes by cracking or by catalytic reaction .

Refraction The change in direction of light as it passes between two objects of differing densities; the angle of refraction and the angle of incidenceÝ are related by Snellís Law: $n_1*\sin\theta_1 = n_2*\sin\theta_2$

Regioselectivity, regioselective A regioselective reaction is one in which one direction of bond making or breaking occurs preferentially over all other possible directions. Reactions are termed completely (100%) regioselective if the discrimination is complete, or partially (x%), if the product of reaction at one site predominates over the product of reaction at other sites. The discrimination may also semi-quantitatively be referred to as high or low regioselectivity.

(Originally the term was restricted to addition reactions of unsymmetrical reagents to unsymmetrical alkenes.)

In the past, the term "regiospecificity" was proposed for 100% regioselectivity. This terminology is not recommended owing to inconsistency with the terms stereoselectivity and stereospecificity ..

Regulation Refers to control of activity of an enzyme (system) or gene expression.

Relative atomic mass Relative atomic mass (A_r) is the ratio of the average mass per atom of the naturally occurring form of an element to 1/12 of the mass of nuclide ^{12}C. The term atomic weight is synonymous with relative atomic mass.

Relative configuration 1. The configuration of any stereogenic (asymmetric) centre with respect to any other stereogenic centre contained within the same molecular entity. Unlike absolute configuration, relative configuration is reflection-invariant. Relative configuration, distinguishing diastereoisomers, may be denoted by the configurational descriptors R*, R* (or l) and R*, S* (or u) meaning, respectively, that the two centres have identical or opposite configurations. For molecules with more than two asymmetric centres the prefix rel- may be used in front of the name of one enantiomer where R and S have been used. If any centres have known absolute configuration then only R* and S* can be used for the relative configuration. See also ??alpha????beta? 1. and 3..

2. Two different molecules Xabcd and Xabce, may be said to have the same relative configurations if e takes the position of d in the tetrahedral arrangement of ligands around X (i.e. the pyramidal fragments Xabc are superposable). By the same token the enantiomer of Xabce may be said to have the opposite relative configuration to Xabcd. The terms may be applied to chiral molecular entities with central atoms other than carbon but are limited to cases where the two related molecules differ in a single ligand.

Both definitions can be generalised to include stereogenic units other than asymmetric centres.

Relative density The relative density, rd, of a substance is the ratio of the density of the substance to the density of some reference substance. For liquids or solids, the reference substance is water. This quantity was formerly called Specific Gravity.

Relative error Relative uncertainty. Compare with absolute error.

The uncertainty in a measurement compared to the size of the measurement. For example, if three replicate weights for an object are 2.00 g, 2.05 g, and 1.95 g, the absolute error can be expressed as ± 0.05 g and the relative error is ± 0.05 g / 2.00 g = 0.025 = 2.5%.

Relative humidity This is the ratio between the actual water vapor content of the atmosphere and the maximum water vapor content possible at that given temperature. If the temperature of a given parcel of air rises, the amount of moisture it can hold increases and as such its relative humidity decreases. If there is no change in temperature but the moisture content decreases then the relative humidity will again decrease because the ratio of actual water vapor present is less than the maximum amount the air could hold. Conversely, if the water content does not change but the temperature falls, the relative humidity increases until saturation and possible precipitation occurs.

Relative molecular Mass Relative molecular mass (i.e. molecular weight) is the ratio of the average mass per molecule of the naturally occurring form of an element or compound to the 1 / 12 part of the mass of the carbon-12 atom. It is equal to the sum of the relative atomic masses of all the atoms that comprise a molecule.

Relative risk (i) The ratio of the risk of disease or death among the exposed to the risk among the unexposed; this usage is synonymous with risk ratio; (ii) alternatively, the ratio of the cumulative incidence rate in the exposed to the cumulative incidence rate in the unexposed, i.e., the cumulative incidence ratio, and (iii) the term "relative risk" has also been used synonymously with "odds ratio" and, in some biostatistical articles, has been used for the ratio of forces of morbidity. The use of the term "relative risk" for several different quantities arises from the fact that for "rare" diseases (e.g., most cancers) all the quantities approximate one another. For common occurrences (e.g., neonatal mortality in infants under 1500g birth weight), the approximations do not hold .

Relaxation If a system is disturbed from its state of equilibrium it relaxes to that state, and the process is referred to as relaxation.

Rem Rem (rem) (roentgen equivalent man) is a non-SI unit of dose equivalent employed in radioprotection (rem = 10^{-2} Sv).

Renal elimination Excretion of a substance through the kidneys

Reorganization energy In a one-electron transfer reaction

$A + D\ A\cdot^- + D\cdot^+$

the reorganization energy is the energy required for all structural adjustments (in the reactants and in the surrounding solvent molecules) which are needed in order that A and D assume the configuration required for the transfer of the electron.

Replicate sampling The act of taking several samples concurrently under comparable conditions (WHO, 1979).

Replication During the course of an experiment or survey, replication is the determination of a value more than once, so as to obtain a better estimation of the variation. Replication should be distinguished from repetition by the fact that replication of an experiment denotes repeated determinations carried out, as far as possible, at one place and one period of time. The successive determinations, including the first, are called replicates (ISO, 1977).

Reproductive effects The adverse effects of a chemical on any aspects of reproduction in an organism (WHO, 1979).

Residence time The amount of time a substance can remain in a reservoir.Ý The reservoir can be aquatic, atmospheric, or terrestrial.Ý In atmospheric chemistry an example of reservoir time would be the amount of time NOx can remain in the atmosphere before being extracted by deposition or chemical reactions.

Residue 1. The substances left after an evaporation or distillation.

2. A recognizable molecular fragment embedded in a larger molecule; for example, amino acid "residues" within a protein.

Resin Insoluble polymeric material which allows ready separation from liquid phase materials by filtration; can be used to carry library members (i.e. solid support) or reagents, or to trap excess reagents or reaction by-products.

ResistanceResistance (R) is electric potential difference divided by current when there is no electromotive force in the conductor. This definition applies to direct current. More generally, resistance is defined as the real part of impedance.

Resonance Description of the ground state of a molecule with delocalized electrons as an average of several Lewis structures. The actual ground state doesn't switch rapidly between the separate structures: it is an average.

Resonance effect Mesomeric effect.

If electron density at a particular point in a molecule is higher or lower than what you'd expect from a single Lewis structure, and various canonical structures can be drawn to show how electron delocalization will explain the discrepancy, the difference in electron density is called a "resonance effect" or "mesomeric effect".

Resonance energy The difference in potential energy between the actual molecular entity and the contributing structure of lowest potential energy. The resonance energy cannot be measured, but only estimated, since contributing structures are not observable molecular entities. See resonance.

Resonance raman spectroscopy A spectroscopic technique increasingly used in bioinorganic chemistry for characterization and assignment of vibrations directly connected with a chromophore, as well for the assignment of the chromophore. The excitation frequency is applied close to the absorption maximum of the chromophore. Particularly useful for deeply coloured species.

Resonance structures Structures used to better depict the location of pi and nonbonding electrons on a molecule. A molecule looks like a hybrid of all resonance structures.

Response The proportion of an exposed population with an effect or the proportion of a group of individuals that demonstrate a defined effect in a given time (e.g., death) (WHO, 1979).

Restricted rotation See free rotation, hindered rotation, restricted rotation.

Re-synthesis Preparation of individual members or pools of a combinatorial library, normally to follow up on some property of interest identified in initial screening, and often in larger scale and/or greater purity than the original preparation.

Retention of configuration See Walden inversion.

Retention The amount of substance that is left of the deposited amount after a certain time. If the retention follows a course in relation to time that is a first-order process, it may be described in terms of biological half-life (WHO, 1979).

Retrospective study A research design which is used to test etiologic hypotheses in which inferences about exposure to the putative causal factor(s) are derived from data relating to characteristics of the persons under study or to events or experiences in their past. The essential feature is that some of the persons under study have the disease or other outcome condition of interest, and their characteristics and past experiences are compared with those of other, unaffected persons. Persons who differ in the severity of the disease may also be compared

Return stroke A short-lived, high amplitude discharge of electricity that accompanies a cloud-to-ground lightning event. The bright lightning flash that is visible to the human eye is actually composed of extremely rapid electric discharges called strokes. The return stoke follows a downward extending leader, or conductive path of ionized air. The electrons in the cloud flow down this conductive pathway toward the surface of the earth. As electrons continually migrate down the path, electrons remaining higher up on the path in the cloud begin to consecutively move down the channel to the surface. Since the path of electron flow is progressively lengthened upward, the discharge of the elevated electrons high in the cloud to a lower place in the cloud and then down the pathway to the ground is called the return stroke.

Reverse osmosis Reverse osmosis is a process of obtaining pure water from saline water, as in a descalination unit. Pure water and the salt are separated by a semi-permeable membrane, by raising the pressure of the salt water above the osmotic pressure. Then, the solvent, water, is able to pass through the membrane, leaving the solute (i.e. the dissolved salts) behind.

Reversible (A process or reaction that can be reversed by an infinitesimally small change in conditions. For example, ice and water coexist at 1 atm and 0°C; a very slight temperature increase causes the ice to melt; a tiny temperature decrease causes the water to freeze. Melting or freezing under these conditions can be considered reversible. Reversible processes are infinitesimally close to equilibrium.) reversible process; reversible reaction. Compare with irreversible and irreversible process.

Reynolds number Reynolds number (Re) is a dimensionless quantity used in fluid mechanics, defined by

$Re = \rho\, vl/\eta$

where ρ is density, v is velocity, l is length, and η is viscosity.

Rf value An Rf value of a substance is the ratio of the distance that the substance moves in a chromatographic separation, to the distance that the solvent moves.

RheologyRheology is the study of the deformation and flow of materials. Has important bearing on the behavior of viscous liquids in plastic molding.

Ribonucleic acids (RNA) Linear polymer molecules composed of a chain of ribose units linked between positions 3 and 5 by phosphodiester groups. The bases adenine or guanine (via atom N-9) or uracil or cytosine (via atom N-1), respectively, areattached to ribose at its atom C-1 by β-N-glycosidic bonds (see nucleotides). The three most important types of RNAs in the cell are: messenger RNA (mRNA), transfer RNA (tRNA), ribosomal RNA (rRNA).

Ribonucleic acids Ribonucleic acids (RNA) is naturally occurring polyribonucleotides that is concerned with protein synthesis. Four types are recognized: messenger RNA (mRNA), ribosomal RNA (rRNA), transfer RNA (tRNA), and viral RNA.

Ribonucleotide reductases Enzymes which catalyze the reduction of ribonucleotide diphosphates or triphosphates to the corresponding deoxyribonucleotides by a radical-dependent reaction. The enzyme of animal, yeast and aerobic E. coli cells contains an oxo-bridged dinuclear iron center and a tyrosyl radical cation, and uses thioredoxin, a thiol-containing protein, as reductant. At least three other ribonucleotide reductases are known from bacteria, containing, respectively, an iron-sulfur cluster with a glycyl radical, adenosyl cobalamin, and a dinuclear manganese cluster.

Ribosomes A subcellular unit composed of specific rRNA and proteins that are responsible for the translation of mRNA into protein synthesis.

Ribulose-1, 5-bisphosphate carboxylase/ oxygenase (Rubisco) A magnesium-dependent enzyme , the primary enzyme of carbon dioxide fixation in plants and autotrophic bacteria. It catalyzes the synthesis of 3-phospho-D-glycerate from ribulose bisphosphate and also the oxidation of ribulose bisphosphate by O_2 to 3-phospho-D-glycerate and 2-phosphoglycolate.

Richard dean baker (b.1913) Born in Hot Springs, South Dakota, and studied chemistry at the South Dakota School of Mines and Technology in Rapid City before taking his Ph.D. at Iowa State University in Ames. After working for US Gypsum, he joined the Los Alamos Science Laboratory, where he worked on the chemistry of uranium and plutonium.

Rieske iron-sulfur protein An iron-sulfur protein of the mitochondrial respiratory chain, in which the (2Fe-2S) cluster is coordinated to two sulfur ligands from cysteine and two imidazole ligands from histidine. The term is also applied to similar proteins isolated from photosynthetic organisms and microorganisms, and other proteins containing (2Fe-2S)clusters with similar coordination.

Ring A ring is an arrangement of atoms in a molecule to form a closed chain. Ring compounds may consist of

an alicyclic ring structure (e.g. cyclohexane),

an aromatic ring structure (e.g. benzene), or

a fused ring structure (e.g. naphthalene).

Ring closure Ring closures are chemical reactions in which one part of a molecule reacts with another part of the same molecule to form a ring structure. Examples of ring closures include the formation of lactams and lactones.

Ring reversal (or ring inversion) The interconversion by rotation about single bonds (coupled with angle deformation in the transition state) of cyclic conformers having equivalent ring shapes (e.g. chair ??chair) though not necessarily equivalent spatial positions of substituents (e.g. equatorial ??axial)

Risk The probability that an event will occur, e.g., that an individual will become ill or die within a stated period of time or age. Also, a nontechnical term encompassing a variety of measures of the probability of a (generally) unfavourable outcome (Last, 1988). Risk should not be confused with the term "hazard". Risk is most correctly applied to predicted or actual frequency of occurrence of an adverse effect of a chemical or other hazard.

Risk assessment A combination of hazard identification, quantification of risk resulting from a specific use or occurrence of a chemical, taking into account the possible harmful effects on individual people or society of using the chemical in the amount and manner proposed and all the possible routes of exposure. Quantification ideally requires the establishment of dose-effect and dose-response relationships in likely target individuals and populations. Compare "risk evaluation".

Risk characterization The outcome of hazards identification and risk estimation applied to a specific use or occurrence of an environmental health hazard (e.g., a chemical compound). The assessment requires quantitative data on the human exposure in the specific situation. The end product is a quantitative statement about the proportion of affected people in a target population (WHO, 1988).

Risk estimation The quantification of dose-effect and dose-response relationships for a given environmental agent, showing the probability and nature of the health effects of exposure to the agent (WHO, 1988).

Risk evaluation Risk evaluation involves the establishment of qualitative or quantitative relationship between risks and benefits, involving the complex process of determining the significance of identified hazards and estimated risks to those organisms or people concerned with or affected by them.

Risk management The managerial, decision-making and control process to deal with those environmental agents for which risk evaluation has indicated that the risk is too high (WHO, 1988).

Risk marker (synonym: risk indicator) An attribute that is associated with an increased probability of occurrence of a disease or other specified outcome and that can be used as an indicator of this increased risk. Not necessarily a causal factor (Last, 1988).

Risk monitoring The process of following up decisions and actions within risk management in order to check whether the aims of reduced exposure and risk are achieved (WHO, 1988).

Ritchie equation The linear free energy relation

$$\log k_N = \log k_0 + N_+$$

applied to the reactions between nucleophiles and certain large and relatively stable organic cations, e.g. arenediazonium, triarylmethyl, and aryltropylium cations in various solvents. k_N is the rate constant for reaction of a given cation with a given nucleophilic system (i.e. given nucleophile in a given solvent). k_0 is the rate constant for the same cation with water in water, and N_+ is a parameter which is characteristic of the nucleophilic system and independent of the cation. A surprising feature of the equation is the absence of a coefficient of N_+, characteristic of the substrate (cf. the s in the Swain-Scott equation), even though values of N_+ vary over 13 log units. The equation thus involves a gigantic breakdown of the reactivity-selectivity principle.

RNA (Ribonucleic acid) Compare with nucleic acid and nucleotide.

A nucleic acid with D-ribose as the sugar component in its nucleotides.

Roald hoffmann (1937) Born in Zloczow, Poland. Went to Columbia and Harvard Universities after narrowly escaping from the Nazis. Collaborated with R. B. Woodward on the conservation of orbital symmetry, hence the Woodward-Hoffmann rules. Awarded the Nobel prize in 1981.

Robotic system Automated device where materials are transferred by the physical movement of a delivery device relative to the ultimate receptacle, or vice versa.

Rochelle's salt Rochelle's Salt, (i.e. Sodium Potassium Tartate), KNaC4H4O6.4H20, is a colourless crystalline salt.

Salt

A salt is the product of the neutralisation of an acid by a base. Salts are usually ionic solids.

Rodenticide A chemical used to kill rodents (rats).

Roger adams (1889-1971) Born in Boston and took his Ph.D. at Harvard before studying under Diels and Willstätter in Germany. Joined the faculty at the University of Illinois at Urbana in 1916, which he converted into the leading centre of organic chemistry in America. Also forged close links between academia and industry, notably with Du Pont. He also established the series Organic Reactions. His name is associated with chemical warfare agent Adamsite (diphenylamine chloroarsine) and the Adams catalyst (platinum or palladium oxide, used for the hydrogenation of carbon-carbon double bonds).

Rotamer One of a set of conformers arising from restricted rotation about one single bond.

Rotational barrier In a rotation of groups about a bond, the potential energy barrier between two adjacent minima of the molecular entity as a function of the torsion angle.

Rotatory power The quantitative measure of optical activity.

Rubisco See ribulose-1,5-bisphosphate carboxylase/oxygenase.

Rubredoxin An iron-sulfur protein without acid-labile sulfur, in which an iron center is coordinated by four sulfur-containing ligands, usually cysteine. The function, where known, is as an electron carrier.

Rubrerythrin A protein assumed to contain both a rubredoxin-like iron center and a hemerythrin-like dinuclear iron center.

Rudolf fischer (1881-1957) Born in Berlin, he studied at Strassburg and Berlin, where he took his doctorate under Emil Fischer in 1903. He took out pioneering patents for colour photography in 1911-1913 and founded his own photographic film company in 1927.

Rule of 90 ó The systematic numbering system originally created by the Du Pont Chemical Company in the early 1930s, which is now the standard method in the naming of chlorofluorocarbons (CFCs). Chlorofluorocarbons, such as dichlorodifluoromethane (CCl_2F_2), and trichlorofluorocarbon (CCl_3F), were given the nicknames CFC-12 and CFC-11, respectively, by Du Pont so that they could be remembered more easily. In order to derive the chemical structure of a chlorofluorocarbon from its common name, the rule of 90 states that you would add 90 to the number at the end of the name. In the case of CFC-12, add 90+12 to get 102. The hundreds place, the one in this example, represents the number of carbon atoms in the molecule. The tens place, the 0, represents the number of hydrogen atoms in the molecule, and the ones place represents the number of fluorine atoms in the molecule. According to the rule of 90, CFC-12 has 1 carbon atom, 0 hydrogen atoms, and 2 fluorine atoms. The final aspect of the rule of 90 pertains to the necessity of carbon to form four bonds. If, after assigning hydrogen and fluorine atoms to it, the carbon atom still requires more bonds to reach the required four, those bonds will be to chlorine atoms. So far, the one carbon atom of CFC-12 has only formed two bonds, both with fluorine atoms. Thus the remaining two bonds go to chlorine atoms.

Rules of five: lipinski's rules Set of criteria for predicting the oral bioavailability of a compound on the basis of simple molecular features (molecular weight, c Log P, numbers of hydrogen-bond donors and acceptors). Often used to profile a library or virtual library with respect to the proportion of drug-like members which it contains.

Runaway greenhouse effect Runaway greenhouse effect is that describes and explains the difference in the surface temperatures between the Earth and Venus. On Venus, the loss of liquid water from evaporation caused the buildup carbon dioxide. Since CO2 is a greenhouse gas, it absorbs the infrared radiation reflected from surface, which heats the atmosphere and prevents the removal of CO2.Ý This phenomenon is caused by a high atmospheric concentration of greenhouse gases, which in turn cause overheating of the planet. The runaway effect causes an acceleration in the warming of the planet as the surface.

Rusticyanin An electron transferprotein, containing a type 1 copper site, from the periplasm of the iron-oxidizing bacterium Thiobacillus ferrooxidans.

S

S cis conformation A conformation in which the two double bonds of a conjugated diene are on the same side of the carbon-carbon single bond that connects them. The required conformation for the Diels-Alder reaction.

S trans conformation The conformation in which the two double bonds of a conjugated diene are on opposite sided of the carbon-carbon single bond that connects them.

S.T.P. (Standard temperature (273.15 K) and pressure (101 325 Pa)) is the standard conditions used as a basis for calculations involving quantities that vary with temperature and pressure. These conditions are used when comparing the properties of gases.

Sacrificial hyperconjugation See hyperconjugation.

Safety (Of a drug or other chemical substance for human health) The extent to which a chemical substance may be used in thc amounts necessary for intended purposes with a minimum risk of adverse health effects (WHO, 1979).

Safety factors (In food additives and contaminants) A factor applied to the no-observed-effect level to derive acceptable daily intake (ADI) (the no-observed-effect level is divided by the safety factor to calculate the ADI). The value of the safety factor depends on the nature of the toxic effect, the size and type of population to be protected, and the quality of the toxicological information available (WHO, 1987).

Safety-catch linker A linker which is cleaved by performing two different reactions instead of the normal single step, thus providing greater control over the timing of compound release.

Salifiable Capable of reacting with an acid to form a salt. Lavoisier classified lime, baryta, alumina, and silica as "salifiable earths".

Salinity Salinity (S) is a parameter used in oceanography to describe the concentration of dissolved salts in seawater. The salinity of normal seawater is 35 parts salt per 1000 parts water.

Salt Salt is an ionic compound formed by the reaction of an acid and a base. The reaction of sodium hidroxide to hydrochloric acid give sodium chloride

$$NaOH + HCl \rightleftarrows NaCl + H_2O$$

Salt bridge Salt bridge is a permeable material soaked in a salt solution that allows ions to be transferred from one container to another. The salt solution remains unchanged during this transfer.

Salt effect See kinetic electrolyte effect.

Sample One or more items taken from a population and intended to provide information on the population, and possibly to serve as a basis for a decision on the population or on the process which had produced it (ISO, 1977).

Sampling The procedure used to constitute a sample (ISO, 1977).

Sampling error Part of the total estimation error of a parameter due to the random nature of the sample (ISO, 1977).

Saponification Saponify; saponified; saponifies.

The hydrolysis of esters using hot sodium hydroxide solution to produce the salt of a carboxylic acid. Usually saponification refers to the hydrolysis of esters of fatty acids to manufacture soaps.

Satellite Satellites are electronic devices placed into orbit to monitor and examine regions which we may not be able observe from earth. These are used routinely to monitor atmospheric ozone, ultraviolet radiation, and many other constituents. Satellite altimetry and radar interferometry are used to measure magnetic field intensity and track Antarctic ice flow.

SATP (Standard ambient pressure and temperature). Compare with STP and standard state.

Used to describe a substance at standard pressure and a temperature of 25°C (298.15 K).

Saturated compound A saturated compound is one in which all the atoms are linked by single bonds. Saturated compounds react with other compounds by undergoing substitution reactions. If a saturated compound undergoes an elimination reaction, an unsaturated compound is formed.

Saturated fat Compare with unsaturated fat.

A lipid that contains no carbon-carbon double bonds. Animal fats like butter and lard are composed of saturated fat. Saturated fats tend to be waxy or greasy solids.

Saturated solution Saturated solution is a solution that holds the maximum possible amount of dissolved material. When saturated, the rate of dissolving solid and that of recrystallisation solid are the same, and a condition of equilibrium is reached. The amount of material in solution varies with the temperature; cold solutions can hold less dissolved solid material than hot solutions. Gases are more soluble in cold liquids than in hot liquids.

Saturation transfer A term used in nuclear magnetic resonance. When a nucleus is strongly irradiated, its spin population may partly be transferred to another nucleus by an exchange process..

Sawhorse projection A perspective formula indicating the spatial arrangement of bonds on two adjacent carbon atoms. The bond between the two atoms is represented by a diagonal line, the left-hand bottom end of which locates the atom nearer the observer and the right-hand top end the atom that is further away. In general a Newman or zig-zag projection is preferred.

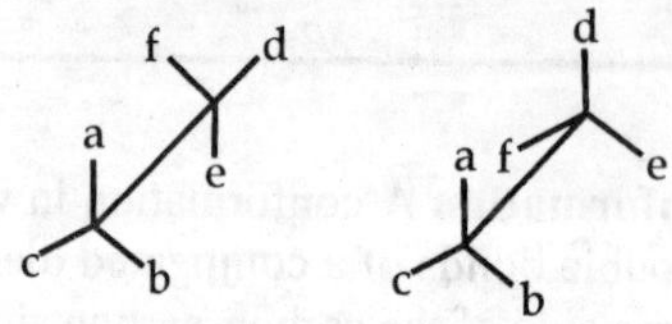

(staggered) (eclipsed)

Saytzeff rule Dehydrohalogenation of secondary- and tertiary-alkyl halides proceeds by the preferential removal of the -hydrogen from the carbon that has the smallest number of hydrogens. Originally formulated by A. Saytzeff (Zaitsev) to generalize the orientation in - elimination reactions of alkyl halides, this rule has been extended and modified, as follows: When two or more olefins can be produced in an elimination reaction, the thermodynamically most stable alkene will predominate. Exceptions to the Saytzeff rule are exemplified by the Hofmann rule. Saytzeff (1875). See also Markownikoff rule.

Scaffold Core portion of a molecule common to all members of a combinatorial library.

Scavenger A substance that reacts with (or otherwise removes) a trace component (as in the scavenging of trace metal ions) or traps a reactive reaction intermediate.

Scavenger resin Solid-supported reagent which will react with undesired materials (such as excess reagents) and remove them from solution.

Schiff base Schiff base is a class of compounds derived by chemical reaction (condesation) of aldehydes or ketones with aromatic amines, for example

RNH_2 + R'CHO RN:CHR' + H_2O

They are named after the German chemist Hugo Schiff (1834-1915).

Schiff bases Imines bearing a hydrocarbyl group on the nitrogen atom: R_2C=NR' (R' H).

Schrödinger equation Schrödinger equation is the basic equation of wave mechanics which, for systems not dependent on time, takes the form:

$$\nabla^2\psi + \frac{8\pi^2 m}{h^2}(E - V)\psi = C$$

where Ψ is the wavefunction, V is the potential energy expressed as a function of the spatial coordinates, E its total energy, ∇^2 is the Laplacian operator, h is Planck's constant, and m is the mass.

Scientific method A research method that involves:

1 Making an observation 2

Developing a hypothesis 3

Experimentation

4 Testing and retesting (usually by many other scientists) 5. Developing a theory

6 If theory is proven beyond a doubt (e.g.: the law of gravity) a law may result (rarely does the scientific method reach this degree of acceptance as fact)

Scientific notation Exponential notation.

A system for reporting very small or very large numbers by writing the number as a decimal number between 1 and 10, multiplied by a power of 10. For example, 602000000000000000000000 is written in scientific notation as 6.02×10^{23}. 0.000323 is written in scientific notation as 3.23×10^{-4}.

S-cis, s-trans The spatial arrangement of two conjugated double bonds about the intervening single bond is described as s- cis if synperiplanar and s-trans if antiperiplanar. This term should not be applied to other systems such as N-alkyl amides (use E/Z or sp/ap).

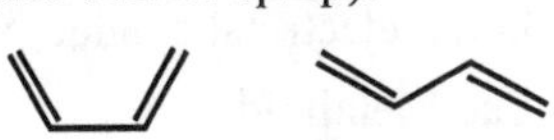

s-cis s-trans

Scrambling See isotopic scrambling.

Screening The presumptive identification of unrecognized disease or defect by the application of tests, examinations, or other procedures which can be applied rapidly. A screening test is not intended to be diagnostic. Persons with positive or suspicious findings must be referred to their physicians for diagnosis and necessary treatment.

Screening is an initial examination only, and positive responders require a second, diagnostic examination. Thc initiative for screening usually comes from the investigator or the person or agency providing care rather than from a patient with a complaint. Screening is usually concerned with chronic illness and aims to detect disease not yet under medical care .

Sea level This is the average level of the ocean over the entire earth. Tidal fluctuation is taken into account when determining sea level. Mean Sea Level (MSL) is used to reference the height of structures above ground level and depth below the ocean's surface. Atmospheric pressure is also referenced from sea level. At sea level, 760mm (29.92 inches) of mercury is normally displaced by the pressure of the air column above that point on the earth.

Second (s) The second (s) is the base unit of time in the SI system of units, defined as the duration of 9,192,631,770 cycles of the radiation associated with a certain color of light emitted by the cesium atom.

Second ionization energy (IE,IP) second ionization potential. Compare with first ionization energy, adiabatic ionization energy, vertical ionization energy, electronegativity, and electron affinity.

The energy needed to remove an electron from an isolated +1 ion. The third ionization energy would be the energy required to remove an electron from an isolated +2 ion, and so on.

Second law Second law of thermodynamics.

The second law states that every spontaneous process causes a net increase in the entropy of the universe. Many alternative statements are possible, including: Heat cannot be converted to work via an isothermal cycle. Heat cannot be converted to work with 100% efficiency. Heat cannot flow from a cold object to a warmer object without doing outside work.

Second order reaction Compare with zero order reaction and first order reaction.

A reaction with a rate law that is proportional to either the concentration of a reactant squared, or the product of concentrations of two reactants.

Secondary alcohol A secondary alcohol is one in which the hydroxyl group (-OH) is attached to a secondary carbon atom (i.e. a carbon atom which has one hydrogen atoms attached to it).

Secondary cell A secondary cell is an electrochemical cell for the production of electricity. After use the secondary cell can be recharged by passing an electric current through the cell in the opposite direction to that delivered during use. Examples of a secondary cell include the lead-acid battery used in cars, and the nickel-cadmium cells used in portable electrical devices.

Secondary kinetic electrolyte effect See kinetic electrolyte effect.

Secondary kinetic isotope effect See isotope effect.

Secondary pollutant See pollutant.

Secondary standard A secondary standard is a working standard material which is used in the laboratory as a working standard, and which has been calibrated against a primary standard material of known composition.

Secondary structure Level of structural organization in proteins described by the folding of the polypeptide chain into structural motifs such as alpha helices and beta sheets, which involve hydrogen bonding of backbone atoms. Secondary structure is also formed in nucleic acids, especially in single-stranded RNA's by internal base pairing.

Secondary treatment Secondary treatment involves the biochemical oxidation of a sewage or trade effluent to meet the standards before its discharge to the aquatic environment.

Sediment Particles carried by water or air that are deposited in other areas or elevation are called sediment. Usually associated with water flow, sediment will accumulate at the mouth of a river or stream as it empties into a larger, slower moving body of water. Sediment deposition is also effected by particle size.

Sedimentation Separation of a dense material (usually a solid) from a less dense material (usually a liquid) by allowing the denser material to settle out of the mixture.

Selectivity The discrimination shown by a reagent in competitive attack on two or more substrates or on two or more positions in the same substrate. It is quantitatively expressed by ratios of rate constants of the competing reactions, or by the decadic logarithms of such ratios.

Selectivity factor A quantitative representation of selectivity in aromatic substitution reactions (usually electrophilic, for monosubstituted benzene derivatives). If the partial rate factor, f, expresses the reactivity of a specified position in the aromatic compound PhX relative to that of a single position in benzene, then the selectivity factor S_f (expressing discrimination between p- and m-positions in PhX) is defined as

$$S_f = \lg (f_p^X/f_m^X).$$

Selenides Selenides are compounds having the structure RSeR (R not equal to H). They are thus selenium analogues of ethers. Also used for metal salts of H_2Se.

SELEX "systematic evolution of ligands by exponential enrichment" Process for identifying aptamers by iterative enrichment of oligonucleotide mixtures with respect to their ability to bind a target.

Semiconductor Semiconductor is a material in which the highest occupied energy band (valence band) is completely filled with electrons at T = 0 K, and the energy gap to the next highest band (conduction band) ranges from 0 to 4 or 5 eV. With increasing temperature electrons are excited into the conduction band, leading to an increase in the electrical conductivity.

Semimetal See Metalloid

Semipermeable membrane A membrane that allows some but not all of the components in a mixture to pass through it. Semipermeable membranes are used in dialysis.

Sensitivity (In chemical analysis) For a simple procedure, the slope of the analytical calibration curve, i.e., the differential of the measure (x) with respect to concentration (c) (i.e., dx/dc). The greater the value of this derivative, the greater the sensitivity. Sensitivity should not be confused with limit of detection .

Sensitivity and specificity (Of a screening test) Sensitivity is the proportion of truly diseased persons in the screened population who are identified as diseased by the screening test. Sensitivity is a measure of the probability of correctly diagnosing a case, or the probability that any given case will be identified by the test (synonym: true positive rate).

Specificity is the proportion of truly nondiseased persons who are so identified by the screening test. It is a measure of the probability of correctly identifying a nondiseased person with a screening test (synonym: true negative rate). The relationships are shown in the following fourfold table, in which the letters a, b, c, and d represent the quantities specified below the table.

Sensitization This term is applied to the exposure to a substance (allergen) which provokes a response in the immune system such that disease symptoms will ensue on subsequent encounters with the same substance.

Separation Separation is required to isolate pure compounds from mixtures and is a prerequisite to the analysis of mixtures. In chemical manufacturing plants, separation is normally achieved by distillation, sublimination, crystallisation, or by filtration. In analytical chemistry, separation is normally achieved by chromatographic methods.

Sequence The order of neighbouring amino acids in a protein or the purine and pyrimidine bases in RNA or DNA.

Sequence rules See priority.

Sequestration-enabling reagent Reagent which converts undesired by-products or residual starting materials into a form which may more easily be removed from the reaction mixture by, for example, solid phase extraction or other phase switch.

Sergei nikolaeviè reformatski (1860-1934) Born in Borisoglebskoe, near Ivanovo (which later became a centre of the Soviet petrochemical industry with the help of Zelinski), and studied chemistry under Aleksandr Michailoviè Sayteff or Zajcev (1841-1910) at Kazan, before travelling to Heidelberg and Leipzig. He took his doctorate at the University of Warsaw. Reformatski was Professor of Organic Chemistry at the University of Kiev for over forty years and was also associated with the development of the Soviet synthetic rubber industry in the 1930s. His synthesis of ?-hydroxy esters (1887) has remained popular.

Serine (Ser, $HOCH_2CH(NH_2)COOH$)

A naturally occuring amino acid with an hydroxyl group on its side chain.

Serine protease A family of proteases, characterized by a serine amino acid at its active site.

Shape of ionic compound Ionic compounds exist in a crystalline form in the solid state. For example, the electrostatic attractive force which exist between the positively charged sodium ion and the negatively charged chloride ion cause these oppositely charged ions to arrange themselves as close as possible in a crystal lattice. The exact structure of the unit cell of the crystal structure will depend on the size and charge on each of the ions.

In the case of sodium chloride the lattice structure is a centered cube. In this type of lattice, a given face of the lattice has a sodium ion at its center and four sodium ions at the corners of the face, a chloride ion is at the center of the face with four chloride ions at the corners of the face and four sodium ions spaced equally between the chloride ions.

As the forces between ions in a crystal structure is considerable, the crystal lattice is a very stable structure. The lattice is not easily broken by vibration of the ions. A great deal of energy (i.e. a high temperature), is needed to overcome these forces. Hence, ionic compound has a high melting point.

Shape of simple covalent compounds The shape of a simple covalent compounds is related to the number of electron pairs in the outer shell of the central atom of the molecule. Each electron pair repels all others and the electron pairs arrange themselves in space so that they are as far apart as possible.

Shell Compare with subshell.

A set of electrons with the same principal quantum number. The number of electrons permitted in a shell is equal to $2n^2$. A shell contains n^2 orbitals, and n subshells.

Shells Where the electrons generally stay. There are 4 types of electron shells: s, p, d and f shells.

Shielding In the context of NMR spectroscopy shielding is the effect of the electron shells of the observed and the neighbouring nuclei on the external magnetic field. The external field induces circulations in the electron cloud. The resulting magnetic moment is oriented in the opposite direction to the external field, so that the local field at the central nucleus is weakened, although it may be strengthened at other nuclei (deshielding).

The phenomenon is the origin of the structural dependence of the resonance frequencies of the nuclei. See also chemical shift.

Short term exposure limit (STEL) According to American Conference of Government Industrial Hygienists, this is the time-weighted average (TWA) airborne concentration to which workers may be exposed for periods up to 15 minutes, which no more than 4 such excursions per day and at least 60 minutes between them.

Short term memory Short term memory is a mechanism for storing temporary information, such as where you parked your car or numbers in a simple arithmetic problems.

SI (Systeme Internationale); International System. Le Systéme Internationale (SI) is a system of units introduced to remove barriers to international trade, based on the older metric system. It is now used in science and technical communications worldwide.

SI unit Stands for Systeme International d'Unites, a international system which established a uniform set of measurement units.

Siderophore Generic term for Fe(III)-complexing compounds released into the medium by bacteria for the purpose of scavenging iron.

Sidestream smoke The smoke that is given off from the cigarette (pipe, cigar) between puffs and is not directly inhaled by the smoker.

SiemensSiemens (S) is the SI derived unit of of electrical conductance of a conductor in which a current of one ampere is produced by an electrical potential difference of one volt ($S = A'''V^{-1} = \Omega^{-1}$). The unit is named after the German scientist Ernst Werner von Siemens (1816-1892).

Sievert Sievert (Sv) is the SI derived unit of dose equivalent when the absorbed dose of ionizing radiation multiplied by the stipulated dimensionless factors is 1 J/kg. It is named after the Swedish physicst Rolf Sievert (1896-1966).

Sigma bond (Bond) Compare with pi bond.

In the valence bond theory, a sigma bond is a valence bond that is symmetrical around the imaginary line between the bonded atoms. Most single bonds are sigma bonds.

Sigma bond (a.k.a. s bond) A bond in which electrons are located between the nuclei of the bonding atoms. Single bonds are sigma bonds.

Sigma, pi (,) The terms are symmetry designations, pi molecular orbitals being antisymmetric with respect to a defining plane containing at least one atom (e.g. the molecular plane of ethene), and sigma molecular orbitals symmetric with respect to the same plane. In practice the terms are used both in this rigorous sense (for orbitals encompassing the entire molecule) and also for localized two-centre orbitals or bonds, and it is necessary to make a clear distinction between the two usages.

In the case of two-centre bonds, a pi bond has a nodal plane that includes the internuclear bond axis, whereas a sigma bond has no such nodal plane. (A delta bond in organometallic or inorganic molecular species has two nodes.)

Radicals are classified by analogy into sigma and pi radicals.

Such two-centre orbitals may take part in molecular orbitals of sigma or pi symmetry. For example, the methyl group in propene contains three C-H bonds, each of which is of local sigma symmetry (i.e. without a nodal plane including the internuclear axis), but these three "sigma bonds" can in turn be combined to form a set of group orbitals one of which has pi symmetry with respect to the principal molecular plane and can accordingly interact with the two-centre orbital of pi symmetry (pi bond) of the double-bonded carbon atoms, to form a molecular orbital of pi symmetry.

Such an interaction between the CH_3 group and the double bond is an example of what is called hyperconjugation. This cannot rigorously be described as "sigma-pi conjugation" since sigma and pi here refer to different defining planes, and interaction between orbitals of different symmetries (with respect to the same defining plane) is forbidden.

Sigmatropic rearrangement A molecular rearrangement that involves both the creation of a new sigma bond between atoms previously not directly linked and the breaking of an existing sigma bond. There is normally a concurrent relocation of pi bonds in the molecule concerned, but the total number of pi and sigma bonds does not change. The term was originally restricted to intramolecular pericyclic reactions, and many authors use it with this connotation. It is, however, also applied in a more general, purely structural, sense.

If such reactions are intramolecular, their transition state may be visualized as an association of two fragments connected at their termini by two partial sigma bonds, one being broken and the other being formed as, for example, the two allyl fragments in (a'). Considering only atoms within the (real or hypothetical) cyclic array undergoing reorganization, if the numbers of these in the two fragments are designated i and j, then the rearrangement is said to be a sigmatropic change of order (i,j) (conventionally (i) <= (j)). Thus the rearrangement (a) is of order (3,3), whilst reaction (b) is a (1,5)sigmatropic shift of hydrogen. (N.B.: By convention square brackets (...) here refer to numbers of atoms, in contrast with current usage in the context of cycloaddition.)

The descriptors a and s (antarafacial and suprafacial) may also be annexed to the numbers i and j; (b) is then described as a (1s,5s) sigmatropic rearrangement, since it is suprafacial with respect both to the hydrogen atom and to the pentadienyl system: The prefix "homo" (meaning one extra atom, interrupting conjugation - cf. "homoaromaticity") has frequently been applied to sigmatropic rearrangements, but is misleading.

Silanes Silanes are saturated silicon hydrides, analogues of the alkanes (compounds of the general formula Si_nH_{2n+2}).

Silicate 1. A negatively charged ion containing silicon and oxygen, usually SiO_3^{-2}, $Si_2O_7^{-6}$, and $Si_3O_7^{-2}$.

2. A compound containing positively charged metal ions combined with negatively charged ions made of silicon and oxygen.

Silicones Silicones are polymeric or oligomeric siloxanes, usually considered unbranched, of general formula $(-OSiR_2-)_n$ (R not equal to H).

Silicosis Damage to the lungs caused by exposure to materials containing crystalline silicon dioxide which is found in three different forms: quatrz, tridymite and cristobalite.

Siloxanes Siloxanes are saturated silicon-oxygen hydrides with unbranched or branched chains of alternating silicon and oxygen atoms (-Si-O-Si-).

Silylene 1. Generic name for H_2Si: and substitution derivatives thereof, containing an electrically neutral bivalent silicon atom with two non-bonding electrons. (The definition is analogous to that given for carbene.)

2. The silanediyl group (H_2Si), analogous to the methylene group (H_2C).

Single bond When an electron pair is shared by two different elements.

Single displacement Single replacement reaction; single displacement reaction; single replacement. Compare with double displacement.

A reaction of the form A + BC = B + AC. For example, zinc displaces hydrogen from hydrochloric acid in the following reaction: $Zn(s) + 2\ HCl(aq) = ZnCl_2(aq) + H_2(g)$.

Single-electron transfer mechanism (SET) A reaction mechanism characterized by the transfer of a single electron between the species occurring on the reaction coordinate of one of the elementary steps.

Single-step reaction A reaction that proceeds through a single transition state.

Singlet Describes an NMR signal consisting of only one peak.

Sink This is a natural removal process for pollutants in the troposphere such as dissolution and removal via rain.

Siroheme A heme-like prosthetic group found in a class of enzymes that catalyze the six-electron reduction of sulfite and nitrite to sulfide and ammonia, respectively.

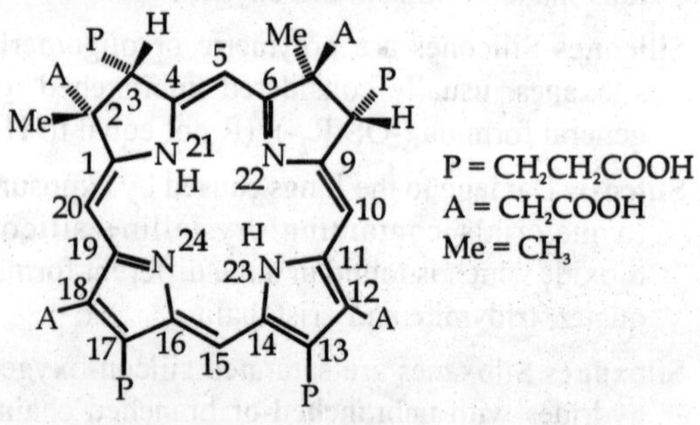

Site isolation Property of solid supports, whereby functional groups are separated from each other by the polymeric framework, and thus, while they may be physically in close proximity, reduced levels of reaction between sites may be observed. Also referred to as pseudo-dilution.

Site-directed mutagenesis See mutagenesis.

Skew See torsion angle and chair, boat, twist.

Slaked lime Slaked lime (i.e. calcium hydroxide), is prepared by adding water to freshly prepared quicklime.

Slater-type orbital An approximate atomic orbital that attempts to allow for electron-electron repulsion by scaling the nuclear charge for each orbital.

Smog This is a term used to describe a mixture of smoke and fog. Smog occurs when high concentrations of moisture is combined with smoke (often containing oxides of sulfur and nitrogen) in the presence of high temperatures or thermal inversions and the absence of wind. These conditions cause polluted air to stagnate over industrial areas and can create a respiratory health hazard. Large coastal industrial centers with surrounding high ground are more prone to smog. There is often a diurnal (over a day) variation in the process of smog formation because one of the necessary components for its formation is sunlight.

Smoke It is the product of incomplete combustion and consisting mostly of carbon and small liquid and solid particles. Smoke is primarily caused by the burning of wood, coal, gasoline, or other biomass. Smoke particles are usually less than 1.0 micrometers in size. Wind intensity, particle size, air temperature, and atmospheric pressure all effect how smoke behaves in the environment.

S_N1 reaction A first order substitution reaction that goes through a carbocation intermediate.

S_N2 reaction A second order substitution reaction that takes place in one step and has no intermediates.

Soap Soap is a mixture of the sodium salt of the fatty acids, which is obtained on hydrolysis of fats with sodium hydroxide. Glycerol is also obtained in this reaction.

SOD See superoxide dismutase.

Sodium thiosulphate Sodium thiosulphate, $Na_2S_2O_3$, is a salt used in photography for the fixing of photographic plates.

Soft acid See hard acid.

Soft base See hard base.

Sol Sols are dispersions of small solid particles in a liquid. The particles may be macromolecules or may be clusters of small molecules.

Lyophobic sols are those in which there is no affinity between the dispersed phase and the liquid (e.g. silver chloride dispersed in water). Lyophobic sols are inherently unstable, in time the particles aggregate, and form a precipitate. Lyiophilic sols, on the other hand, are more like true solutions in which the solute molecules are large and have an affinity for the solvent (e.g. starch in water). Association colloids are systems in which the dispersed phase consists of clusters of molecules that have lyophobic and lyophilic parts (e.g. soap in water).

Solar radiation This is electromagnetic radiation (light energy) emitted by the sun. This energy is transmitted through space in the units of electromagnetic energy called photons. The strength of the solar energy that reaches our outer atmosphere is called the solar constant and has a value of approximately 2.0 calories per minute per square centimeter. Roughly one-half (1.0 cal min^{-1} cm^{-2}) of all the solar radiation that enters the upper atmosphere is available at the earth's surface. Of the fifty percent loss, approximately 32 percent is reflected back into space by either clouds, dust, or aerosols; the remaining 18 percent is absorbed by clouds, dust, water, and atmospheric gases. The short high-energy wavelengths that are absorbed by the Earth are re-radiated at longer, lower energy wavelengths mostly in the IR wavelength region. These lower-energy wavelengths are, in part, prevented from escaping into space by the greenhouse gases in the atmosphere. Solar radiation is the fuel source that provides a mechanism for all biological processes on earth. Without solar radiation, life on earth would not exist. (This statement, of course, neglects the contribution to the earth's surface temperature from radioactive decay in the earth.)

Solar wind The outflow of charged particles from the solar corona into space. Because of the high temperature of the particles of the corona (mostly protons and electrons), they are moving at speeds higher than the solar escape velocity. At the orbit of the Earth, these particles are moving at about 500 km/sec. Some of these particles are captured by the magnetic fields of the planets, forming their magnetospheres.

Solid A solid is a relatively dense, rigid state of matter, with a definite volume and shape. Molecules in solids are often packed close together in regularly repeating patterns, and vibrate around fixed positions.

Solid phase extraction Method for sample purification, whereby either the desired or undesired components of a mixture have preferential affinity for a solid material. Adding the mixture to the solid material then allows facile separation of the desired material by filtration.

Solid support Insoluble, functionalized, polymeric material to which library members or reagents may be attached (often via a linker) allowing them to be readily separated (by filtration, centrifugation, etc.) from excess reagents, soluble reaction by-products or solvents.

Solubility Solubility is the quantity of solute that dissolves a given quantity of solvent to form a saturated solution, and is expressed in kilograms per meter cubed, moles per kilogram of solvent, etc. The solubility of a substance in a given solvent depends on temperature. No simple general relation exists between solubility and temperature, so each case must be investigated separately. Polar solutes are more soluble in polar solvents.

Solubility of gases A general property of gases is that they diffuse to fill the volume in which they are contained (i.e. they have neither fixed shape or fixed volume). Thus, gases which do not react with each other, are infinitely soluble in each other in all proportions, due to this power of diffusion. This behaviour is described by Graham's Law and by Dalton's Law of Partial Pressures.

Similarly, when a gas dissolves in a liquid with which it does not react, its behaviour, it is uniformly distributed throughout the volume of the solvent and its behaviour is described by Henry's Law. In general, the solubility of a gas which is only sparingly soluble in a solvent, decreases with increasing temperature. Hydrogen, nitrogen and oxygen are non-polar and are only sparingly soluble in water. In the case of these three gases, the gases continue to exist as covalent molecules in solution and there is no significant alteration to the structure of the molecules of gas.

Solubility product constant Solubility product constant (K_{sp}) (or the solubility product) is the product of the molar concentrations of the constituent ions, each raised to the power of its stoichiometric coefficient in the equilibrium equation. For instance, if a compound A_aB_b is in equilibrium with its solution

$$A_aB_b(s) \rightleftharpoons aA^+ + bB$$

the solubility product is given by

$$K_{sp}(A_aB_b) = (A^+)^{a'''}(B^-)^b$$

Solubilizing group A group or substructure on a molecule that increases the molecule's solubility. Solubilizing groups usually make the molecule they are attached to ionic or polar. For example, hydrocarbon chains can be made water-soluble by attaching a carboxylic acid group to the molecule.

Soluble A substance is soluble in a liquid, if it dissolves in that liquid. A liquid can also dissolve in another liquid is it is miscible with the liquid.

Soluble salt An ionic compound that dissolves in a solvent (usually water).

Soluble support An attachment, common to all library members, which renders the library components soluble under conditions for library synthesis, but which can be readily separated from most other soluble components when desired by some simple physical process. This process has been termed liquid phase chemistry. Examples of soluble supports include linear polymers such as poly(ethylene glycol), dendrimers or fuorinated compounds which selectively partition into fluorine-rich solvents.

Solute A substance dissolved in a sòlvent to make a solution.

Solutions A solution is a mixture of a liquid, called as the solvent, with a gas, liquid or solid known as the solute.

Solutions are homogeneous mixtures, and usually consist of a mixture of a solid (i.e. the solute) and a liquid (i.e. the solvent). When a solution if formed, the molecules of the solute are evenly distributed amongst the molecules of the solvent. In a solution, there are interactions between the molecules of the solute and the molecules solvent, and this interaction is known as solvation.

In this case, the solid is called the solute, is said to have dissolved in the liquid. The liquid is called the solvent.

If a solid can dissolve in a given liquid, the solid is said to be soluble in that liquid. If the solid cannot dissolve in a liquid, that solid is insoluble in that liquid.

For example, when sugar is dissolved in water, the sugar is the solute, the water is the solvent and the resultant mixture is the solution.

The amount of a solute which a solvent can hold in solution is dependent on the temperature. When a solution contains as much solute as the solvent can hold at a particular temperature, the solution is said to be a saturated solution at that temperature.

An unsaturated solution is one in which more solute will go into solution.

A supersaturated solution is one in which too much solute has been dissolved in a solvent. As a general rule, hot solvents can dissolve more solute than the same quantity of cold solvent. A hot saturated solution forms a supersaturated solution when cooled to a lower temperature. Because the solvent is unable to hold all the solute in solution at the lower temperature, some of solute precipitates out in the form of crystals. This process is called crystallisation.

Crystallisation is a method used to purify solids. If one takes a solid containing a small amount of impurities and dissolves it in a suitable solvent to form a saturated solution at elevated temperature, the impurities are likely to remain in solution when the solution is cooled and the solid crystallises out.

Solutions of electrolytes Solutions of electrolytes are the electrically conducting solutions formed when salts are dissolved in water. The high electrical conductivity of the solution is due to the presence of ions from the dissociation of the salt.

Solvation Any stabilizing interaction of a solute (or solute moiety) and the solvent or a similar interaction of solvent with groups of an insoluble material (i.e., the ionic groups of an ion-exchange resin). Such interactions generally involve electrostatic forces and van der Waals forces, as well as chemically more specific effects such as hydrogen bond formation.

Solvatochromic relationship A linear free-energy relationship based on solvatochromism.

Solvatochromism The (pronounced) change in position and sometimes intensity of an electronic absorption or emission band, accompanying a change in the polarity of the medium. Negative (positive) solvatochromism corresponds to a hypsochromic (bathochromic) shift with increasing solvent polarity.

Solvent A solvent is a liquid that dissolves another substance or substances to form a solution.

Solvent extraction Solvent extraction is a method for separating mixtures by exploiting differences in the solubilities of the components. For example, a coffee machine extracts the soluble components of ground coffee with water, and leaves the insoluble components behind. The sample is shaken or mixed with solvent (or with two immiscible solvents) to effect the separation. The "like dissolves like" is a useful guide for selecting solvents to use in the extraction. Nonpolar substances are usually successfully extracted into nonpolar solvents like hexane or methylene chloride. Polar and ionic substances are often extracted with water.

Solvolysis Generally, reaction with a solvent, or with a lyonium ion or lyate ion, involving the rupture of one or more bonds in the reacting solute. More specifically the term is used for substitution, elimination and fragmentation reactions in which a solvent species is the nucleophile ("alcoholysis" if the solvent is an alcohol, etc.).

Solvolysis Reaction with a solvent involving the rupture of one or more bonds in the reacting solute.

Solvophobicity parameter A solvent parameter defined by

$$S_p = 1 - M/M(\text{hexadecane})$$

derived from the Gibbs energy of transfer ($_tG^o$) of a series of solutes from water to numerous aqueous-organic mixtures and to pure solvents:

$$_tG^o(\text{to solvent}) = MR_T + D$$

where R_T is a solute parameter, and M and D characterize the solvent. The M values are used to define a solvent solvophobic effect so that S_p values are scaled from unity (water) to zero (hexadecane).

SOMO A singly occupied molecular orbital (such as the half-filled HOMO of a radical).

Sonication Irradiation with (often ultra)sound waves, e.g. to increase the rate of a reaction or to prepare vesicles in mixtures of surfactants and water.

Soret band A very strong absorption band in the blue region of the optical absorption spectrum of a heme protein.

Sorption A class of processes by which one material is taken up by another. Absorption is refers to the process of the penetration of one material into another; adsorption to the action of one material being collected on another's surface.

Sort and combine Use of directed sorting to facilitate library assembly. Related to pool/split protocol, but more commonly applied to macroscopic solid supports (such as pins and related carriers) where each library member is found on only one, or a small number of carriers.

Sp A hybrid orbital made by mixing one s orbital and one p orbital.

Sp² A hybrid orbital madc by mixing one s orbital and two p orbitals.

Sp³ A hybrid orbital made by mixing one s orbital and three p orbitals.

Space formula Synonym for stereochemical formula.

Spatially addressable Having the ability to identify at least part of the structure of a library component or pool by noting its physical location in an array.

Special salt effect The initial steep rate increase observed in the kinetic electrolyte effect on certain solvolysis reactions, upon addition of some non-common ion salts, especially $LiClO_4$.

Speciation Refers to the chemical form or compound in which an element occurs in both non-living and living systems. It may also refer to the quantitative distribution of an element. In biology, it refers to the origination of a new species.

Species See chemical species.

Specific catalysis The acceleration of a reaction by a unique catalyst, rather than by a family of related substances. The term is most commonly used in connection with specific hydrogen-ion or hydroxide-ion (lyonium ion or lyate ion) catalysis. See also general acid catalysis, general base catalysis, pseudo-catalysis.

Specific chromatographic methods Gas chromatography (GC) Applied to volatile organic compounds. The mobile phase is a gas and the stationary phase is usually a liquid on a solid support or sometimes a solid adsorbent.

High-performance liquid chromatography (HPLC) A variation of liquid chromatography that utilizes high-pressure pumps to increase the efficiency of the separation.

Liquid chromatography (LC) Used to separate analytes in solution including metal ions and organic compounds. The mobile phase is a solvent and the stationary phase is a liquid on a solid support, a solid, or an ion-exchange resin.

Size-exclusion chromatography (SEC) Also called gel-permeation chromatography (GPC), the mobile phase is a solvent and the stationary phase is a packing of porous particles.

Thin-layer chromatography (TLC) A simple and rapid method to monitor the extent of a reaction or to check the purity of organic compounds. The mobile phase is a solvent and the stationary phase is a solid adsorbent on a flat support.

Chromatography Theory(h) The underlying principles that determine chromatographic separations are dynamic behaviors that depends on partitioning and mass transport. These phenomena are too complex to model directly, and chromatography theory consists of empirical relationships to describe chromatographic columns and the separation of peaks in chromatograms.

Specific equilibrium constants The equilbrium constant has specific names for several classes of reactions: 1. Gas-phase reactions that use units of partial pressure: K_p

2. Dissociation of water: dissociation constant of water, K_w

2. Dissociation of acids: acid dissociation constant, K_a 3. Reaction of bases with water: base hydrolysis constant, K_b

3. Solubility of precipitates: solubility product, K_{sp}

4. Formation of complexes: formation constant, K_f

Flame-Ionization Detector (FID)(h) An FID consists of a hydrogen/air flame and a collector plate. The effluent from the GC column passes through the flame, which breaks down organic molecules and produces ions. The ions are collected on a biased electrode and produce an electrical signal. The FID is extremely sensitive with a large dynamic range. Its only disadvantage is that it destroys the sample.

Molecular Fluorescence Spectroscopy(h) Molecular fluorescence is the optical emission from molecules that have been excited to higher energy levels by absorption of electromagnetic radiation. The main advantage of fluorescence detection compared to absorption measurements is the greater sensitivity achievable because the fluorescence signal has in principle a zero background. Analytical applications include quantitative measurements of molecules in solution and fluorescence detection in liquid chromatography. The theory of quantitative fluorescence measurements is given in a separate document.

Specific gravity The ratio of the density of a substance to the density of a reference substance. Specific Gravity does not have any units. When water is used as the reference material, specific gravity equals density (without units).

Specific Gravity (SG) = (Density of substance g/ml)/(Density of reference substance g/ml)

Notice that g/ml cancels out so that there are no units associated with specific gravity. Specific Gravity then becomes an abstract number.

Specific heat The ratio of heat capacity of a substance to the heat capacity of water. Another way of looking at it is the quantity of heat required to raise the temperature of one gram of a substance by one degree centigrade or one degree Kelvin (1 J/gC = 1 J/gK).

Specific quantity Specific quantity is often convenient to express an extensive quantity (e.g., volume, enthalpy, heat capacity, etc.) as the actual value divided by mass. The resulting quantity is called specific volume, specific enthalpy, etc.

Specific volume Compare with density. The volume of a unit mass of substance. For example, the specific volume of water at 4°C is 1.00000 mL/g. Specific volume is the reciprocal of density.

Specificity (in chemical analysis) The degree to which a given analytical procedure detects a specified component but not other components that may be present in the sample (WHO, 1979).

Spectator ion A spectator ion is an ion that appears as both a reactant and a product in an ionic equation. For example, in the ionic equation

$Ag^{+}(aq) + NO_3^{-}(aq) + Na^{+}(aq) + Cl^{-}(aq) = AgCl(s) + Na^{+}(aq) + NO_3^{-}(aq)$ the sodium and nitrate ions are spectator ions.

Spectator mechanism A pre-association mechanism in which one of the molecular entities, C, is already present in an encounter pair with A during formation of B from A, but does not assist the formation of B, e.g.

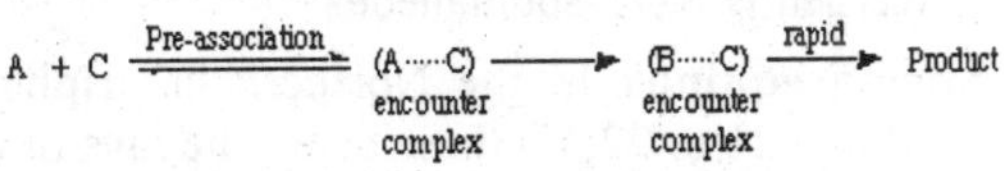

The formation of B from A may itself be a bimolecular reaction with some other reagent. Since C does not assist the formation of A, it is described as being present as a spectator, and hence such a mechanism is sometimes referred to as a spectator mechanism.

Spectrophotometer An instrument for measuring the amount of light absorbed by a sample.

Spectrophotometry Spectrophotometric.

Determination of the concentration of a material in a sample by measurement of the amount of light the sample absorbs.

Spectroscope Spectrometer.

An instrument for measuring the spectrum of light or radiation.

Spectroscopy Spectrometry; spectroscopic.

Spectroscopy is analysis of the interaction between electromagnetic radiation and matter. Different types of radiation interact in characteristic ways with different samples of matter; the interaction is often unique and serves as a diagnostic "fingerprint" for the presence of a particular material in a sample. Spectroscopy is also a sensitive quantitative technique that can determine trace concentrations of substances.

Spectrum 1. A sequence of colors produced by passing light through a prism or diffraction grating.

2. A range of wavelengths of electromagnetic radiation.

3. A plot that shows how some intensity-related property of a beam of radiation or particles depends on another property that is related to dispersal of the beam by a prism, a magnet, or some other device. For example, a plot of light absorbance vs. wavelength is an absorption spectrum; a plot of ion abundance vs. mass is a mass spectrum.

Spectrum of hydrogen The spectrum of hydrogen is obtained by an electrical discharge in a Geissler Tube. The spectrum contains four bright lines, due to atomic hydrogen and is used in calibrating spectroscopes and refractomeres. The principal lines in the spectrum of hydrogen are :

a red line, H, at 656.2 nm (i.e. Fraunhofer's C),

a blue line, H, at 434.0 nm,

a greenish-blue line, H, at 486.1 nm,

(i.e. Fraunhofer's F) and

an indigo line, H, at 410.2 nm.

Spin Electrons have an intrinsic angular momentum that is similar to what would be observed if they were spinning. Electron spin is sometimes called a "twoness" property because it can have two values, referred to as "spin up" and "spin down". Nuclei can have spins of their own.

Spin density The unpaired electron density at a position of interest, usually at carbon, in a radical. It is often measured experimentally by electron paramagnetic resonance (EPR, ESR (electron spin resonance)) spectroscopy through hyperfine coupling constants of the atom or an attached hydrogen.

Spin label A stable paramagnetic group (typically a nitryl radical) that is attached to a part of a molecular entity whose microscopic environment is of interest and may be revealed by the electron spin resonance (ESR) spectrum of the spin label.

When a simple paramagnetic molecular entity is used in this way without covalent attachment to the molecular entity of interest it is frequently referred to as a "spin probe".

Spin pair (Paired spins; electron pair; paired electrons). Compare with unpaired spin.

Two electrons with opposite spins, usually occupying the same orbital.

Spin quantum number This spin quantum number, s relates to the spin on an electron. The spin may have one of two values, s = + 0.5. or s = - 0.5. Each sub-orbital can accommodate two electrons, and the two electrons in each sub-orbital must have opposite spins. Allowing two electrons in each sub-orbital and the magnetic quantum, ml explains the number of electrons which can be accommodated in each sub-orbital.

Spin trapping In certain reactions in solution a transient radical will interact with a diamagnetic reagent to form a more persistent radical. The product radical accumulates to a concentration where detection and, frequently, identification are possible by EPR/ESR spectroscopy. The key reaction is usually one of attachment; the diamagnetic reagent is said to be a "spin trap", and the persistent product radical is then the "spin adduct". The procedure is referred to as spin trapping, and is used for monitoring reactions involving the intermediacy of reactive radicals at concentrations too low for direct observation. Typical spin traps are C-nitroso compounds and nitrones, to which reactive radicals will rapidly add to form nitryl radicals. A quantitative development, in which essentially all reactive radicals generated in a particular system are intercepted, has been referred to as "spin counting". Spin trapping has also been adapted to the interception of radicals generated in both gaseous and solid phases. In these cases the spin adduct is in practice transferred to a liquid solution for observation in order to facilitate interpretation of the EPR/ESR spectra of the radicals obtained.

Spin-orbit coupling The interaction of the electron spin magnetic moment with the magnetic moment due to the orbital motion of the electron.

Spin-spin coupling The interaction between the spin magnetic moments of different electrons and/or nuclei. In NMR spectroscopy it gives rise to multiplet patterns, and cross-peaks in two-dimensional NMR spectra. Between electron and nuclear spins this is termed the nuclear hyperfine interaction. Between electron spins it gives rise to relaxation effects and splitting of the EPR spectrum.

Spontaneous Spontaneity; spontaneous process; spontaneous reaction.

A spontaneous process occurs because of internal forces; no external forces are required to keep the process going, although external forces may be required to get the process started. For example, the burning of wood is spontaneous once the fire is started. The combination of water and carbon dioxide to reform the wood and oxygen is NOT spontaneous!

Spring equinox In the Northern hemisphere, March 21 or 22; on these dates, the rays of the sun are striking vertically at noon at the equator (0 degrees latitude), for the earth is in such a position in its orbit that the axis is tilted neither toward nor away from the sun. For latitudes above the equator, the dates are earlier then these dates.

Stability constant An equilibrium constant that expresses the propensity of a species to form from its component parts. The larger the stability constant, the more stable is the species. The stability constant (formation constant) is the reciprocal of the instability constant (dissociation constant).

Stabilizer A substance that makes a mixture more stable. Antioxidants and antiozonants are examples of stabilizers; stabilizers are added to paints to prevent the components of the mixture from separating over time.

Stable Stable is a term describing a system in a state of equilibrium corresponding to a local minimum of the appropriate thermodynamic potential for the specified constraints on the system. Stability cannot be defined in an absolute sense, but if several states are in principle accessible to the system under given conditions, that with the lowest potential is called the stable state, while the other states are described as metastable. Unstable states are not at a local minimum. Transitions between metastable and stable states occur at rates which depend on the magnitude of the appropriate activation energy barriers which separate them.

Staggered conformation Conformation about a carbon-carbon single bond in which bonds off of one carbon are at a maximum distance apart from bonds coming off of an adjacent carbon (60°dg apart when viewed in a Newman projection).

Standard electrode potential Standard electrode potential (E°) (standard reduction potentialis) are defined by measuring the potential relative to a standard hydrogen electrode using 1 mol solution at 25 °C. The convention is to designate the cell so that the oxidized form is written first. For example,

$Pt(s)|H_2(g)H^+(aq)|Zn^{2+}(aq)|Zn(s)$

The e.m.f. of this cell is -0.76 V and the standard electrode potential of the $Zn^{2+}|Zn$ half cell is -0.76 V.

Standard enthalpy change (H°) standard enthalpy. Compare with enthalpy change.

A change in enthalpy associated with a reaction or transformation involving substances in their standard states.

Standard enthalpy of formation ($H_f°$) standard heat of formation; heat of formation; enthalpy of formation.

The change in enthalpy when one mole of compound is formed from its elements in their most stable form and in their standard states.

Standard enthalpy of reaction ($H_{rxn}°$) standard heat of reaction.

A change in enthalpy associated with a reaction involving substances in their standard states.

Standard entropy of reaction ($S_{rxn}°$) entropy of reaction.

A change in entropy associated with a reaction involving substances in their standard states. A superscript circle (°) distinguishes standard enthalpy changes from enthalpy changes which involve reactants and products that are not in their standard states.

Standard heat of combustion The standard heat of combustion is the heat evolved when one gram mole of a substance is burned in oxygen.

Standard hydrogen electrode Standard hydrogen electrode is a system in which hydrogen ion and gaseous hydrogen are present in their standard states. The convention is to designate the cell so that the standard hydrogen electrode is written first.

$H_2(g) \rightleftarrows 2H^+(aq) + 2e^-$

The electrode is used as a reference (of zero) for the values of other standard electrode potentials.

Standard mean ocean water Standard mean ocean water (SMOW) is a standard sample of pure water of accurately known isotopic composition which is maintained by the International Atomic Energy Agency. It is used for precise calibration of density and isotopic composition measurements.

Standard molar entropy (S°) The entropy of one mole of a substance in its standard state.

Standard molar volume The volume of 1 mole of an ideal gas at STP, equal to 22.414 liters.

Standard pressure (P° or $P^\ominus$)

Standard pressure is a pressure of 1 bar. Before 1982, the standard pressure was 1 atm (1 atm = 1.01325 bar).

Standard reduction potential (E°)

The voltage associated with a reduction process at standard state. The reduction potential of $2H^+(aq, 1M) + 2\ e^- \ H_2(g, 1\ atm)$ is taken as exactly zero volts.

Standard reference electrode A standard reference electrode is required in all electrochemical measurements.

In the case of the determination of standard reduction potentials, the standard reference electrode is the hydrogen half-cell.

In the case of pH measurement, the standard reference electrode is the calomel electrode.

Standard solution A standard solution is one which contains a known weight of reagent in a definite volume of solution.

Standard state ($^\circ$ or $^\ominus$) A set of conditions defined to allow convenient comparison of thermodynamic properties. The standard state for a gas is the the state of the pure substance in the gaseous phase at the standard pressure, with the gas behaving ideally. The standard state for liquids and solids is the state of the most stable form of the substance at the standard pressure. Temperature is not included in the definition of standard state and must be specified, but when not given a temperature of 25°C is usually implied.

Standard temperature and pressure (STP) The standard conditions of 273.15 K and 1 atm (0°C at sea level). used as a baseline for calculations involving quantities that vary with temperature and pressure.

Standardization A set of techniques used to remove as far as possible the effects of differences in age or other confounding variables, when comparing two or more populations. The common method uses weighted averaging of rates specific for age, sex, or some other potential confounding variable(s) according to some specified distribution of these variables. There are two main methods, as follows:

Direct method: The specific rates in a study population are averaged, using as weights the distribution of a specified standard population. The directly standardized rate represents what the crude rate would have been in the study population if that population had the same distribution as the standard population with respect to the variable(s) for which the adjustment or standardization was carried out.

Indirect method: This is used to compare study populations for which the specific rates are either statistically unstable or unknown. The specific rates in the standard population are averaged, using as weights the distribution of the study population. The ratio of the crude rate for the study population to the weighted average so obtained is the standardized mortality (or morbidity) ratio, or SMR. The indirectly standardized rate itself is the product of the SMR and the crude rate for the standard population .

Standardized mortality (morbidity) ratio (SMR) The ratio of the number of events observed in the study group or population to the number of deaths expected if the study population had the same specific rates as the standard population, multiplied by 100

Standards Standards are materials containing a known concentration of an analyte. They provide a reference to determine unknown concentrations or to calibrate analytical instruments.

The accuracy of an analytical measurement is how close a result comes to the true value. Determining the accuracy of a measurement usually requires calibration of the analytical method with a known standard. This is often done with standards of several concentrations to make a calibration or working curve.

A primary standard is a reagent that is extremely pure, stable, has no waters of hydration, and has a high molecular weight.

A secondary standard is a standard that is prepared in the laboratory for a specific analysis.

It is usually standardized against a primary standard.

Starch A polysaccharide used by plants to stockpile glucose molecules. The most common forms are amylose and amylopectin.

State function A property that depends only on the condition or "state" of the system, and not on the path used to obtain the current conditions. Energy, enthalpy, temperature, volume, pressure, and temperature are examples of state functions; heat and work are examples of non-state functions.

State of matter There are three common states of matter: gases, liquids, and solids. States of matter differ in the way the molecules are arranged at the molecular level, but not in the structure of the molecules themselves. Other states (the plasma and Bose-Einstein condensate states) are uncommon on Earth.

State property A state property is a quantity that is independent of how the substance was prepared. Examples of state properties are altitude, pressure, volume, temperature and internal energy.

Stationary phase A stationary phase is a substance that shows different affinities for different components in a sample mixture in a separation of the mixture by chromatography. The mobile phase (a solution containing the sample) flows over or through the stationary phase to effect the separation.

Stationary state 1. In quantum mechanics: A state that does not evolve with time.

2. In kinetics: See steady state.

Steady state If during the course of a chemical reaction the concentration of an intermediate remains constant, the intermediate is said to be in a steady state. In a static system a reaction intermediate reaches a steady-state if the processes leading to its formation and those removing it are approximately in balance. The steady-state hypothesis leads to a great simplification in reaching an expression for the overall rate of a composite reaction interms of the rate constants for the individual elementary steps. Care must be taken to apply the steady-state hypothesis only to appropriate reaction intermediates. An intermediate such as an atom or a free radical, present at low concentrations, can usually be taken to obey the hypothesis during the main course of the reaction. In a flow system a steady-state may be established even for intermediates present at relatively high concentrations.

Steam cracking Steam cracking is the industrial process using steam at high temperature to break down high molecular weight hydrocarbons into smaller molecules.

Steam reforming Steam reforming is a process used to convert methane derived from natural gas into a mixture of carbon monoxide and hydrogen, which is then used to synthesise organic chemicals. The reaction occurs at about 900 degC using a nickel catalyst.

$CH4 + H2O \Rightarrow CO + 3H2$

Steel An alloy of top"iron and top"carbon. Steel contains anywhere between 0.2% carbon (for soft wire and sheet steel) and 1.5% carbon (for cutting tools), with small amounts of many other elements often present.

Stellacyanin An electron transferprotein, containing a type 1 copper site, isolated from exudates of the Japanese lacquer tree.

Stepwise reaction A chemical reaction with at least one reaction intermediate and involving at least two consecutive elementary reactions.

Steradian (STR) A solid angle with vertex at the center of a sphere of radius r that encompasses an area of r^2 on the surface of the sphere.

Stereochemical formula (Stereoformula) A three-dimensional view of a molecule either as such or in a projection.

Stereochemical Refers to the three-dimensional view of a molecule either as such or in a projection.

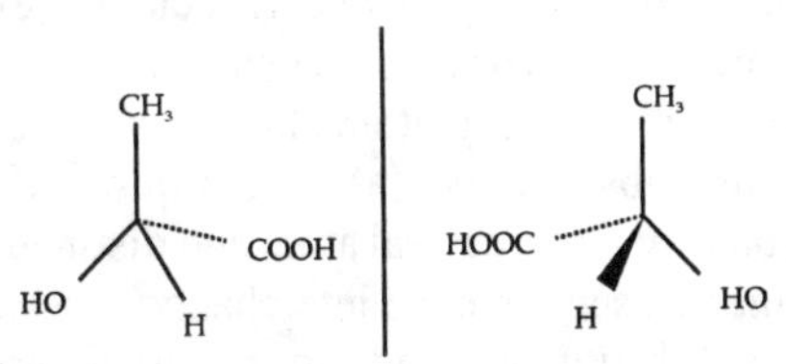

Stereochemistry Stereochemistry is the study of how the properties of a compound are affected by the spatial positions of groups within its molecules. Stereochemistry is also concerned with determining the structure of stereoisomers.

Stereoconvergence The predominant formation of the same stereoisomer or stereoisomer mixture of a reaction product when two different stereoisomers of the reactant are used in the same reaction. When that product involved in the reaction is one enantiomer the result has been called enantioconvergence.

Stereodescriptor A prefix to specify configuration (absolute or relative) or conformation. For example R,S; r,s; P,M; Re,Si; E,Z; ap,sp; etc.

Stereoelectronic control Control of the nature of the products of a chemical reaction (or of its rate) by stereoelectronic factors. The term is usually applied in the framework of an orbital approximation. The variations of molecular orbital energies with relative nuclear geometry (along a reaction coordinate) are then seen as consequences of variations in basis-orbital overlaps.

Stereoelectronic Pertaining to the dependence of the properties (especially the energy) of a molecular entity in a particular electronic state (or of a transition state) on relative nuclear geometry. The electronic ground state is usually considered, but the term can apply to excited states as well. Stereoelectronic effects are ascribed to the different alignment of electronic orbitals in different arrangements of nuclear geometry.

Stereogenic unit (Or stereogen or stereoelement) A grouping within a molecular entity that may be considered a focus of stereoisomerism. At least one of these must be present in every enantiomer (though the presence of stereogenic units does not conversely require the corresponding chemical species to be chiral). Three basic types are recognised for molecular entities involving atoms having not more than four substituents: (a) A grouping of atoms consisting of a central atom and distinguishable ligands, such that the interchange of any two of the substituents leads to a stereoisomer. An asymmetric atom (chirality centre) is the traditional example of this stereogenic unit. (b) A chain of four non-coplanar atoms (or rigid groups) in a stable conformation, such that an imaginary or real (restricted) rotation (with a change of sign of the torsion angle) about the central bond leads to a stereoisomer. (c) A grouping of atoms consisting of a double bond with substituents which give rise to cis-trans isomerism.

Stereoheterotopic Either enantiotopic or diastereotopic. (In contrast the term constitutionally heterotopic has been used to describe identical groups located in constitutionally non-equivalent positions.)

Stereoisomer Molecules with the same atoms and bond structure, but different three dimensional arrangements of atoms. For example, the CH_3 groups in $CH_3CH{=}CHCH_3$ can be placed on the same side of the double bond in one stereoisomer and on opposite sides in another.

Stereoisomeric Isomerism due to difference in the spatial arrangement of atoms without any difference in connectivity or bond multiplicity between the isomers.

Stereoisomerism Isomerism due to differences in the spatial arrangement of atoms without any differences in connectivity or bond multiplicity between the isomers.

Stereoisomerism, planar See planar chirality.

Stereoisomers Isomers that possess identical constitution, but which differ in the arrangement of their atoms in space. See enantiomer, diastereoisomer, cis-trans isomers.

Stereomutation A change of configuration at a stereogenic unit brought about by physical or chemical means.

Stereoselective synthesis A chemical reaction (or reaction sequence) in which one or more new elements of chirality are formed in a substrate molecule and which produces the stereoisomeric (enantiomeric or diastereoisomeric) products in unequal amounts. Traditionally called asymmetric synthesis. See also stereoselectivity.

Stereoselectivity, stereoselective Stereoselectivity is the preferential formation in a chemical reaction of one stereoisomer over another. When the stereoisomers are enantiomers, the phenomenon is called enantioselectivity and is quantitatively expressed by the enantiomer excess; when they are diastereoisomers, it is called diastereoselectivity and is quantitatively expressed by the diastereomer excess. Reactions are termed (100%) stereoselective if the discrimination is complete or partially (x%) stereoselective if one product predominates. The discrimination may also be referred to semiquantitatively as high or low stereoselectivity.

Stereospecificity, stereospecific 1. A reaction is termed stereospecific if starting materials differing only in their configuration are converted into stereoisomeric products. According to this definition, a stereospecific process is necessarily stereoselective but not all stereoselective processes are stereospecific. Stereospecificity may be total (100%) or partial. The term is also applied to situations where reaction can be performed with only one stereoisomer. For example the exclusive formation of trans-1,2-dibromocyclohexane upon bromination of cyclohexene is a stereospecific process, although the analogous reaction with (E)-cyclohexene has not been performed.

2. The term has also been applied to describe a reaction of very high stereoselectivity, but this usage is unnecessary and is discouraged.

Steric effect The effect on a chemical or physical property (structure, rate or equilibrium constant) upon introduction of substituents having different steric requirements. The steric effect in a reaction is ascribed to the difference in steric energy between, on the one hand, reactants and, on the other hand, a transition state, (or products). A steric effect on a rate process may result in a rate increase ("steric acceleration") or a decrease ("steric retardation"). (The adjective "steric" is not to be confused with stereochemical.)

Steric effects arise from contributions ascribed to strain as the sum of 1. non-bonded repulsions, 2. bond angle strain, and 3. bond stretches or compressions.

For the purpose of correlation analysis or linear free-energy relations various scales of steric parameters have been proposed, notably A values, Taft's E_S and Charton's scales.

In a reactant molecule RY and an appropriate reference molecule R^0Y, the "primary steric effect" of R is the direct result of differences in compressions which occur because R differs from R^0 in the vicinity of the reaction centre Y. A "secondary steric effect" involves the differential moderation of electron delocalization by non-bonded compressions.

Some authors make a distinction between "steric" effects attributed to van der Waals repulsions alone, and "strain" effects, attributed to deviations of bond angles from "ideal" values. See Taft equation, van der Waals forces.

Steric hindrance Term referring to the way that atoms can shield a site by getting in the way of approach of a reactant.

Steric isotope effect See isotope effect, steric.

Steric strain 1. See strain

2. See angle strain.

3. See eclipsing strain.

4. See transannular strain.

Steric approach control Control of stereoselectivity of a reaction by steric hindrance towards attack of the reagent, which is directed to the less hindered face of the molecule. Partial bond making is strong enough at the transition state for steric control to take place. This suggests that the transition state should not be close to products.

Stochastic effect Effect for which the probability of occurrence depends on the absorbed dose. Hereditary effects and cancer induced by radiation are considered to be stochastic effects. The term "stochastic" indicates that the occurrence of effects so named would be random. This means that, even for an individual, there is no threshold of dose below which the effect will not appear, but the chance of experiencing the effect increases with increasing dose (WHO, 1979).

Stoichiometric coefficient The coefficients given before substances in a balanced chemical equation. For example, the stoichiometric coefficient of carbon dioxide in the following reaction is 4:

$2\ C_2H_6(g) + 7\ O_2(g)\ 4\ CO_2 + 6\ H_2O$

Stoichiometric number Stoichiometric number (ν) is the number appearing before the symbol for each compound in the equation for a chemical reaction. By convention, it is negative for reactants and positive for products.

Stoichiometry 1. Ratios of atoms in a compound. 2. Ratios of moles of compounds in a reaction. 3. A branch of chemistry that quantitatively relates amounts of elements and compounds involved in chemical reactions, based on the law of conservation of mass and the law of definite proportions.

Stoichiometry of electrolysis The stoichiometry of electrolysis is described by Faraday's Laws of Electrolysis.

Stopped flow A technique for following the kinetics of reactions in solution (usually in the millisecond time range) in which two reactant solutions are rapidly mixed by being forced through a mixing chamber. The flow of the mixed solution along a uniform tube is then suddenly arrested. At a fixed position along the tube the solution is monitored (as a function of time following the stoppage of the flow) by some method with a rapid response (e.g. photometry). See mixing control.

Strain Strain is present in a molecular entity or transition structure if the energy is enhanced because of unfavourable bond lengths, bond angles, or dihedral angles ("torsional strain") relative to a standard.

It is quantitatively defined as the standard enthalpy of a structure relative to a strainless structure (real or hypothetical) made up from the same atoms with the same types of bonding. (The enthalpy of formation of cyclopropane is 53.6 kJ mol^{-1}, whereas the enthalpy of formation based on three "normal" methylene groups, from acyclic models, is -62 kJ mol^{-1}. On this basis cyclopropane is destabilized by ca. 115 kJ mol^{-1} of strain energy.)

Stratification The process of or result of separating a sample into several subsamples according to specified criteria such as age groups, socioeconomic status, etc. The effect of confounding variables may be controlled by stratifying the analysis of results. For example, lung cancer is known to be associated with smoking. To examine the possible association between urban atmospheric pollution and lung cancer, controlling for smoking, the population may be divided into strata according to smoking status. The association between air pollution and cancer can then be appraised separately within each stratum. Stratification is used not only to control for confounding effects but also as a way of detecting modifying effects. In this example, stratification makes it possible to examine the effect of smoking on the association between atmospheric pollution and lung cancer.

Stratified sampling Of a population which can be divided into different subpopulations (called strata), sampling carried out in such a way that specific proportions of the sample are drawn from the different strata (ISO, 1977).

Stratocumulus clouds Low altitude gray colored clouds composed of water droplets that have a patchy appearance. Each cloud patch consists of a rounded mass. This cloud has a somewhat uniform base and normally covers the entire sky. Between the patches blue sky can be seen.

Stratopause The boundary between the stratosphere and the mesosphere. It occurs at a height in the atmosphere of approximately 50 kilometers; however this depends on latitude. The atmosphere is characterized by a decrease in pressure with respect to increased altitude. More importantly, regions within the atmosphere like the troposphere, stratosphere, and mesosphere are distinguishable because of distinct temperature gradients with relatively well-defined starting and ending points. The stratopause is the highest portion of the stratosphere, with a temperature of approximately 0 degrees Celsius; the stratopause can also be described as the warmest region between the mesosphere and the stratosphere.

Stratosphere The thermal atmospheric region of the atmosphere between the troposphere and the mesosphere. The lower boundary of the stratospheric region is marked by the tropopause and begins at approximately 13 kilometers; however, this altitude of the troposphere depends on latitude. The upper limit of the stratosphere is marked by the stratopause at approximately 50 kilometers. The stratosphere is characterized by relatively stable temperatures (between -80 and -50 degrees Celsius) in the lower regions, and begins warming near 20 kilometers, reaching its maximum temperature of approximately 0 degree Celsius at the stratopause. Stratospheric chemistry is of particular interest to scientists because ozone, the principal substance that shields the earth from incoming solar ultraviolet radiation, is found in the stratosphere. It should also be noted that wind currents in the stratosphere are primarily horizontal in nature.

Stratosphere/troposphere exchange The exchange between the stratosphere and the troposphere is extremely important. It is caused by upwelling in the troposphere and when troposphere species are pushed into the stratosphere. It starts in the tropics and moves toward the poles. If certain ozone destroying species are brought into the stratosphere from the troposphere when they mix damage to the ozone layer and global warming can result.

Stratospheric clouds These are clouds found in the stratospheric region of the atmosphere. Unlikc the moisture laden troposphere, the stratosphere is almost moisture free. A lack of moisture means that cloud formation in the stratosphere is a relatively uncommon occurrence. Because of the very low temperatures, nacreous clouds, a type of stratospheric cloud, occur during winter conditions at the higher latitudes. A second type of stratospheric cloud, called a polar stratospheric cloud (PSC), occurs in the polar regions during winter and early spring. Polar stratospheric clouds are of particular interest to scientists studying the Antarctic Ozone Hole. Polar stratospheric clouds form under extremely cold conditions, whereby nitric acid, water vapor and other trace chemicals freeze to form ice crystals. PSCs provide a solid surface on which chlorine-containing reservoir compounds can gather, and on this surface the chemical reactions involved in the depletion of ozone are greatly increased.

Stratospheric ozone This is also called the ozone layer, where ozone concentrations are as high as 10 parts per million, and is a vitally important region of the atmosphere. This layer of ozone is located approximately 20-50 kilometers above the earth's surface. Stratospheric ozone is important because it prevents most of the high-energy ultraviolet solar radiation from reaching the earth's surface. Photodissociation, a photochemical process, is responsible for the formation of the protective ozone layer in the stratosphere. In the upper atmosphere, diatomic oxygen absorbs high-energy ultraviolet radiation. The absorption of radiation causes the diatomic oxygen molecule to break forming two oxygen radicals. The oxygen radical can then recombine with other diatomic oxygen molecules to form triatomic oxygen, or ozone. In the middle regions of the stratosphere, ozone is found in concentrations as high as 10 parts per million. Ozone can also form in the lower portions of the troposphere, due to anthropogenic activity and by a completely different mechanism. Without the protective stratospheric ozone layer in the upper atmosphere, life (as we know it) on earth would not be possible.

Stripping Stripping is a technique for removing volatile components in a mixture by bubbling a stream of an chemically unreactive gas (like nitrogen) through the sample, and then 'scrubbing' the nitrogen through a solution or solid adsorbent that can recover the volatile materials.

Strong acids and bases Strong acids and bases are acids and bases which are completely dissociated in solution.

Strong acid Compare with weak acid. A strong acid is an acid that completely dissociates into hydrogen ions and anions in solution. Strong acids are strong electrolytes. There are only six common strong acids: HCl (hydrochloric acid), HBr (hydrobromic acid), HI (hydroiodic acid), H_2SO_4 (sulfuric acid), $HClO_4$ (perchloric acid), and HNO_3 (nitric acid).

Strong base A strong base is an base that completely dissociates into ions in solution. Strong bases are strong electrolytes. The most common strong bases are alkali metal and alkaline earth metal hydroxides.

Strong electrolyte A strong electrolyte is a substance which undergoes almost complete dissociation into ions in solution. Solution of a strong electrolyte consist almost entirely of positively charged ions and negatively charged ions, which are stabilised in solution by solvation (i.e. by being associated with aggregates of molecules of the solvent). Examples of strong electrolytes include almost all salts, the caustic alkalis and the mineral acids in dilute solutions.

Strong ligand Strong field ligand. Compare with weak ligand. A ligand that causes a large crystal field splitting which results in a low-spin complex.

Structural formula Compare with molecular formula and empirical formula. A structural formula is a diagram that shows how the atoms in a molecule are bonded together. Atoms are represented by their element symbols and covalent bonds are represented by lines. The symbol for carbon is often not drawn. Most structural formulas don't show the actual shape of the molecule (they're like floor plans that show the layout but not the 3D shape of a house).

Structure activity relationship (SAR) An analysis which defines the relationship between the structure of a molecule and its ability to affect a biological system.

Structure of crystals The structure of crystals is described in terms of the geometry of a unit cell of the crystal structure.

Styrene Chemical formula C_8H_8, also commonly called vinyl benzene. Styrene is an aromatic hydrocarbon originating from petroleum and natural gas by-products. It is used to manufacture resilient, lightweight packaging materials and fiberglass. Styrene is extremely volatile in air and evaporates quickly; as a result, it does not remain in the atmosphere for long periods of time.

Subacute toxicity test An animal experiment serving to study the effects produced by the test material when administered in repeated doses (or continuously in food, drinking water) over a period of up to about 90 days (WHO, 1979).

Subjacent orbital The next-to-highest occupied molecular orbital ("NHOMO", also called "HOMO^{-}") and the second lowest unoccupied molecular orbital (SLUMO). Subjacent orbitals are sometimes found to play an important role in the interpretation of molecular interactions according to the frontier orbital approach. .

Subjective environment (Synonym:perceived environment) The environment as it is perceived by persons living in it, e.g., eye irritation caused by air pollution, or pleasure arising from good housing conditions (WHO, 1979).

Sublevel One part of a level, each of which can hold different numbers of electrons.

Sub-library A subset of a combinatorial library physically separated from the rest of the library, generally with one or more fixed building blocks.

Sublimation Sublimate; sublimating. Conversion of a solid directly into a gas, without first melting into a liquid.

Sub-monomer synthesis Process resulting in an oligomer in which each monomer residue is formed from two or more building blocks. This approach has been used for peptoid synthesis.

Subshell Sublevel. A set of electrons with the same azimuthal quantum number. The number of electrons permitted in a subshell is equal to 2 + 1.

Substituent An atom or group of bonded atoms that can be considered to have replaced a hydrogen atom (or two hydrogen atoms in the special case of bivalent groups) in a parent molecular entity (real or hypothetical).

Substituent electronegativity See electronegativity.

Substitution reaction A reaction, elementary or stepwise, in which one atom or group in a molecular entity is replaced by another atom or group. For example,

$CH_3Cl + OH^- CH_3OH + Cl^-$

Substrates 1. A chemical species of particular interest, the reaction of which with some other chemical reagent is under observation (e.g. a compound that is transformed under the influence of a catalyst).

2. The chemical entity whose conversion to a product or products is catalyzed by an enzymes.

3. A solution or dry mixture containing all ingredients which are necessary for the growth of a microbial culture or for product formation.

4. A component in the nutrient medium, supplying the organisms with carbon (C-substrate), nitrogen (N-substrate) etc.

Subunit An individual polypeptide chain in a protein, containing more than one polypeptide chain. Different types of subunits are frequently designated by , α , β , γ etc.

Successor complex The radical ion pair which forms by the transfer of an electron from the donor D to the acceptor A after these species have diffused together to form the precursor or encounter complex:

$A + D (A D) (A^{\cdot -} D^{\cdot +})$

sSugar A carbohydrate with a characteristically sweet taste. Sugars are classified as monosaccharides, disaccharides, or trisaccharides.

Suicide inhibition See mechanism-based inhibition.

Sulfate (SO_4^{2-}) sulphate. 1. The SO_4^{2-} ion, formed by reaction of sulfuric acid with a base.

2. A compound containing the SO_4^{2-} ion.

Sulfate anion (SO_4^{2-}) The sulfate anion in the atmosphere is the result of the oxidation of sulfur dioxide (SO_2). The sulfate anion is water soluble and is removed from the atmosphere through precipitation.

Sulfite (SO_3^{2-}) sulphite. 1. The SO_3^{2-} ion, formed by reaction of sulfurous acid with a base.

2. A compound containing the SO_3^{2-} ion.

Sulfite reductase Enzymes which catalyze the reduction of sulfite to sulfide. All known enzymes of this type contain siroheme and iron-sulfur clusters.

Sulfur dioxide This is a colorless gas consisting of a single sulfur atom and two oxygen atoms. Sulfur dioxide is a major primary pollutant in the atmosphere originating mostly from coal fired power plants and other fossil fuels combustion. In the atmosphere, sulfur dioxide is usually oxidized by ozone and hydrogen peroxide to form sulfur trioxide, a secondary pollutant. Sulfur trioxide, similar to sulfur dioxide, is extremely soluble in water. If these sulfur oxides are present in the atmosphere when condensation occurs, droplets of sulfuric acid (acid rain) are formed. Volcanic eruptions provide a natural source of sulfur dioxide in the atmosphere. However the real problem associated with the production of sulfuric acid in the atmosphere is not with the sulfur dioxide expelled by volcanoes. Anthropogenic production of sulfur dioxide, caused by the burning of fossil fuels, is largely responsible for damage caused by acid rain.

Sulfur hexafluoride (SF6) is an inert gas categorized as hydrofluorocarbons. It is used as an insulator in circuit breakers and is used to produce other elements such aluminum and magnesium. Sulfur hexafluoride absorbs thermal infrared radiation and could increase global warming as its concentration in the atmosphere increases.

Sulfuric acid (H_2SO_4) oil of vitriol; sulphuric acid. An oily, corrosive liquid that acts as a strong acid when dissolved in water. Sulfuric acid has so many industrial uses that sulfuric acid production was once used as an index of industrial productivity. Salts of sulfuric acids are called sulfates.

Sulfurous acid (H_2SO_3) sulphurous acid. A colorless liquid that acts as a weak acid when dissolved in water, sometimes used as a bleach. Salts of sulfurous acid are called sulfites.

Summer solstice In the Northern Hemisphere occurs on about June 21 of each year. The summer solstice is the longest day of the year and marks the beginning of the summer season. If the earth's axis were not tilted in relation to the plane of earth's orbit around the sun (the ecliptic) there would be no seasonal changes on earth. But because the earth's axis is tilted in relationship to its plane of orbit around the sun (~23.5 degrees) and assumes a constant direction—with the North Pole pointing towards the North Star, seasons occur, and daylight hours can fluctuate in length. Each day prior to the summer solstice the sun appears farther north in the sky; but at the summer solstice, the sun reaches its most northerly position in the sky, directly above the Tropic of Cancer, 23.5 degrees north latitude. Following the summer solstice the sun moves farther south in the sky, and on about September 23 it is directly overhead at the equator (the fall equinox).

Sunspots A dark area, which is strongly magnetic, that occurs in the area of the photosphere on the disk of the sun.

Sunspot cycle This cycle, lasting an average period of 11 years, involves recurring increases and decreases in the number of sunspots that occur on the sun's surface.

Superacid A medium having a high acidity, generally greater than that of 100 wt.-% sulfuric acid. The common superacids are made by dissolving a powerful Lewis acid (e.g. SbF_5) in a suitable Brønsted acid, such as HF or HSO_3F. (An equimolar mixture of HSO_3F and SbF_5 is known by the trade name "magic acid".)

In a biochemical context "superacid catalysis" is sometimes used to denote catalysis by metal ions analogous to catalysis by hydrogen ions.

By analogy, a compound having a very high basicity, such as lithium diisopropylamide, is called a "superbase".

Superbase See superacid.

Superconductivity The ability of certain materials to carry an electric current with zero electrical resistance.

SuperconductorSuperconductor is a material that experiences a nearly total loss of electrical resistivity below a critical temperature T_c.

Supercooling Supercooled; supercool. Liquids at temperatures below their normal freezing points are said to be "supercooled".

Supercritical fluid A fluid state that occurs when the pressure and temperature exceed the substance's critical pressure and critical temperature. Supercritical fluids fill their containers like gases but dissolve substances like liquids, which makes them very useful as solvents. Their density and other properties are intermediate between gases and liquids.

Superfluid Superfluid is a fluid with near-zero viscosity and extremely high thermal conductivity. Liquid helium exhibits these properties below 2.186□K (the λ point).

Superhyperfine See electron paramagnetic resonance spectroscopy.

Superoxide Superoxide ion.A binary compound containing oxygen in the -½ oxidation state. For example, KO2 is potassium superoxide, an ionic compound containing the superoxide ion, O2-.

Superoxide dismutases (SOD) Enzymes which catalyze the dismutation reaction of superoxide anion to dihydrogen peroxide and dioxygen. The enzymes have active sites containing either copper and zinc (Cu/Zn-superoxide dismutase), or iron (Fe-superoxide dismutase), or manganese (Mn-superoxide dismutase).

Superposability The ability to bring two particular stereochemical formulae (or models) into coincidence (or to be exactly superposable in space, and for the corresponding molecular entities or objects to become exact replicas of each other) by no more than translation and rigid rotation.

Supersaturated solution Supersaturated. A supersaturated solution has concentration of solute that is higher than its solubility. A crystal of solute dropped into a supersaturated solution grows; excess solute is deposited out of the solution until the concentration falls to the equilibrium solubility.

Supramolecular chemistry This is defined as the chemistry of molecular assemblies and of the intermolecular bond, as "chemistry beyond the molecule", bearing on the organized entities of higher complexity that result from the association of two or more chemical species held together by intermolecular forces. Thus, supramolecular chemistry may be considered to represent a generalized coordination chemistry extending beyond the coordination of transition elements by organic and inorganic ligands to the bonding of all kinds of substrates; cationic, anionic and neutral species of either inorganic, organic or biological nature.

Supramolecule A system of two or more molecular entities held together and organized by means of intermolecular (noncovalent) binding interactions.

Surface tension The work required to expand the surface of a liquid by unit area.

Surfactant A material that spreads along a surface, changing the properties of the surface. For example, soap spreads over a water surface and lowers its surface tension.

Surroundings Compare with system. In thermodynamics, the surroundings refer to the universe outside the system.

Surveillance Ongoing scrutiny, generally using methods distinguished by their practicability, uniformity, and frequently their rapidity, rather than by complete accuracy. Its main purpose is to detect changes in trend or distribution in order to initiate investigative or control measures .

Suspension A heterogenous mixture in which droplets or particles are suspended in a liquid.

Swain-lupton equation A dual parameter approach to the correlation analysis of substituent effects, which involves a field constant (F) and a resonance constant (R). The original treatment was modified later. Swain and Lupton (1968); Swain, unger, rosenquist and Swain (1983).

The procedure has been considerably applied, but also much criticized .

Swain-scott equation The linear free-energy relation of the form

$$\lg(k/k_0) = sn$$

applied to the variation of reactivity of a given electrophilic substrate towards a series of nucleophilic reagents. n is characteristic of the reagent (i.e. a measure of its nucleophilicity) and s is characteristic of the substrate (i.e. a measure of its sensitivity to the nucleophilicity of the reagent). A scale of n values is based on the rate coefficients k for the reaction of methyl bromide with nucleophiles in water at 25 ^{o}C, s being defined as 1.00 for these reactions and n being defined as 0.00 for the hydrolysis of methyl bromide. (Other scales have been devised.) Swain and Scott (1953).

Symbiosis The term was originally applied to describe the maximum flocking of either hard or soft ligands in the same complexes. For hydrocarbon molecules, symbiosis implies that those containing a maximum number of C-H bonds (e.g. CH_4) or C-C bonds (e.g. Me_4C) are the most stable.

Symproportionation Synonymous with comproportionation.

Syn addition A reaction in which two groups of a reagent X-Y add on the same face of a carbon-carbon double bond.

Synchronization (Principle of nonperfect synchronization) This principle applies to reactions in which there is a lack of synchronization between bond formation or bond rupture and other primitive changes that affect the stability of products and reactants, such as resonance, solvation, electrostatic, hydrogen bonding and polarizability effects. The principle states that a product-stabilizing factor whose development lags behind bond changes at the transition state, or a reactant-stabilizing factor whose loss is ahead of bond changes at the transition state, increases the intrinsic barrier and decreases the "intrinsic rate constant" of a reaction. For a product-stabilizing factor whose development is ahead of bond changes, or reactant factors whose loss lags behind bond changes, the opposite relations hold. The reverse effects are observable for factors that destabilize a reactant or product.

Synchronous A concerted process in which the primitive changes concerned (generally bond rupture and bond formation) have progressed to the same extent at the transition state is said to be synchronous. The term figuratively implies a more or less synchronized progress of the changes. However, the progress of the bonding change (or other primitive change) has not been defined quantitatively in terms of a single parameter applicable to different bonds or different bonding changes. The concept is therefore in general only qualitatively descriptive and does not admit an exact definition except in the case of concerted processes involving changes in two identical bonds.

Synergistic effect A synergistic effect is any effect of two chemicals acting together which is greater than a simple sum of their effects when acting alone.

Synthase An enzyme that catalyzes a reaction in which a particular molecule is synthesized, not necessarily by formation of a bond between two molecules (contrast synthetase).

Synthesis Synthesize; synthetic reaction. Compare with decomposition.

Formation of a complex product from simpler reactants. For example, water can be synthesized from oxygen and hydrogen gas: $H_2(g) + ½O_2 (g) H_2O$.

Synthetase See ligase.

Synthetic Synthetic material; artificial; artificial substance.

A substance manufactured by chemical synthesis.

System Compare with surroundings.

In thermodynamics, the system is the part of the universe that is of interest.

Systematic error Determinate error. Compare with random error, gross error and mistake. Systematic errors have an identifiable cause and affect the accuracy of results.

Systemic toxicity This term is applied when a substance affects a system in the organism other than and often distant from the site of application or exposure.

Systems analysis The analysis of an existing or proposed system in order to find optimal solutions for achieving the objectives of the system. The method owes much to the development of computer logic, and computerized methods of data control are used in the systems analysis when necessary (WHO, 1979).

T

Taft equation Various equations are associated with R.W. Taft, but the term is most often used to designate the family of equations that emerged from Taft's analysis of the reactivities of aliphatic esters, and which involved the polar substituent constant * and the steric substituent constant E_S

$$\lg k = \lg k_O + ^{**} + E_S$$

or the one-parameter forms applicable when the role of either the polar term or the steric term may be neglected. Nowadays * is usually replaced by the related constant $_I$.

Tag a) One of a set of surrogate analytes which are used in a decoding process; b) pendant function which allows a molecule to be selected from a mixture.

Target (Of environmental pollution) A human being or any organism, organ, tissue, cell, resource, or any constituent of the environment, living or not, that is subject to the activity of a pollutant or other chemical or physical activity or other agent (WHO, 1979).

Target organ(s) Organ(s) in which the toxic injury manifests itself in terms of dysfunction or overt disease (WHO, 1979).

Target population (i) The collection of individuals, items, measurements, etc., about which we want to make inferences. The term is sometimes used to indicate the population from which a sample is drawn and sometimes to denote any "reference" population about which inferences are required; (ii) The group of persons for whom an intervention is planned.

Tautomer A structure formed by facile motion of a hydrogen from one site to another within the same molecule.

Tautomeric effect See electromeric effect.

Tautomerization The isomerization by which tautomers are interconverted. It is a heterolytic molecular re-arrangement and is frequently very rapid. See tautomerism.

Tautomers Molecules that differ in the placement of a hydrogen and double bonds and are easily interconvertible. Keto and enol forms are tautomers.

Tea-bag A type of reaction vessel consisting of a porous mesh which encloses the resin but allows passage of reagents and solvents when immersed in an appropriate secondary container. Several tea-bags may be treated in a single vessel without mixing of the enclosed resins; manipulation of multiple tea-bags allows preparation of libraries by pool/split or directed sorting techniques.

Tectonic activity The process of the formation, movement, interaction, and destruction of parts of the earth's crust on a generally large scale. This can be used to explain seismicity, volcanism, and mountain building.

Tectonics The structural arrangement of rocks in any crust of a planet. Used to understand rock structures.

Tele-substitution A substitution reaction in which the entering group takes up a position more than one atom away from the atom to which the leaving group was attached.

Telomerization The formation of an addition oligomer, having uniform end groups X'...X", by a chain reaction in which a chain transfer limits the length of the polymer ("telomer") produced. An example is the polymerization of styrene in bromotrichloromethane solution (X' = CCl_3, X" = Br), where Cl_3C. radicals are formed in the initiation step to produce $Cl_3C(CH_2CHPh)_nBr$, with n greater than 1 and often less than ca. 10.

Temperature The thermal state of matter with reference to its ability to transfer heat to other matter. Temperature is distinguished from heat in that heat is the energy that is transferred between matter by radiation, conduction and/or convection. The three most common scales for measuring temperature are centigrade, Fahrenheit and Kelvin.

Temperature inversion The abnormal reversal of temperature in the troposphere caused by the entrapment of urban air pollution. Under normal circumstances, air in the troposphere is cold at high altitudes and warm at low altitudes, or near the earthIs surface. When air pollutants, such as SO_2 or NO_x, in urban areas get too concentrated, smog is produced. Since cold air underlies the warmer air at high altitudes, the air does not mix well which can lead to human health disturbances in people with respiratory illnesses such as asthma.

Temporary acceptable daily intake Used when data are sufficient to conclude that use of the substance is safe over the relatively short period of time required to generate and evaluate further safety data, but are insufficient to conclude that use of the substance is safe over a lifetime. A higher-than-normal safety factor is used when establishing a temporary ADI and an expiration date is established by which time appropriate data to resolve the safety issue should be available (WHO, 1987).

Temporary hardness Temporary hardness is that fraction of the hardness of water that is due to dissolved calcium bicarbonate.

Temporary hardness is removed on boiling, when the calcium bicarbonate is converted to calcium carbonate.

Temporary maximum residue limit A temporary maximum residue limit is a maximum residue limit established for a specified, limited period when (i) only a temporary or conditional acceptable daily intake has been established for the pesticide concerned, or (ii) although an acceptable daily intake has been established, the residue data are inadequate for firm maximum residue recommendations (WHO, 1976).

Tenta gel Trademark of Rapp Polymere GmbH, Tubingen, Germany. Beaded solid support with a crosslinked polystyrene core and grafted linear poly(ethylene glycol) (PEG) chains with terminal functional groups.

Teratogen This is the descriptor applied to any substance that can cause non-heritable birth defects.

Teratogenicity The property (or potential) to produce structural malformations or defects in an embryo or fetus (WHO, 1987).

Term Each compound or element in a chemical equation.

Terminal 1. The end of a polymer molecule.

2. A point at which electrical connections can easily be made or broken.

Terminal reaction A reaction that ends a cycle or chain of other chemical reactions.

Termination Termination is the final step in a free radical mechanism that results in the stopping of the free radical reaction. Normally, termination occurs only after a significant number of propagation steps in the free radical process have occurred.

Terminus The end of a polymer molecule.

Termolecular reaction A reaction containing three particles. In the atmosphere an example of a termolecular reaction is one in which a molecule acquires the excess energy released by a reaction so that the products created do not convert back to the original state of the reactants. The reaction of molecular oxygen with an oxygen atom (radical) to produce ozone is a good example of an important atmospheric termolecular reaction. In the atmosphere this third body is most often molecular nitrogen because statistically this is the most likely next collision since atmospheric N_2 = 78% by volume.

Terpene A naturally occurring organic compound, of the general empirical formula, $C_{10}H_{16}$, biologically built from a naturally occurring "monomer" called isoprene, C_5H_8, which is found as a volatile oil in plants. A clearly important—and olifactorily obvious—example is alpha-terpinene found in lemon oil.

Tertbutyl alcohol a highly flammable, volatile, colorless liquid which has a camphor-like odor. It is miscible in esters, and aromatic and aliphatic hydrocarbons, and soluble in water. Based on its vapor pressure, tert-butyl alcohol will exist mainly in the vapor phase in the atmosphere. It is most likely to react with the hydroxyl (OH) radical at this state. The reaction products are formaldehyde and acetone.

Tertiary alcohol Tertiary alcohols are aliphatic alcohols in which the hydroxyl group (-OH) is attached to a tertiary carbon atom (i.e. a carbon atom which has no hydrogen atoms attached directly to it).

Tertiary period This extended from about 65 million years ago to about 2.5 million years ago.

Tertiary structure The overall three-dimensional structure of a biopolymer. For proteins this involves the side chain interactions and packing of secondary structure motifs. For nucleic acids this may be the packing of stem-loops or supercoiling of double helices.

Tertiary treatment Tertiary treatment is the polishing process to which a sewage effluent is subjected after it has received secondary treatment in order to reduce further its nutrient content.

Tesla Tesla (T) is the SI derived unit of magnetic flux density. The tesla is magnetic flux density of a magnetic flux of one weber per square metre ($T = Wb/m^2$). The unit is named after the Croatian scientist Nikola Tesla (1857-1943), equal to $V\ s/m^2$.

Tetradecane a light, oily hydrocarbon with a melting point of 5.89*f*C and a boiling point of 253.7*f*C. As a gas-phase pollutant it is inactive as a human contact allergen.

Tetragonal decomposition A reaction where a single compound breaks down into simpler compounds.

Tetrahedral A molecular shape that results when there are four bonds and no lone pairs around the central atom in the molecule. The atoms bonded to the central atom lie at the corners of a tetrahedron with 109.5° angles between them. The ammonium ion (NH_4^+) has a tetrahedral molecular geometry.

Tetrahedral intermediate A reaction intermediate in which the bond arrangement around an initially double-bonded carbon atom (typically a carbonyl carbon atom) has been transformed from trigonal to tetrahedral. For example, aldol in the condensation reaction of acetaldehyde (but most tetrahedral intermediates have a more fleeting existence).

Tetrahedron See coordination.

Tetrahydrofolates Reduced folate derivatives that contain additional hydrogen atoms in positions 5, 6, 7 and 8. Tetrahydrofolates are the carriers of activated one carbon units and are important in the biosynthesis of amino acids and precursors needed for DNA synthesis.

Thalassemia A chronic inherited disease characterized by defective synthesis of hemoglobin. Defective synthesis of the α chain of hemoglobin is called α -thalassemia and defective synthesis of the β chain ofhemoglobin is called β-thalassemia. Thalassemias result in anemia that can be severe and are found more frequently in areas where malaria is endemic.

The amount of a substance The amount of a substance is measured in moles. One mole is the quantity of the substance (i.e. the weight of the substance in grams) that is present in numerically the same as its molecular weight.

The periodic table The periodic table is the structured arrangement of the elements in a chart that accentuates the relationships between the chemical properties of the different elements. The structure of the modern version of the periodic table is explained by the electronic configuration of the elements.

Thematic libraries synthesis Over the past several years, combinatorial chemistry has gradually realigned itself with changing business needs. In many organizations diversity-driven library production intended to broadly cover druglike chemical space has to a large extent been replaced by thematic as well as project-directed libraries. Thematic libraries are particularly useful for a platform target-based approach to drug discovery because the control of a common biochemical theme (e.g., the use of enzyme inhibitors, peptidomimetics of receptor ligands, etc.) can often cross over providing leads in several therapeutic areas.

Theodor curtius (1857-1928) Born in Duisburg in the Ruhr, and originally intended to study music, but switched to chemistry under the influence of Bunsen at Heidelberg and Kolbe at Leipzig. After working with Baeyer at Munich, he became the director of analytical chemistry at Erlangen, before taking up a Chair in Chemistry at Kiel in 1890. After a brief appointment at Bonn, Curtius succeeded Victor Meyer as Professor of Chemistry at his old university, Heidelberg, in 1898. He published the Curtius rearrangement in 1890 and also discovered diazoacetic acid, hydrazine and hydrazoic acid. He also composed music and sang in concerts.

Theodor wagner-jauregg (b.1903) Born in Vienna, the son of Leopold Wagner-Jauregg, the Nobel Laureate for medicine in 1927. He studied chemistry at the Universities of Vienna and Munich, before working with Richard Kuhn (1900-1967) in Zurich. He taught at Heidelberg before taking up a position at the Georg-Speyer-Haus research institute in Frankfurt in 1936. After the war, Wagner-Jauregg was transferred by the US Army to the United States, one of the few German chemists to co-operate with "Operation Paperclip". He worked in the US Chemical Corps at Edgewood Arsenal in Maryland. In 1955, he went to Switzerland to become research director of a small pharmaceutical company. The Wagner-Jauregg reaction (1930), the addition of maleic anhydride to 1,1-diarylethenes, is fairly obscure, and is usually considered to be a special case of the Diels-Alder reaction.

Theoretical yield Theoretical yield is the maximum quantity of a product that could be formed in a chemical reaction if all the limiting reactant reacted to form products (distinguished from actual yield).

Theories of catalysis The theories of catalysis explain the influence of the catalysts on the rate of a reaction by describing the detailed mechanism by which the catalyst is involved in the steps of the chemical reaction.

Theory Experimental measurements are usually made in terms of transmittance (T), which is defined as:

$$T=\frac{P}{P_0}$$

where P is the power of light after it passes through the sample and P_0 is the initial light power. The relation between A and T is: A = –

$$\log(T) = -\log\left(\frac{P}{P_0}\right)$$

In analytical applications we often want to measure the concentration of an analyte independent of the effects of reflection, solvent absorption, or other interferences. The figure to the right shows the two transmittance measurements that are necessary to use absorption to determine the concentration of an analyte in solution. *Absorption of light by a sample*

Modern absorption instruments can usually display the data as either transmittance, %-

transmittance, or absorbance. An unknown concentration of an analyte can be determined by measuring the amount of light that a sample absorbs and applying Beer's law. If the absorptivity coefficient is not known, the unknown concentration can be determined using a working curve of absorbance versus concentration derived from standards.

Therapeutic index For a substance used to alleviate disease, pain or injury the therapeutic index is the ratio between toxic and therapeutic doses (the higher the ratio, the greater the safety of the therapeutic dose).

Thermal Pertaining to heat.

Thermal conductivity Thermal conductivity (Λ) is rate of heat flow divided by area and by temperature gradient.

Thermal energy energy an object possesses by virtue of its temperature. For example, 1 g of water at 15°C has 4.184 J more energy than 1 g of water at 14°C.

Thermal pollution Thermal pollution is the increase in temperature of natural waters resulting from the discharge to these waters of hot effluents from industrial and power plants. The higher temperatures reduce the concentration of dissolved oxygen in the receiving waters and promote eutophication.

Thermionic emission The emission of electrons or ions by a hot object. For example, the filament in a mass spectrometer spits out energetic electrons that ionize atoms and molecules in the sample.

Thermistor A device that senses temperature changes by using a resistor with an electrical resistance that falls when temperature rises.

Thermochemical equation An compact equation representing a chemical reaction that describes both the stoichiometry and the energetics of the reaction. For example, the thermochemical equation $CH_4(g) + 2\,O_2(g)\,CO_2(g) + 2\,H_2O(g)$, H = -2220 kJ means "When one mole of gaseous CH_4 is burned in two moles of oxygen gas, one mole of CO_2 gas and 2 moles of steam are produced, and 2220 kilojoules of heat are released."

Thermochemistry In every chemical reaction, there is a difference between the intrinsic energy in the reactants and the intrinsic energy in the products. This energy, which is either adsorbed or released during the course of the reaction, is the heat of reaction.

The heat of reaction (i.e. the change in the energy during the course of a reaction) is expressed as dH, (i.e. "delta-H"), which has positive values for endothermic reactions and negative values for exothermic reactions .

The unit of heat is the joule. Another unit of heat which is widely used in thermochemistry, the calorie (i.e. the amount of heat required to raise 1 g of water through 10 C) which is equal to 4.18 joules.

Thermocouple A device that senses temperature changes by using a pair of joined wires made of dissimilar metals that produces a voltage that changes with temperature.

Thermodynamics The science of heat and temperature and of the laws governing the conversion of heat into mechanical, electrical, or chemical energy.

Thermodynamic control (of product composition) The term characterizes conditions that lead to reaction products in a proportion governed by the equilibrium constant for their interconversion and/or for the interconversion of reaction intermediates formed in or after the rate-limiting step. (Some workers prefer to describe this phenomenon as "equilibrium control".)

Thermodynamic equilibrium A system is at thermodynamic equilibrium if the energy it gains from its surroundings is exactly balanced by the energy it loses, no matter how much time is allowed to pass.

Thermodynamic laws Thermodynamic laws are the foundation of the science of thermodynamics:

First law: The internal energy of an isolated system is constant; if energy is supplied to the system in the form of heat dq and work dw, then the change in energy dU = dq + dw.

Second law: No process is possible in which the only result is the transfer of heat from a reservoir and its complete conversion to work.

Third law: The entropy of a perfect crystal approaches zero as the thermodynamic temperature approaches zero.

Thermodynamic product The reaction product with the lowest energy..

Thermoelectron An electron emitted by a very hot object.

Thermolysin A calcium- and zinc-containing neutral protease isolated from certain bacteria.

Thermolysis The uncatalysed cleavage of one or more covalent bonds resulting from exposure of a compound to a raised temperature, or a process in which such cleavage is an essential part.

Thermometer An instrument for measuring temperature.

Thermometry The science of temperature measurement.

Thermoplastic Compare with thermosetting. A polymer that softens or melts on heating, and becomes rigid again on cooling. Thermoplastic polymer chains are not cross-linked. Polystyrene is a thermoplastic.

Thermoplastic material Thermoplastic materials are polymeric materials which can be repeatedly softened by heating and hardened again on cooling.

Thermosetting material Thermosetting plastics are those rigid plastic which are molded into shape at the time of manufacture (i.e. at the time of polymerisation) and they an extensively cross-linked structure. These plastics cannot be remolded to another shape as their rigidity is due to the highly cross-linked structure formed between the molecules at the time of polymerisation."

Thermosphere Thermosphere is the layer of the earth's atmosphere extending from the top of the mesosphere (80 km - 90 km above the surface) to about 500 km. It is characterized by a rapid increase in temperature with increasing altitude up to about 200 km, followed by a leveling off in the 300 km - 500 km region.

Thin layer chromatography (TLC) Compare with chromatography. A technique for separating components in a mixture on the basis of their differing polarities. A spot of sample is placed on a flat sheet coated with silica and then carried along by a solvent that soaks the sheet. Different components will move different distances over the surface. TLC is a useful screening technique in clinical chemistry; for example, it can be used to detect the presence of drugs in urine.

Thio- A prefix that means, "replace an oxygen with sulfur". For example, sulfate ion is SO42-; thiosulfate ion is S2O32-. Cyanate ion is OCN-; thiocyanate ion is SCN-.

Thiol A molecule containing an SH group. Also a functional group.

Third law Third law of thermodynamics. The entropy of a perfect crystal is zero at absolute zero.

Thixotropic fluid Thixotropy. A liquid that becomes less viscous when stirred. Paint and printing inks are thixotropic fluids; they are formulated so that they flow more freely when brushed or rolled.

-amino-ß%3b-hydroxybutyric%20acid">threonine. ($CH_3CH(OH)CH(NH_2)COOH$) -amino-ß-hydroxybutyric acid.

A naturally occuring amino acid with a hydroxyl group on its side chain. Closely related to serine.

Three dimensional quantitative structure activity relationship (3D-QSAR) The analysis of the quantitative relationship between the biological activity of a set of compounds and their spatial properties using statistical methods.

Three way safety bulb The three way safety bulb is used for pipeting. The attachment is placed over the mouth of the pipet. The air (A) valve is used to empty the bulb of air. The suction (S) valve is used to draw liquid into the pipet. The empty (E) valve is used to drain liquid out of a pipet.

Threshold limit value (TLV) This is a quideline value defined by the by the American Conference of Government Industrial Hygienists to establish the airborne concentration of a potentially toxic substance to which it is believed that healthy working adults may be exposed safely through a 40 hour working week and a full working life. This concentration is measured as a time-weighted average concentration. They are developed only as a quidelines to assist in the control of health hazards and are not developed for use as legal standards.

Thylakoids Enclosed membrane structures inside chloroplasts and photosynthetic bacteria.

Tight-ion pair See ion pair.

Time weighted average (TWA) exposure This is a regulatory value defining the concentration of a substance to which a person is exposed in ambient air divided by the total time of observation. For occupational exposure a working shift of eight hours is commonly used as the averaging time.

Titrant The substance that quantitatively reacts with the analyte in a titration. The titrant is usually a standard solution added carefully to the analyte until the reaction is complete. The amount of analyte is calculated from the volume of titrant required for complete reaction.

Titration A procedure for determining the amount of some unknown substance (the analyte) by quantitative reaction with a measured volume of a solution of precisely known concentration (the titrant).

Titration curve A plot that summarizes data collected in a titration. A linear titration curve plots moles of analyte (or, some quantity proportional to moles of analyte) on the Y axis, and the volume of titrant added on the X axis. Nonlinear plots use the log of the concentration of the analyte instead. Nonlinear titration curves are often used for neutralization titrations (pH vs. mL NaOH solution). Logs are used to exaggerate the rate of change of concentration on the plot, so that the endpoint can be determined from the point of maximal slope.

Tolerance Tolerance is the ability to experience exposure to potentially harmful amounts of a substance without showing an adverse effect. An adaptive state characterized by diminished responses to the same dose of a chemical (WHO, 1979).

Toluene A highly volatile and water-soluble benzene derivative that is highly toxic to most forms of life. Its estimated lifespan in air is 60 hours in the case of a photochemical reaction with hydroxyl radical and is eventually oxidized to carbon dioxide. Toluene is used in a variety of industries including the manufacture of TNT, pharmaceuticals, rubbers, and other benzene derivatives, and is also a fuel additive and a paint solvent. Toluene also reaches the atmosphere from natural sources such as coal tar, mineral oils, and forest fires.

TOMS (total ozone mapping spectrometer) An instrument that measures the total ozone in a vertical column of atmosphere. Measurements are made by comparing the amount of backscattered ultraviolet radiation with the amount of incoming solar energy. These changes in this ratio, the earthís albedo, at specific wavelengths can be used to determine the amount of ozone present. 4 satellites create global maps daily of the earthís ozone measurements.

Tonne Tonne (t) is an alternative name for megagram (1000 kg).

Topomerisation The identity reaction leading to exchange of the positions of identical ligands. The indistinguishable molecular entities involved are called topomers. For example in the reaction below the two identical ligating atoms N_a and N_b are interchanged by rotation about the C-aryl bond but may be identified by n.m.r. spectroscopy. This is a degenerate isomerisation.

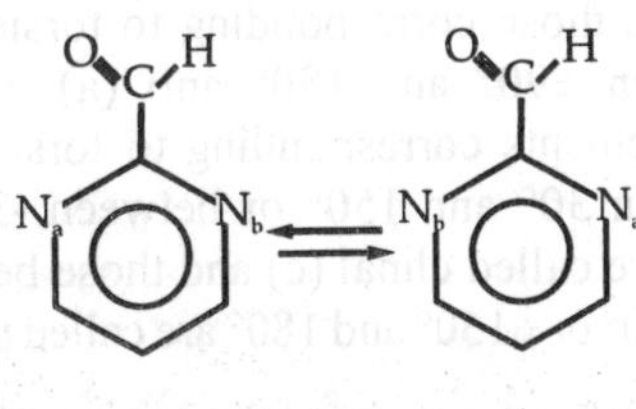

Torquoselectivity The term refers to the preference for "inward" or "outward" rotation of substituents in conrotatory or disrotatory electrocyclic ring opening reactions.

Torr Torr; mm Hg. Compare with barometer and pressure. A unit of pressure, defined so that 760 Torr is exactly 1 atmosphere. A Torr is equivalent to 1 mm Hg on barometer readings taken at 0°C; at other temperatures, the conversion from mm Hg to Torr is approximately p(Torr) = p(mm Hg) × (1 - 1.8×10^{-4}t), where t is in °C.

Torsion angle In a chain of atoms A-B-C-D, the dihedral angle between the plane containing the atoms A,B,C and that containing B,C,D). In a Newman projection the torsion angle is the angle (having an absolute value between 0° and 180°) between bonds to two specified (fiducial) groups, one from the atom nearer (proximal) to the observer and the other from the further (distal) atom. The torsion angle between groups A and D is then considered to be positive if the proximal bond A-B is rotated in a clockwise direction through less than 180° in order that it may eclipse the distal bond C-D: a negative torsion angle requires rotation in the opposite sense.

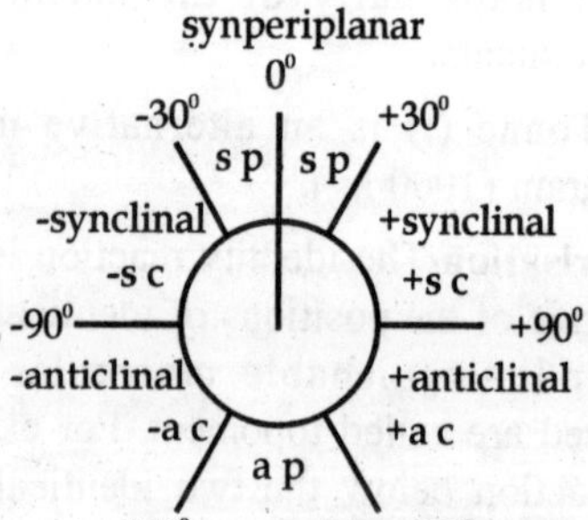

Stereochemical arrangements corresponding to torsion angles between 0° and ±90° are called syn (s), those corresponding to torsion angles between ±90° and 180° anti (a). Similarly, arrangements corresponding to torsion angles between 30° and 150° or between -30° and -150° are called clinal (c) and those between 0° and ±30° or ±150° and 180° are called periplanar (p). The two types of terms can be combined so as to define four ranges of torsion angle; 0° to ±30° synperiplanar (sp); 30° to 90° and -30° to -90° synclinal (sc); 90° to 150°, and -90° to -150° anticlinal (ac); ±150° to 180° antiperiplanar (ap). The synperiplanar conformation is also known as the syn- or cis-conformation; antiperiplanar as anti or trans and synclinal as gauche or skew. For macromolecular usage the symbols T, C, G^+, G^-, A^+ and A^- are recommended (ap, sp, +sc, -sc +ac and -ac respectively).

Torsional stereoisomers Stereoisomers that can be interconverted (actually or conceptually) by torsion about a bond axis. This includes E,Z-isomers of alkenes, atropisomers and rotamers.

Total column ozone The total amount of ozone that is found in a column of air above the earth from the surface to the top of the atmosphere. The majority of this amount is typically found in the stratosphere.

Total diet studies Studies undertaken to show the range and amount of various foodstuffs in the typical diet or to estimate the total amount of a specific substance (e.g., pesticide residue, vitamin, or food contaminant) in a typical diet (WHO, 1979).

Total diet study A total diet study is a study designed to establish the pattern of pesticide residue intake by a person consuming a defined diet (WHO, 1979).

Toxicokinetics A term with the same meaning as chemobiokinetics for substances not used as drugs (WHO, 1979).

Toxicology The study of poisons, including identification, isolation, biological effects, mechanism of action, and development of antidotes.

Toxicometry A combination of investigation methods and techniques for making a quantitative assessment of toxicity and hazards of poisons.

Trace elements Elements required for physiological functions in very small amounts that vary for different organisms.Included among the trace elements are Co, Cu, F, Fe, I, Mn, Mo, Ni, Se, V, W, and Zn. Excess mineral intake may produce toxic symptoms.

Trace gases These are gases in the atmosphere that do not occur in large quantities but are significant to life on Earth or are important constituents of the chemical cycles in the atmosphere.

Traceless linker Type of linker which leaves no residue on the compound after cleavage.

Trade winds Wind systems occupying most of the tropics, which blow from 30 degrees North and South toward the equator; winds are northeasterly in the Northern Hemisphere and southeasterly in the Southern Hemisphere.

Trajectory The curve or a path that an object makes while in space. Trajectory is usually a set of points on the path taken.

Trans In inorganic nomenclature a structural prefix designating two groups directly across a central atom from each other (not generally recommended for precise nomenclature purposes of complicated systems).

Transactinides All elements that follow the Actinides. As with all elements larger than uranium (U), these elements can only be produced in nuclear reactors or particle accelerators. While some of these elements of the transactinides are commonly lumped with metals, not enough information is available for this site to positively conclude whether elements 116 and 118 should be included with the metals or nonmetals. Indeed it may be completely in appropriate to lump these elements with either the metals or nonmetals. Only time will tell what the scientific community will ultimately decide.

Transannular strain In medium-sized ring compounds, the strain due to repulsive nonbonded interactions between substituents or hydrogen atoms attached to non-adjacent ring atoms.

Transcription The process by which the genetic information encoded in a linear sequence of nucleotides in one strand of DNA is copied into an exactly complementary sequence of RNA.

Transduction 1. The transfer of genetic information from one bacterium to another by means of a transducing bacteriophage. When the phage is grown on the first host, a fragment of the host DNA can be incorporated into the phage particles. This foreign DNA can be transferred to the second host upon infection with progeny phage from the first experiment.

2. In cell biology the transduction of a signal (mechanical signal, hormone, etc.) to cells or tissues summarizes the chain of events between the primary reception of the signal and the final response (change in growth and/or metabolism) of the target cells or tissues. Inorganic substances (e.g., calcium ions) are frequently involved in the transduction of signals.

Transferability Transferability assumes invariance of properties, associated conceptually with an atom or a fragment present in a variety of molecules. The property, such as electronegativity, nucleophilicity, NMR chemical shift, etc. is held as retaining a similar value in all these occurrences.

Transferase An enzyme of EC class 2, which catalyzes the transfer of a group from one substrate to another.

Transference number Different ions carry different fractions of the current because different ions move at different speeds under the same potential gradient. The transference number of an ion is the fraction of the total current that is carried by that ion during electrolysis.

In general, a cation and an anion differ in the amount of current they can carry during electrolysis.

Transferrin An iron-transport protein of blood plasma, which comprises two similar iron-binding domains with high affinity for Fe(III). Similar proteins are found in milk (lactoferrin) and eggs (ovotransferrin).

Transformation The conversion of a substrate into a particular product, irrespective of reagents or mechanisms involved. For example, the transformation of aniline ($C_6H_5NH_2$) into N-phenylacetamide ($C_6H_5NHCOCH_3$) may be effected by use of acetyl chloride or acetic anhydride or ketene. A transformation is distinct from a reaction, the full description of which would state or imply all the reactants and all the products.

Transient (chemical) species Relating to a short-lived reaction intermediate. It can be defined only in relation to a time scale fixed by the experimental conditions and the limitations of the technique employed in the detection of the intermediate. The term is a relative one.

Transient species are sometimes also said to be "metastable". However, this latter term should be avoided, because it relates a thermodynamic term to a kinetic property, although most transients are also thermodynamically unstable with respect to reactants and products.

Transition coordinate The reaction coordinate at the transition state corresponding to a vibration with an imaginary frequency. Motion along it in the two opposite senses leads towards the reactants or towards the products. See also reaction coordinate, transition state.

Transition state analogue A substrate designed to mimic the properties or the geometry of the transition state of reaction.

Transition state In theories describing elementary reactions it is usually assumed that there is a transition state of more positive molar Gibbs energy between the reactants and the products through which an assembly of atoms (initially composing the molecular entities of the reactants) must pass on going from reactants to products in either direction. In the formalism of "transition state theory" the transition state of an elementary reaction is that set of states (each characterized by its own geometry and energy) which an assembly of atoms, when randomly placed there, would have an equal probability of forming the reactants or of forming the products of that elementary reaction. The transition state is characterized by one and only one imaginary frequency. The assembly of atoms at the transition state has been called an activated complex.

It may be noted that the calculations of reaction rates by the transition state method and based on calculated potential-energy surfaces refer to the potential energy maximum at the saddle point, as this is the only point for which the requisite separability of transition state coordinates may be assumed. The ratio of the number of assemblies of atoms that pass through to the products to the number of those that reach the saddle point from the reactants can be less than unity, and this fraction is the "transmission coefficient" . (There are also reactions, such as the gas-phase colligation of simple radicals, that do not require "activation" and which therefore do not involve a transition state.).

Transition structure A saddle point on a potential-energy surface. It has one negative force constant in the harmonic force constant matrix.

Translation The unidirectional process that takes place on the ribosomes whereby the genetic information present in an mRNA is converted into a corresponding sequence of amino acids in a protein.

Transmission electron microscopy (TEM) Transmission electron microscopy (TEM) images the electrons that pass through a sample. Since electrons interact strongly with matter, electrons are attenuated as they pass through a solid and require the samples to be prepared in very thin sections, less than 100 nm. Specialized electron microscopes can use electron energies on the order of a MeV to study samples as thick as 1 μm or more. Materials samples are or for particles dispersed as a monolayer on a support grid. Biological samples are embedded in a solid matrix and microtomed. In all cases, the sample is placed on some type of support grid. Either the diffraction pattern or the image of the sample is observed on a phosphor screen below the sample, and can be recorded with film.

Transmittance Transmittance (τ) is ratio of the radiant or luminous flux at a given wavelength that is transmitted to that of the incident radiation. Also called transmission factor.

Transoid Conformation (usage strongly discouraged) See cisoid for use of transoid as a stereodescriptor.

Transuranium element Transuranic; transuranic element. An element with an atomic number higher than 92 (uranium's atomic number). Transuranium elements are unstable and occur in extremely low concentrations (if at all) in nature. Most are made artificially.

Trapping The interception of a reactive molecule or reaction intermediate so that it is removed from the system or converted into a more stable form for study or identification.

Traugott sandmeyer (1854-1922) Born in Wettingen near Zurich, and lived in the Zurich area for nearly all of his life. He trained as a precision instrument-maker, but became interested in chemistry. Self-educated in chemistry, he carried out chemical experiments in his kitchen. In 1881, he became a lecture assistant to Victor Meyer (1848-1897). He followed Meyer to Göttingen in 1886, but soon returned to Zurich and worked for Arthur Hantzsch (1857-1935). Sandmeyer joined Geigy as a research scientist in 1888, and eventually became a director of the firm. He discovered the decomposition of aryl diazonium chlorides to chloroarenes in the presence of copper (I) chloride in 1884. He also worked on the triphenylmethane dyes and the synthesis of isatin. Many years before, he had suggested to Victor Meyer an impurity in commerical benzene was responsible for the isatin reaction with sulphuric acid, thereby paving the way for Meyer's discovery of thiophen.

Treatment of drinking Water The treatment of drinking water involves the purification of raw waters to render them suitable for human consumption. This involves the removal of 1. the colloidal matter which is responsible for colour, and

2. the suspended solids responsible for turbidity.

Subsequently, this processed water has its pH adjusted, to make it less corrosive to the metals of the distribution pipes. Finally, fluoride is added to protect teeth from decay and the water is disinfected with chlorine before being put into the distribution system as drinking water.

Treatment of sewage The treatment of sewage is the purification of sewage and may involve 1. primary treatment (i.e. the removal of suspended solids),

2. secondary treatment (i.e. the removal of dissolved organic matter) and

3. tertiary treatment (i.e. the removal of nutrients)

before it is discharged to the aquatic environment.

Triads The triads are the three groups of elements, each of three metals, in Group VII Elements of the periodic table.

Trichloroethylene (TCE) C_2HCl_3, This is an industrial solvent used primarily for vapor degreasing and cold cleaning. It is an extraction solvent for greases, oils, fats, waxes, and tars, and is used by the textile processing industry to sour cotton, wool, and other fabrics. Exposure to moderate amounts of trichloroethylene may cause various health effects such as headaches, vertigo, visual disturbance, tremors, and dizziness. Death may occur from inhalation of large amounts. People located near or downwind of sites where TCE may be in operation might be exposed to higher levels. It is a carcinogen.

Trichlorofluoromethane (Freon[Æ]-11) One of the principal greenhouse gases, a gas with absorption bands in the infrared portion of the spectrum. There is extensive evidence showing that a class of synthetic compounds, the chlorofluorocarbons (CFCs), are responsible for the destruction of the ozone layer. CFCs are molecules that contain one or more atoms of both chlorine and fluorine. In September of 1992, the ozone hole over Antarctica was the largest ever recorded and was almost THREE times as large as the area of the U.S. Because CFCs are so unreactive, they do not break down when released into the air in the troposphere where they are spilled. In time, air currents and diffusion carry them into the stratosphere, where, under the influence of UV radiation, they release chlorine radicals that initiate the destruction of ozone. Data collected by NASA have shown conclusively, that there is an inverse relationship between ozone concentration and the chlorine monoxide radical in the stratosphere; ClO is formed by chlorine atom attack on O_3. CFCs are very useful inert, nontoxic, nonflammable compounds that had been used for years as coolants and as spray can propellants for aerosol forms of hair sprays and deodorants. They had been unsurpassed as solvents for cleaning electronic microcircuits. Commercially, the most important CFCs are the halogenated methanes, Freon-11 (trichlorofluoromethane) and Freon-12 (dichlorodifluoromethane). Over 50% of asthma inhalers contain chlorofluorocarbns as the solvent and gaseous propellent. These CFCs have relatively recently been prohibited in all products except in those medicinal inhaler dispensers for asthmatics and a few other limited exceptions. In Finland alone there are over a million medicinal dispensers that disperse freons into the atmosphere. This is equal to the freon concentration of the cooling devices in approximately l00,000 refrigerators.

Triglyceride A triglyceride is an ester of glycerol and three fatty acids. Most animal fats are composed primarily of triglycerides. In the structures below, the fatty acids attached to the glycerol are represented by 'R'. The fatty acids can be the same or different.

Trigonal bipyramidal Trigonal bipyramidal shape; trigonal bipyramidal geometry. A molecular shape that results when there are five bonds and no lone pairs on the central atom in the molecule. Three of the bonds are arranged along the atom's equator, with 120° angles between them; the other two are placed at the atom's axis. Axial bonds are at right angles to the equatorial bonds. The PCl_5 molecule has a trigonal bipyramidal molecular geometry.

Trigonal planar A molecular shape that results when there are three bonds and no lone pairs around the central atom in the molecule. The pairs are arranged along the central atom's equator, with 120° angles between them. The carbonate ion (CO_3^{2-}) has a trigonal planar geometry.

Trigonal pyramidal A molecular shape that results when there are three bonds and one lone pair on the central atom in the molecule. NH_3 is a trigonal pyramidal molecule.

Triple bond A covalent bond that involves 3 bonding pairs. In the valence bond theory, one of the bonds in a triple bond is a sigma bond and the other two are pi bonds. For example, the central bond in acetylene is a triple bond: H-CC-H.

Triple point Triple point is the point in p,T space where the solid, liquid, and gas phases of a substance are in thermodynamic equilibrium.

Triplet Describes an NMR signal split into three peaks.

Tritium A radioisotope of hydrogen with two neutrons and one proton in its nucleus.

Trophic level The amount of energy, in terms of food, that an organism needs. Organisms capable of utilizing inorganic chemicals, e.g., plants, or food of low energy content are said to be on a low trophic level whereas, for example, predator species needing food of high energy content are said to be on a high trophic level. The trophic level, thus, indicates the level of the organism in the food chain (WHO, 1979).

Tropopause The area where the temperature in the troposphere no longer decreases, indicating the boundary between the troposphere and the stratosphere.

Troposphere The lowest layer of the Earth's atmosphere, ranging from the ground to the base of the stratosphere with a altitude of 10-15 kilometers depending on the latitude. This is where all weather occurs. The word troposphere is derived from the Greek words tropos—turning and spharia—sphere.

Trueness Compare with accuracy. Trueness is the closeness of an average measurement to a "true" value, while accuracy is the the closeness of a single measurement to the true value.

-amino-ß%3b-indolylpropionic%20acid">

Tryptophan Trp; -amino-ß-indolylpropionic acid. A naturally occuring amino acid with a heterocyclic group on its side chain.

T-shape A molecular shape that results when there are 3 bonds and 2 lone pairs around the central atom in the molecule. The atoms bonded to the central atom lie at the ends of a "T" with 90° angles between them. ICl_3 has a T-shaped molecular geometry.

Tub conformation A conformation (of symmetry group D_{2d}) of an eight-membered ring in which the four atoms forming one pair of diametrically opposite bonds in the ring lie in one plane and all other ring atoms lie to one side of that plane. It is analogous to the boat conformation of cyclohexane.

Tumour (neoplasm) This term describes any growth of tissue forming an abnormal mass. Cells of a benign tumour will not spread and will not cause cancer. Cells of a malignant tumour can spread through the body and cause cancer.

Tunnelling The process by which a particle or a set of particles crosses a barrier on its potential energy surface without having the energy required to surmount this barrier. Since the rate of tunnelling decreases with increasing reduced mass, it is significant in the context of isotope effects of hydrogen isotopes.

Tyndall effect Light passing through a colloid is scattered by suspended particles. The light beam becomes clearly visible; this phenomenon is called the Tyndall effect. For example, car headlight beams can be seen in fog, but the beams are invisible in clear air.

Type 1,2,3 copper Different classes of copper-binding sites in proteins, classified by their spectroscopic properties as Cu(II). In type 1, or blue copper centers the copper is coordinated to at least two imidazole nitrogens from histidine and one sulfur from cysteine. They are characterized by small copper hyperfine couplings and a strong visible absorption in the Cu(II) state. In type 2, or non-blue copper sites, the copper is mainly bound to imidazole nitrogens from histidine. Type 3 copper centers comprise two spin-coupled copper ions, bound to imidazole nitrogens.

Tyrosinase A copper protein containing an antiferromagnetically coupled dinuclear copper unit (type 3 like site) which oxygenates the tyrosine group to catechol and further oxidizes this to the quinone.

U

UARS (upper atmosphere research satellite) A satellite deployed by NASA to study the chemistry in the upper atmosphere of Earth.

Ulrich hoffmann (1900-1970) Born in Chemnitz and studied chemistry at the Technische Hochschule in Dresden. At I. G. Farben in Ludwgishafen in the 1920s, he worked with Walter Reppe on the industrial synthesis of butadiene. Became joint managing director of Chemical Werke Hüls (1938-1945). He was dismissed from CW Hüls by the British and later worked for Degussa.

Ulrich hofmann (1903-1986) Born in Munich, son of Karl Andreas Hofmann. Studied under his father at the Technische Hochschule, Charlottenburg. Professor of inorganic chemistry at several number of universities, starting with Rostock in 1936 and culminating with Heidelberg in 1960.

Ultraviolet light Ultraviolet; ultraviolet radiation; ultraviolet region; UV.

Electromagnetic radiation with wavelength longer than that of x-rays but shorter than that of visible light. Ultraviolet light can break some chemical bonds and cause cell damage.

Umpolung Any process by which the normal alternating donor and acceptor reactivity pattern of a chain, which is due to the presence of O or N heteroatoms, is interchanged. Reactivity umpolung is most often achieved by temporary exchange of heteroatoms (N, O) by others, such as P, S, and Se.

The original meaning of the term has been extended to the reversal of any commonly accepted reactivity pattern. For example, reaction of R-CCX (X = halide) as a synthon for "R-CC$^+$ (i.e. electrophilic acetylene) is an umpolung of the normal more common acetylide, R-CC$^-$ (i.e. nucleophilic) reactivity.

Unbiased library Library prepared from building blocks and scaffold chosen without bias towards a particular target.

Uncertainty principle Heisenberg's uncertainty principle; Heisenberg principle; indeterminancy; indeterminancy principle.

The exact momentum and exact location of a particle cannot be specified. Werner Heisenberg stated that the product of uncertainties in location and momentum measurements can never be smaller than h/4, where h is Planck's constant.

Unified atomic mass unit Unified atomic mass unit (u or m_u) is a unit of mass used in atomic, molecular, and nuclear science, defined as the mass of one atom of ^{12}C divided by 12. Its approximate value is

u = (1.660 565±0.000 008 6)'"10^{-27} kg.

Unimolecular reaction A reaction that involves isomerization or decomposition of a single molecule.

Unit A standard for comparison in measurements. For example, the meter is a standard length which may be compared to any object to describe its length.

Unit cell Unit cell is the smallest fragment of the structure of a solid that by repetition can generate the entire structure.

Units of measurement The base units of the SI system are: metre (m), kilogram (kg), second (s), ampere (A), kelvin (K), candela (cd), and mole (mol) (BIPM, 1979).

Universal indicator A universal indicator is an indicator which undergoes several color changes over a wide range of pH. The color is used to "indicate" pH directly. Universal indicators are usually mixtures of several indicators.

Universal library A hypothetical compound collection which will show activity in all assays; a library with useful activity in many assays.

Unpaired spin A single electron occupying an orbital.

Unreactive Failing to react with a specified chemical species under specified conditions. The term should not be used in place of stable, since a relatively more stable species may nevertheless be more reactive than some reference species towards a given reaction partner.

Unsaturated compound An organic compound with molecules containing one or more double bonds.

Unsaturated fat Compare with saturated fat. A lipid containing one or more carbon-carbon double bonds. Unsaturated fats tend to be oily liquids and are obtained from plants.

Unsaturated hydrocarbons Unsaturated hydrocarbons are organic compounds containing double (alkenes) or triple (alkynes) bonds in its molecules.

Unsaturated solution An unsaturated solution is a solution which contains less than the maximum amount of solute that can be dissolved in the solvent at that temperature. If added, more solute will dissolve in an unsaturated solution.

Unstable The opposite of stable, i.e. the chemical species concerned has a higher molar Gibbs energy than some assumed standard. The term should not be used in place of reactive or transient, although more reactive or transient species are frequently also more unstable.

(Very unstable chemical species tend to undergo exothermic unimolecular decompositions. Variations in the structure of the related chemical species of this kind generally affect the energy of the transition states for these decompositions less than they affect the stability of the decomposing chemical species. Low stability may therefore parallel a relatively high rate of unimolecular decomposition.)

Upfield See chemical shift.

Uptake (synonym: absorption) The entry of a chemical substance into the body, into a cell, or into the body fluids by passage through a membrane or by other means (WHO, 1979).

Uranium-thorium dating An absolute dating technique which uses the properties of the radioactive half-life of Uranium-238 and Thorium-230. The half-life of uranium-238 is 4.5 billion years, that is, in that many years half of the original amount is still uranium-238 while the other half has lost nucleons to form a different element which is more stable.

Urban air pollution Pollution that occurs in cities from the burning of fossil fuels and the emissions of hydrocarbons. Urban air pollution can be in the form of ozone, smog, acid rain, carbon monoxide, particulate matter (PM-10), and nitrogen oxides.

Urban heat island effect- Urban heat island effect is the condition in which a city is relatively warmer than the surrounding areas due to the heat generated by the good conductors such as concrete, cars, and city buildings and the lack of vegetation.

Urease A nickel enzyme, urea amidohydrolase ,that catalyzes the hydrolysis of urea to ammonia and carbon dioxide. The active site comprises two Ni(II) ions, bridged by a carbamate.

Uses hydrogen In the reduction of oxide ores, in the refining of petroleum,

in the production of hydrocarbons from coal, to fill balloons and airships, as it is the least dense gas known (i.e. it is lighter than air). Previously, coal gas was often used for the same purpose, as it contains a high percentage of hydrogen. However, because the flammable nature of hydrogen makes it dangerous for such use, this use of hydrogen has been to replaced by helium,

in the Synthesis of ammonia,

as a fuel in Oxy-Hydrogen blowpipes,

for the hardening of vegetable or animal oils (i.e. to convert them into saturated fats which are solids), and

for the hydrogenating petroleum fractions, coal and other organic compounds.

Uses Iodine In medicine, where its solution in alcohol, called "tincture of iodine", is used as a disinfectant, in the manufacture of compounds used in photography (e.g. silver iodide which is a light sensitive material used in film), in the manufacture of dyestuffs and drugs, and as a reagent in analytical chemistry.

Uses Iron An important component of steel alloys.

Iron in its alloys is important in structures.

Iron in essential for life and is an important component of the Hemoglobin in blood.

Uses Lead In the arts, as it can be easily worked and molded, for pipes to carry water, although this use is now diminishing due to its toxicity, in a range of alloys including

Plumbers Alloy (33% lead and 66% tin), Pewter (75% tin and 25% lead), and

Type Metal (75% lead, 5% tin and 20% antimony), in the chemical industry for tanks and pipe work, as it is resistant to attack by sulphuric acid, and in roofing in the form of the sheet metal. Lead, Pb, is a metallic main group element, found in Group IVb of the periodic table.

Atomic Number : 82

Relative Atomic Mass : 207.2

Uses Mercury As the indicating fluid in thermometers, in mirrors, in the manufacture of detonators, and in the preparation of sodium amalgam which is used as a reducing agent in the laboratory.

Uses Nitrogen is used for the manufacture of ammonia in the Haber Process. Nitrogen is also used as the filler gas in electric lamps and in high temperature thermometers. Nitrogen is used to provide an inert atmosphere in metallurgy to prevent oxidation in certain industrial processes.

Uses Oxygen is essential for life and it takes part in processes of combustion, its biological functions in respiration make it important.

Oxygen is sparingly soluble in water, but the small quantity of dissolved oxygen in is essential to the life of fish.

Oxygen gas is used with hydrogen or coal gas in blowpipes and with acetylene in the oxy-acetylene torch for welding and cutting metals.

Oxygen gas is also used in a number of industrial processes.

Medicinally, oxygen gas is used in the treatment of pneumonia and gas poisoning, and it is used as an anesthetic when mixed with nitrous oxide, ether vapour, etc.. Carbon Dioxide is often mixed with the oxygen as this stimulates breathing, and this mixture is also used in cases of poisoning and collapse for restoring respiration.

Liquid oxygen mixed with powdered charcoal has been used as an explosive.

Sulphur

Sulphur, S, is a yellow nonmetallic main group element belonging to the Group VIb of the periodic table.

Atomic Number : 16

Relative Atomic Mass : 32.06

Melting Point : 112.8 degC

Boiling Point : 444.667 degC

Relative Density : 2.07

Although sulphur is in the same group in the periodic table as oxygen, there are more differences in the chemical characteristics of these elements than there are similarities. Thus, while oxygen always displays a valency of two, sulphur displays valences of four (in sulphur dioxide, SO2), six (in thionyl chloride, socl2) and eight (in sulphur hexafluoride, SF8).

In the elemental state, sulphur exists in polymeric forms.

Sulphur is an essential element for living organisms, and is present in some amino-acids.

Uses Sulphur is an important industrial raw material which is used in the manufacture of sulphuric acid, in the manufacture of sulphur dioxide and sodium sulphite which are used for bleaching straw and wood fibers, and for removing lignin from wood pulp for the paper industry, and in the manufacture of organic chemicals which contain sulphur.

The Transition Metals

The Transition Metals are the elements found between the Group IIA Elements and the Group IIB Elements in the periodic table. The GroupIIb are sometimes considered transition elements. The transition elements are also known as the d-block elements, because while the outermost level contains at most two electrons, their next to outermost main levels have incompletely filled d sub-orbitals, which are filled-up progressively on going across the periodic table from 8to 18 electrons. The filling of the d sub-orbitals of the transition elements across a row of the periodic table is not always regular.

Uses Uranium A colourant in pottery and the manufacture of glass.

Uranium is used in the nuclear industry as a fissionable material.

Zinc Zinc, Zn, is a transition metallic element found in Group IIb of the periodic table.

Atomic Number : 30

Relative Atomic Mass : 65.38

Uses Zinc In widespread use in the manufacture of electric cells (i.e. Dry Cells), in the manufacture of galvanized iron sheets, to protect against corrosion of the iron, in the manufacture of die-castings objects (e.g. the carburettor in car engines), in alloys, in medicine (as the Oxide), and as a filler for Rubber and in Paints.

UV radiation Energy that is emitted in the form of electromagnetic waves with a wavelength of 1-380nm, which is composed of UV-A, UV-B, and UV-C light; UV-A light being the longest wavelength and having the smallest energy and UV-C light being the shortest wavelength and having the highest energy. All UV-C light and most UV-B light is filtered out of the atmosphere via the ozone layer. However, UV-A light travels to the surface of the earth.

UV/vis spectrophotometry Ultraviolet/visible spectrophotometry is used to measure the absorption capabilities of certain compounds with reference to wavelengths of light on the ultraviolet and visible light ranges.

UV-A Ultraviolet light that has wavelengths of 320nm to 400nm. The wavelengths of this light are shorter than visible light and are not absorbed in the stratosphere by ozone.

UV-B An ultraviolet wavelength of light between 280 nm and 320 nm whose intensity is increasing at the earth's surface (probably causing increases in skin cancer), This UV increase is because of decreases in statospheric ozone.

UV-C Light that spans the spectrum with wavelengths of 10nm to 290nm. These wavelengths of ultraviolet light are extremely dangerous to human and animal tissue and is totally absorbed in the stratosphere by ozone and molecular oxygen.

V

VacancyVacancy is a missing atom or ion in a crystal lattice.

Vacuum Absolute vaccuum. Compare with partial vaccuum. A volume which contains no matter.

Vadose zone- The unsaturated zone in a soil where chemical processes are at their most active. Its extent is determined partly by the content of the soil water, but it cannot extend beyond the water table, below which voids are completely filled with water.

Valence The maximum number of univalent atoms (originally hydrogen or chlorine atoms) that may combine with an atom of the element under consideration, or with a fragment, or for which an atom of this element can be substituted.

Valence bond In the valence bond theory, a valence bond is a chemical bond formed by overlap of half-filled atomic orbitals on two different atoms.

Valence bond theory VB theory; VB. A theory that explains the shapes of molecules in terms of overlaps between half-filled atomic orbitals, or half filled "hybridized" orbitals (which are a mixture of atomic orbitals).

Valence electron Electrons that can be actively involved in chemical change; usually electrons in the shell with the highest value of n. For example, sodium's ground state electron configuration is $1s^2\ 2s^2\ 2p^6\ 3s^1$; the 3s electron is the only valence electron in the atom. Germanium (Ge) has the ground state electron configuration $1s^2\ 2s^2\ 2p^6\ 3s^2\ 3p^6\ 3d^{10}\ 4s^2\ 4p^2$; the 4s and 4p electrons are the valence electrons.

Valence electron potential (eV) Provides a quantitative indication of an elements reactivity and is based on the charge of the valence electrons and the ionic radius. It is determined by the equation: (-eV) = *kn/r*. '*k*' is a proportionality factor converting Ångstroms into centimeters and expressing the force exerted by the valence electrons in electron volts which is equated to 14.399. '*n*' is the valence. '*r*' is the ionic radius in Ångstroms.

Valence isomer A constitutional isomer interrelated with another by pericyclic reactions. For example, Dewar benzene, prismane and benzvalene are valence isomers of benzene.

Valence shell The shell corresponding to the highest value of principal quantum number in the atom. The valence electrons in this shell are on average farther from the nucleus than other electrons; they are often directly involved in chemical reaction.

Valence shell electron pair repulsion theory (VSEPR-) A model that explains the shapes of molecules by assuming that electron pairs arrange themselves around atoms in a way that minimizes electron-electron repulsions.

-aminoisovaleric%20acid">valine.

$((CH_3)_2CHCH(NH_2)COOH)$ Val; aminoisovaleric acid.

A naturally occuring aliphatic amino acid with a nonpolar isopropyl side chain.

Valence tautomerization The term describes simple reversible and generally rapid isomerizations or degenerate re-arrangements involving the formation and rupture of single and/or double bonds, without migration of atoms or groups.

Valency Valency is the combining power of an element (i.e. the number of bonds which the atom can form with other atoms).

Valency is governed by the number of electrons in the outermost electronic shell of the atoms of that element (i.e. the valency electrons). These valency electrons are given up to other atoms or are received from other atoms to make Ionic Bonds, or the valency electrons are shared with other atoms to make Covalent Bonds. The process of transferring electrons between atoms or of sharing electrons between atoms results in the formation of compounds, where the atoms in the compound achieve the stable electronic configuration of the Nobel Gas elements.

In order to achieve the electronic configuration of the nearest inert gas, which is a stable electronic configuration, atoms may lose, gain or share electrons. When these valency electrons are given up by an atom, or are received by an atom, to make ionic bonds are formed. When these valency electrons are shared with other atoms, covalent bonds are formed. Thus, the electron transfers result in the formation of bonds between atoms, and in the formation of compounds.

In general, the normal valency of an atom is thus equal the smaller of the following two numbers its periodic group number, or eight minus its group number.

However, a number of elements show more than one valency. Thus, the valency displayed by an element in any particular compound depends on its oxidation state in that compound.

Validity, study The degree to which the inference drawn from a study, especially generalizations extending beyond the study sample, are warranted when account is taken of the study methods, the representatives of the study sample, and the nature of the population from which it is drawn. Two varieties of study validity are distinguished: 1. Internal validity: The index and comparison groups are selected and compared in such a manner that the observed differences between them on the dependent variables under study may, apart from sampling error, be attributed only to the hypothesized effect under investigation.

2. External validity (generalizability): A study is externally valid or generalizable if it can produce unbiased inferences regarding a target population (beyond the subjects in the study). This aspect of validity is only meaningful with regard to a specified external target population. For example, the results of a study conducted using only white male subjects might or might not be generalizable to all human males (the target population consisting of all human males). It is not generalizable to females (the target population consisting of all people). The evaluation of generalizability usually involves much more subject-matter judgment than internal validity .

Value (rho-value) A measure of the susceptibility to the influence of substituent groups on the rate constant or equilibrium constant of a particular organic reaction involving a family of related substrates. Defined by Hammett for the effect of ring substituents in meta- and para-positions of aromatic side-chain reactions by the empirical "-equation" of the general form

$lg(kx/k_H) = X$

in which x is a constant characteristic of the substituent X and of its position in the reactant molecule.

More generally (and not only for aromatic series), -values (modified with appropriate subscripts and superscripts) are used to designate the susceptibility of reaction series for families of various organic compounds to any substituent effects, as given by the modified set of -constants in an empirical -correlation.

Reactions with a positive -value are accelerated (or the equilibrium constants of analogous equilibria are increased) by substituents with positive -constants. Since the sign of was defined so that substituents with a positive increase the acidity of benzoic acid, such substituents are generally described as attracting electrons away from the aromatic ring. It follows that reactions with a positive -value are considered to involve a transition state (or reaction product) so that the difference in energy between this state and the reactants is decreased by a reduction in electron density at the reactive site of the substrate. See also Taft equation.

Van der waals' equation Van der Waals' equation is an equation of state for fluids which takes the form:

$$pV_m = RT\left(\frac{1}{V_m - b} - \frac{a}{V_m^2}\right)$$

where p is pressure, V_m is molar volume, T is temperature, R is the molar gas constant, and a and b are characteristic parameters of the substance which describe the effect of attractive and repulsive intermolecular forces.

Van der waal's force The Van der Waal's Force is the attractive force between molecules, which is caused by the induced temporary polarisation of molecules by the dipole moments of molecules.

In complex covalent compounds, an attraction occurs between the protons of one molecule and the electrons of another. As two molecules of the compound approach one another, the electrons in the outer shells of both molecules which have a negative charge begin to repel each other. When the molecules are a certain distance apart, the forces of attraction and repulsion are equal and opposite (i.e. the forces are balanced at that distance).

Graphite, which is a crystalline form of carbon, consists of flat planes of carbon atoms. The carbon atoms in each plane are held together by covalent bonds. However, there are no covalent bonds between the planes, and the only forces between the planes are the weak intermolecular interactions (i.e. the Van der Waal's Force). The planes can slip and slide over each other with ease, a fact which helps to explain why graphite is soft and a good lubricant.

The Van der Waal's Force give rise to intermolecular attractive forces between the molecules of gases, and are responsible for the deviation of the behaviour of real gases from the ideal gas laws.

Van der waals radius Van der Waals radii.

One half the distance between two nonbonded atoms, when attractive and repulsive forces between the atoms are balanced.

Van't hoff equation Van't Hoff equation is the equation expressing the temperature dependence of the equilibrium constant K of a chemical reaction:

$$\frac{d\ln K}{dT} = \frac{\Delta_r H''}{RT^2}$$

where $\Delta_r H°$ is the standard enthalpy of reaction, R the molar gas constant, and T the temperature.

Vapor A substance in the gas state that is below its critical temperature but still suspended in air. It is possible for a vapor to be liquefied by increased pressure. Vapors are usually not visible.

Vapor pressure The pressure exerted when a material in its solid or liquid state is in equilibrium with it's gas state. As this is changes as a function of temperature, vapor pressure must be qualified by the relevant temperature.

Vapor pressure lowering Vapour pressure depression; vapor pressure depression. A colligative property of solutions. The vapor pressure of a solution is always lower than the vapor pressure of the pure solvent; the ratio of solution to pure solvent vapor pressures is approximately equal to the mole fraction of solvent in the solution.

Variable Compare with independent variable and dependent variable. A quantity that can have many possible values. In designing experiments, variables that affect measurements must be identified and controlled. For example, an experiment that measures reaction rates must control temperature, because temperature is a variable that can change the rate of reaction.

Velocity Speed of an object; the change in position over time.

Venus Second planet from the sun; mass 4.872 x 10^{27}g; radius of cloud surface about 6100 Km; radius of solid surface about 6056 Km; surface temperature 737K; temperature of cloud tops 250K; cloud composition more than 75% H_2SO_4 and about 25% water; atmosphere (by volume) 90-95% CO_2 with traces of SO_2, water, CO, H_2SO_4, Ar, Ne, HF, HCl. Atmospheric pressure 92-95 atm.

Vernal equinox A time during the spring season when the sun is directly above the equator which results in an equal amount of time for day and night.

Vertical flux In dry deposition this refers to the rate at which a gas particle is deposited onto another surface without the aid of precipitation. Vertical flux equals concentration of the species multiplied by the deposition velocity.

Vertical ionization energy Compare with adiabatic ionization energy. The energy required to remove an electron from an atom, molecule, or ion in the gas phase without moving any nuclei. The vertical ionization energy is greater than or equal to the adiabatic ionization energy.

Vertical mixing The movement of air in a vertical direction usually caused by differences in temperature and density.

Victor villiger (1868-1934) Born in the Swiss village of Cham am Zuger See in the Aarau, and studied chemistry under Carl Graebe (1841-1927) at the University of Geneva. After his military service, Villiger went to Munich to work with Adolf von Baeyer. Baeyer was very pleased with his new assistant and they worked together for eleven years. In 1904, Villiger decided to become a research chemist at BASF, a firm with close links with Baeyer. At BASF in Ludwigshafen, he initially worked with August Bernthsen (1855-1931) on thiazin and oxazin dyestuffs, and later joined the central research laboratory. His name is associated with the still popular Baeyer-Villiger oxidation of ketones (1899).

Viktor fischer (1875-1943) Born in Vienna and trained as an engineer at the Technische Hochschule there. However, he went into physical chemistry when he worked with the Gesellschaft für Lindes Eismachinen in the 1930. He studied the equilibria and phase diagrams for mixtures including ammonia and water, and carbon monoxide, hydrogen and water.

Vinyl Polyethylene. A polymer made by linking ethylene ($CH_2{=}CH_2$) or substituted ethylene molecules together.

Vinyl bromide a toxic compound that was established as a federal hazardous air pollutant in 1993. Sources of vinyl bromide may include rubber substitution, flame-retarding treatment of acrylics, laminating, and/or preparing films. It is also important to point out that vinyl bromide may be the result of the atmospheric degradation of 1,2-dibromoethane. Based on a typical atmospheric gas phase reaction with a hydroxyl radical, the lifetime of vinyl bromide in the atmosphere is between 1.5 and 2.1 days resulting in the products formaldehyde and formyl bromide.

Vinyl chloride Chemical Formula C2H3Cl: Synonyms: Ethylene, chloro-; Chloroethene; Chloroethylene; Monochloroethylene; Vinyl chloride; Vinyl chloride monomer; Vinyl C monomer; C2H3Cl; Ethylene monochloride; Monochloroethene; Chlorethene; Chlorethylene; Chlorure de vinyle; Cloruro di vinile; Rcra waste number U043; Trovidur; UN 1086; VC; VCM; Vinile; Vinylchlorid; Vinyl chloride, inhibited; Vinyle(chlorure de); Winylu chlorek. This colorless gas is used mainly to produce polyvinylchloride. As there are no natural sources known for this compound, all sources areanthropogenic. The compound reacts with hydroxyl radicals in the atmosphere to yield chlorine radicals, which catalytically destroy ozone.

Virtual library A library which has no physical existence, being constructed solely in electronic form or on paper. The building blocks required for such a library may not exist, and the chemical steps for such a library may not have been tested. These libraries are used in the design and evaluation of possible libraries.

Virtual screening Selection of compounds by evaluating their desirability in a computational model. Also termed in silico screening.

Viscosity Coefficient of viscosity. The resistance a liquid exhibits to flow. Experimentally, the frictional force between two liquid layers moving past each other is proportional to area of the layers and the difference in flow speed between them. The constant of proportionality is called "viscosity" or "coefficient of viscosity", and is given the symbol . The time required for a liquid to drain out of a capillary tube is directly proportional to its viscosity. The poise is a non-SI unit frequently used to express viscosities.

Visibility A measurement of the ability to see and identify objects at different distances.

Visible light Visible light is electromagnetic radiation with a wavelength between 400 and 750 nm.

Vital force The vital force was the entity presumed to confer the properties of life in the organic compounds found in animals and plants. Wohler's synthesis of urea an organic compound, from the inorganic compound, ammonium cyanate discredited the theory of the existence of a vital force.

Vitamin A substance that is critical for proper functioning of a living organism that the organism is unable to produce in sufficient quantities for itself.

Vitamin B-12 See cobalamin.

Volatility The tendency of a material to pass into the gas state at a particular temperature from a solid or liquid state.

Volcanoes Openings in the earth's crust that release molten rock, steam, reduced and oxidized compounds, and ash from within the earth's interior into the atmosphere.

Volt Volt (V) is the SI derived unit of electric potential. One volt is the difference of potential between two points of an electical conductor when a current of 1 ampere flowing between those points dissipates a power of 1 watt. It is named after the Italian physicist Alessandro Volta (1745-1827).

Voltage (V) A measured electric potential, in volts.

Voltage symbol V or E, unit: volt (V) Voltage (also called potential, potential difference, potential drop, or electromotive force - EMF) is the electronic potential energy between two points, and is the driving force that causes charge to flow. One volt is defined as the potential difference that requires one joule of energy to move one coulomb of charge.

Voltage is always measured relative to some other point in a circuit, e.g., the potential across a resistor. Voltage measurements made at a single point in a circuit are made relative to the earth (ground), which is assigned an "absolute" voltage of zero.

Voltaic cell Voltaic cells are the electrochemical cells that are constructed to be a source of electric current to power external devices.

Voltaic pile An early battery consisting of disks of dissimilar metals (usually zinc and copper) separated by moist paper or cloth soaked in an electrolyte solution.

Voltammeter An instrument for measuring voltages and amperages.

Volume (V) 1. The amount of space an object takes up.

2. The amount of space a container can hold. The SI unit of volume is the cubic meter (m^3); the liter is a popular unit of volume in chemistry.

Volume fraction Volume fraction (φ_j) is defined as

$$\phi_A = \frac{V_A}{\sum V_i}$$

where V_j is the volume of the specified component and the V_i are the volumes of all the components of a mixture prior to mixing.

Volume of activation, V A quantity derived from the pressure dependence of the rate constant of a reaction (mainly used for reactions in solution), defined by the equation

V = -RT(ln k/p)T

providing that the rate constants of all reactions (except first-order reactions) are expressed in pressure-independent concentration units, such as mol dm^{-3} at a fixed temperature and pressure.

The volume of activation is interpreted, according to transition state theory as the difference between the partial molar volumes of the transition state (V) and the sums of the partial volumes of the reactants at the same temperature and pressure, i.e.,

$$V = V - (rV_R)$$

where r is the order in the reactant R and V_R its partial molar volume.

Volume percentage ((v/v)%) Volume percentages express the concentration of a component in a mixture or an element in a compound. For example, 95% ethanol by volume contains 95 mL of ethanol in 100 mL of solution (NOT in 100 mL of water!)

Volumetric analysis Volumetric analysis is the quantitative analysis of liquids and gases. In the case of liquids, the concentration of a solution of unknown strength is determined by causing the solution to react with another solution of known concentration. Volumetric analysis is mainly concerned with methods of titrating various pairs of solutions, one against the other, and with the calculations arising from such titration's. In the case of gases, the main technique involves reacting or absorbing gases in graduated containers over mercury, and measuring the volume change during the course of reaction.

Vulcanization Vulcanization of rubber. A process of combining rubber with sulfur or other substances that causes the polymer chains to crosslink, making them stronger and more elastic.

W

Waldemar rudolf johann fischer (1881-1934) Born in Moszczenice in what was then Russian Poland. Took his Ph.D. under Hantzsch at Leipzig in 1903. Became Professor of Analytical Chemistry at Riga.

Walden Inversion Retention of configuration is the preservation of integrity of the spatial arrangement of bonds to an chiral centre during a chemical reaction or transformation. It is also the configurational correlation when a chemical species Xabcd is converted into the chemical species Xabce having the same relative configuration. The configurational change when a chemical species Xabcd (where X is typically carbon), having a tetrahedral arrangement of bonds to X, is converted into the chemical species Xabce having the opposite relative configuration (or when it undergoes an identity reaction in which Xabcd of opposite configuration is produced) is called a Walden inversion or inversion of configuration.

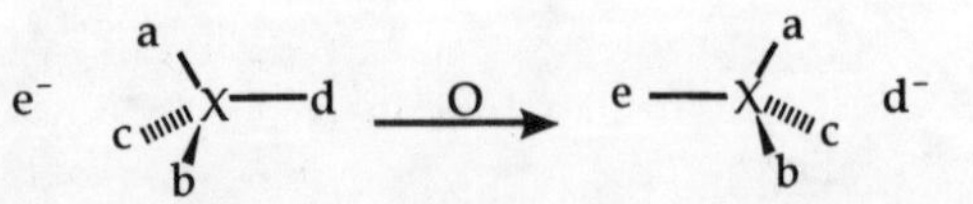

Walker cell A zonal circulation of the atmosphere confined to equatorial regions and driven principally by the oceanic temperature gradient. In the Pacific, air flows westward from the colder, eastern area to the warm, western ocean, where it acquires warmth and moisture and subsequently rises. A return flow aloft and subsidence over the eastern ocean complete the cell.

Water Water consists of two hydrogen atom and one oxygen atom (H_2O). Water covers three-fourths of the Earth's surface, 60-70% of the world's weight, regenerates and redistributes through evaporation and other atmospheric processes. Water is involved in electrical charge separation because it has two types and positions of atoms giving it a net dipole movement. Water vapor also absorbs 17% of solar radiation in the troposphere, thus making it one of the two principal greenhouse gases. Of the solar energy absorbed by the Earth's surface a little more than half goes into latent heat, which is heat absorbed by water because of its transformation from a liquid to a gas. When these molecules condense back into a liquid, usually higher in the atmosphere, they released that energy back into the atmosphere as local warming.

Water cycle The process by which water is transpired and evaporated from the land and water, condensed in the clouds, and precipitated out onto the earth once again to replenish the water in the bodies of water on the earth.

Water gas Blue gas; synthesis gas. A fuel gas used in industrial synthesis of organic chemicals, and in welding, glassmaking, and other high-temperature industrial applications. Water gas made by passing steam over a bed of hot coal or coke. It consists mainly of of carbon monoxide (CO) and hydrogen (H_2), contaminated with small amounts of CO_2, N_2, CH_4, and O_2.

Water hardness Hard water. Compare with water softener. Hard water is water contaminated with compounds of calcium and magnesium. Dissolved iron, manganese, and strontium compounds can also contribute to the "total hardness" of the water, which is usually expressed as ppm $CaCO_3$. Water with a hardness over 80 ppm $CaCO_3$ is often treated with water softeners, since hard water produces scale in hot water pipes and boilers and lowers the effectiveness of detergents.

Water of crystallization Water of hydration. Water that is stoichiometrically bound in a crystal; for example, the waters in copper sulfate pentahydrate.

Water pollution Water pollution is the degradation of the quality of water resources as a result of physical, chemical and/or microbiological contamination. In general, polluted surface waters result when water contains either too little dissolved oxygen, or too high a concentration of toxic substances, such that they cannot support living organisms. The most common causes of water pollution are. 1. the discharge of untreated treated domestic sewages, detergents and industrial trade effluents from point sources to rivers and streams,

2.. the presence of nutrients, fertilisers and pesticides in the diffuse run off from agricultural sources to ground waters and surface waters, and

3. the presence of oils, contaminated suspended solids in the storm water runoff from urban areas to surface waters.

Water resources Water resources are the bodies of water which are available for exploitation for beneficial purposes.

Water softener Soft water; water softening. Compare with water hardness. A material that lowers water hardness when dissolved in water. For example, sodium carbonate ("washing soda") softens water by precipitating Ca^{2+} ions as $CaCO_3$. Zeolites soften water by exchanging Ca^{2+} ions with Na^+ ions.

Water treatment Water treatment is the purification of water, by a physical, chemical or biochemical process, to render it suitable for a specific use.

For example, water must undergoes treatment to render it suitable for use as drinking water.

Water vapor Water present in the atmosphere in gaseous form; the source of all forms of condensation and precipitation. Water vapor, clouds, and carbon dioxide are the main atmospheric components in the exchange of terrestrial radiation in the troposphere, serving as a regulator of planetary temperatures via the greenhouse effect. Approximately 50 percent of the atmosphere's moisture lies within about 1.84 km of the earth's surface, and only a minute fraction of the total occurs above the tropopause.

Watt Watt (W) is the SI derived unit of power. One watt is a power of one joule per second (W = J/s). It is named after the Scottish engineer James Watt (1736-1819).

Wave An oscillating motion that moves outward from the source of some disturbance (ripples running away from a pebble tossed in a pond). Waves transmit the energy of the disturbance away from its source.

Wavefunction A mathematical function that gives the amplitude of a wave as a function of position (and sometimes, as a function of time and/or electron spin). Wavefunctions are used in chemistry to represent the behavior of electrons bound in atoms or molecules.

Wavelength The distance between adjacent peaks (or adjacent troughs) on a wave. Varying the wavelength of light changes its color; varying the wavelength of sound changes its pitch.

Wavelength of light The wavelength of light is the distance between successive crests of the transverse electromagnetic wave. The wavelength of each colour of visible light is different.

Wavenumber The number of wave crests per unit distance. Wavenumber is the reciprocal of wavelength. Wavenumbers are used extensively in infrared spectroscopy, and usually have units of cm^{-1}.

Wave-particle duality The observation that electrons; photons; and other very small entities behave like particles in some experiments and like waves in others.. Compare with quantum .

Wax An ester formed from long-chain fatty acids and alcohols that is usually solid at room temperature.

Weak acid Compare with strong acid.

An acid that only partially dissociates into hydrogen ions and anions in solution. Weak acids are weak electrolytes. Recognize weak acids by learning the six common strong acids; any acid that doesn't appear on the list of strong acids is usually a weak acid.

Weak base Compare with strong base. A base that only partially dissociates into ions in solution. Weak bases are weak electrolytes. top"Ammonia is an example of a weak base; the reaction $NH_3(aq) + H_2O(l) = NH^{4}{}_{-}^{+}(aq) + OH^-(aq)$ is reversible.

Weak electrolyte Compare with strong electrolyte. A weak electrolyte is a solute that incompletely dissociates into ions in solution. For example, acetic acid partially dissociates into acetate ions and hydrogen ions, so that an acetic acid solution contains both molecules and ions. A solution of a weak electrolyte can conduct electricity, but usually not as well as a strong electrolyte because there are fewer ions to carry the charge from one electrode to the other.

Weak ligand Weak field ligand. Compare with strong field ligand. A ligand that causes a small crystal field splitting which results in a high-spin complex.

Weak sun paradox This refers to the fact that our sun has probably increased its luminosity during its lifetime (approximately 3.5 billion years). This assumption is based on astrophysical observations of many other stars. The best scientific conclusions are that when our sun was young it had a luminosity (total energy output) of 70-80% of what it is today; however, there has clearly been (based on geological records and long term evidence for the presence of liquid water) approximately the SAME long term planetary temperature on earth (about 15 degrees C, averaged planet-wide) instead of colder earlier surface temperatures followed later by warmer surface temperatures. This paradox suggests that the planet's surface temperature may have been controlled by atmospheric gas concentrations that varied over time while keeping (via the greenhouse effect) the surface temperature relatively constant (see Gaia hypothesis).

Weber Weber (Wb) is the SI derived unit of magnetic flux. The weber is the magnetic flux which, linking a circuit of one turn, produces in it an electromotive force of one volt as it is reduced to zero at a uniform rate in one second (Wb = V'"s). The unit is named after the German scientist W.E. Weber (1804-1891).

Wedge projection A stereochemical projection, roughly in the mean plane of the molecule, in which bonds are represented by open wedges, tapering off from the nearer atom to the farther atom.

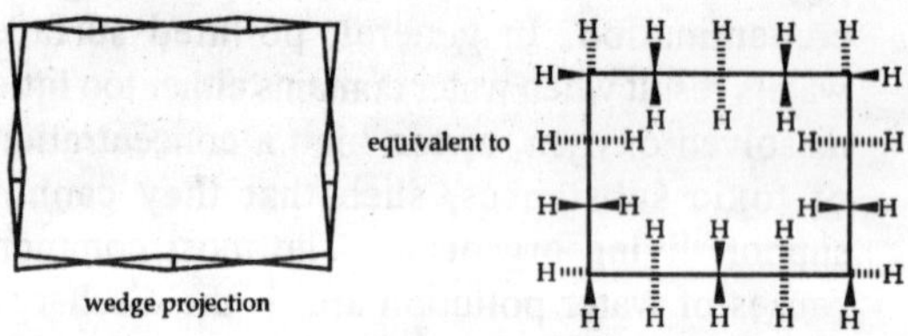

Weight (W) Compare with mass. Weight is the force exerted by an object in a gravitational field. The weight of an object (W) arises from its mass (m):

$$W = mg$$

where g is the acceleration due to gravity (about 9.8 m/s^2 on Earth).

Werner emmanuel bachmann (1901-1951) Born in Detroit, he studied chemistry and chemical engineering at Wayne State University in Detroit and at the University of Michigan in Ann Arbor nearby. He took his doctorate under Gomberg and spent the rest of his career at the University of Michigan, although he often studied overseas. He was interested in physical organic chemistry (rearrangements, free radicals) and organic synthesis. Bachmann was a pioneer in steroid synthesis, and carried out the first total synthesis of a steroidal hormone, equilenin. His name is associated with the Gomberg-Bachmann reaction for the synthesis of diaryl compounds from aryl diazonium chlorides.

Werner reinhold lothar fischer (b.1902) Born in Elberfeld and studied chemistry at the Technische Hochschule in Hanover. He is best known for his purification of rare metals such as hafnium and zirconium. Professor of Inorganic Chemistry at Hanover, he rebuilt the institute after the war.

Wet deposition The deposition of dissolved substances such as airborne particles and gaseous molecules by precipitation. Airborne particles/gaseous molecules in the atmosphere are trapped in water droplets due to falling precipitation (i.e. rain and snow). These substances are then dissolve in the water droplets and are deposited on the surface in which the water droplets come into contact with.

Wetting Wet. Covering with a surface with thin film of liquid. Liquid beads up on a surface if it cannot wet it.

Wienís displacement law The relationship between wavelength and temperature in relation to blackbody radiation. The wavelength decreases as the temperature increases. The common formula to determine this relationship is Plankís Formula.

Wild type The most frequently encountered genotype in natural breeding populations.

Wilhelm bachmann (1885-1933) Born in Kassel, he studied chemistry at the universities of Munich and Göttingen, where he took his doctorate. After a spell as a Privatdozent at Göttingen, under Zsigmondy, he worked for the de Haen company (later Reidel de Haen). He then became Professor of Pure and Applied Colloid Chemistry at the Technische Hochschule in Hanover. At de Haen, Bachmann pioneered the development of collodial graphite. He died in a hunting accident.

William john bailey (1921-1989) Born in East Grand Forks, Minnesota, he studied chemistry at the University of Minnesota. He took his Ph.D. at the University of Illinois at Urbana under Carl "Speed" Marvel. After taking up a position at Wayne State University in Detroit, Bailey moved to the University of Maryland at College Park. He synthesised monomers for polymerisation studies, developed anionic polymerisation and studied the relationship between the structure of polymers and their properties.

William oliver baker (b.1915) Born in Chesterton, Maryland, he studied chemistry at Washington College in Chesterton before taking his Ph.D. under Smyth at Princeton. He joined Bell Laboratories and was soon working on polythene obtained from ICI. He pioneered the application of physical methods to polymers, especially in the context of quality control. He went on to contribute to the development of electronics before taking over the direction of research of Bell Labs.

Wilson baker (b.1900) Born in Runcorn, he studied chemistry at Manchester under Lapworth and Robinson. He taught chemistry at Queen's College, Oxford, before becoming Professor of Organic Chemistry at Bristol. Following in Robinson's footsteps, his early interests were in flower pigments, but he also worked on the chemistry of penicillin during the Second World War. Baker also became interested in non-benzenoid aromatic compounds. He revised Sidgwick's Organic Chemistry of Nitrogen with Thomas Taylor in 1937. His name is associated with the Baker-Venkataraman rearrangement which converts o-acyloxyketones into 1,3 diketones, providing a facile route to chromones and flavones.

Wilson's disease An inherited condition in which copper fails to be excreted in the bile. Copper accumulates progressively in the liver, brain, kidney and red blood cells. As the amount of copper accumulates hemolytic anaemia, chronic liver disease and a neurologic syndrome develop.

Wind rose A diagram that presents information about wind speed and direction for a particular area over a certain amount of time.Ý The diagram is a circle with spokes. The center is 0 and as the wind frequency increases bars extend from the center of the circle to the outside of the circle, in the direction the wind is blowing.Ý Direction can be determined by which way the bar is extending to the outside of the circle because the wind rose is set up like a compass.Ý Information for a windrose is gathered by monitoring stations throughout the area being monitored.Ý By determining wind speed, direction, and frequency with a wind rose, movement of air, and pollutants can be predicted.

Winkler titration The Winkler titration is the back titration used to determine the concentration of Dissolved Oxygen in a sample of water.

Wöhler's synthesis Wöhler's synthesis is a synthesis of urea performed by German chemist Friedrich Wöhler (1800-1882) in 1828. He discovered that urea ($CO(NH_2)_2$) was formed when a solution of ammonium isocyanate (NH_4NCO) was evaporated. At the time it was believed that organic substances such as urea could be made only by living organisms, and its production form an inorganic compound was a notable discovery.

Woodward-hoffmann rules See orbital symmetry.

Work Compare with heat. Work is the energy required to move an object against an opposing force. Work is usually expressed as a force times a displacement. Dropping a stone from a window involves no work, because there is no force opposing the motion (unless you consider air friction...). Pushing against a stone wall involves no work, unless the stone wall actually moves.

X

XANES Acronym for X-ray absorption near edge structure. See extended X-ray absorption fine structure.

Xenobiotic A xenobiotic (Greek, xenos "foreign"; bios "life") is a compound that is foreign to a living organism. Principal xenobiotics include: drugs, carcinogens and various compounds that have been introduced into the environment by artificial means.

Xenon Xenon, Xn, is an inert Nobel Gas element, found in Group 0 Elements of the periodic table.

Atomic Number ; 54

Relative Atomic Mass : 131.30

Discovery

Xenon was discovered by Ramsey and Travers in the final residues from the evaporation of liquidified air.

Occurrence

Xenon is present in the atmosphere in trace amounts (approximately 0.08 ppm).

Extraction

Xenon is extracted from the final residues of liquified air by fractional distillation, from which the major components (i.e. oxygen, nitrogen, carbon oxide, and water vapour) had been removed.

Properties

Xenon is a colourless, tasteless odourless inert gas.

Xenon is very inert chemically due to its electronic configuration.

Uses

Xenon is used as a filler gas in electric lamps used in projectors.

Xenon is used to create a plasma arc, which gives brilliant white light.

Detection and Analysis

Xenon is detected by spectroscopic methods.

X-ray A very high energy form of electromagnetic radiation (though not as high energy as gamma rays). X-rays typically have wavelengths from a few picometers up to 20 nanometers. X-rays easily penetrate soft tissue, which makes them useful in medical imaging and in radiation therapy.

X-ray absorption near edge structure (XANES) See extended X-ray absorption fine structure.

X-ray crystallography Determination of three dimensional arrangement of atoms in a crystal by analysis of x-ray diffraction patterns.

X-ray diffraction pattern Interference patterns created by x-rays as they pass through a solid material. Studying x-ray diffraction patterns gives detailed information on the three-dimensional structure of crystals, surfaces, and atoms.

X-ray photoelectron spectroscopy Technique for determining the elemental composition at a solid surface by measuring the energy of elctrons emitted in response to X-rays of different frequency. Has been applied to solid-phase combinatorial chemistry by incorporating a tracer atom in the linker.

X-ray spectrum x-ray spectra. A set of characteristic x-ray frequencies or wavelengths produced by a substance used as a target in an x-ray tube. Each element has a characteristic x-ray spectrum, and there is a strong correlation between atomic number and the frequencies of certain lines in the x-ray spectrum.

X-ray tube A cathode ray tube that focuses energetic streams of electrons on a metal target, causing the metal to emit x-rays.

Xylene An aromatic hydrocarbon used as a common industrial solvent, also named dimethylbenzene; exists as 3 isomers: m-xylene, o-xylene, and p-xylene. Xylene is usually obtained from petroleum and natural gas distillates. It is used in paint, rubber and leather industries, and also as a cleaning agent or degreaser.

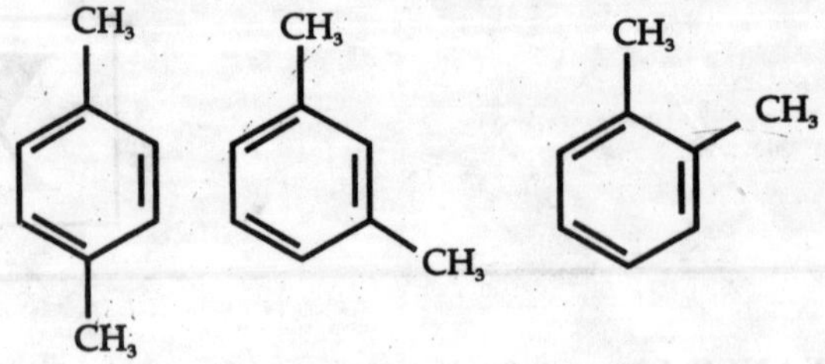

Y

Yield Experimental yield; actual yield. Compare with theoretical yield and percent yield.

The amount of product actually obtained in a chemical reaction.

Ylide A chemical species produced (actually or notationally) by loss of a hydron from an atom directly attached to the central atom of an "onium ion.

Ytterbium Yb. Element 70, atomic weight 173.04, a very rare, malleable metal used in special alloys for X-ray sources.

Yttrium Y. Element 39, atomic weight 88.90585, a dark gray metal that is used in alloys and nuclear technology for its high neutron transparency. Yttrium compounds are used in the manufacture of color tv screens, superconducting ceramics, and fireproof bricks.

Yukawa-tsuno equation A multiparameter extension of the Hammett equation to quantify the role of enhanced resonance effects on the reactivity of meta- and para-substituted benzene derivatives, e.g.

$$\lg k = \lg k_o + + r\ (^+ -))$$

The parameter r gives the enhanced resonance effect on the scale ($^+$-), or ($^-$-), respectively.

Z

Z isomer Isomer in which the two highest-priority substituents are on the same side of a double bond or ring.

Zeeman effect Zeeman effect is the splitting of the lines in a spectrum when the source of the spectrum is exposed to a magnetic field. It was discovered in 1896 by Pieter Zeeman (1865-1943).

Zeolite Zeolite is a natural or synthetic hydrated aluminosilicate with an open three-dimensional crystal structure, in which water molecules are held in cavites in the latice. The water can be driven off by heating and the zeolite can then absorb other molecules of suitable size. Zeolites are used for separating mixtures by selective absorption.

ZeotropeZeotrope is a liquid mixture that shows no maximum or minimum when vapor pressure is plotted against composition at constant temperature.

Zero order reaction Compare with first order reaction and second order reaction. A reaction with a reaction rate that does not change when reactant concentrations change.

Zero point energy A minimum possible energy for an atom or molecule predicted by quantum mechanics. Electrons stay in motion and bonds continue to vibrate even at absolute zero because of zero point energy.

Zero-order reaction Zero-order reaction is a reaction for which the rate of reaction is independent of the concentration of reactants.

Zeta potential Zeta potential (ζ) is the electric potential at the surface of a colloidal particle relative to the potential in the bulk medium at a long distance. Also called electrokinetic potential.

Zeta potential (ζ) is the potential across the interface of all solids and liquids. Specifically, the potential across the diffuse layer of ions surrounding a charged colloidal particle, which is largely responsible for colloidal stability. Also called electrokinetic potential.

Ziegler process Ziegler process is an industrial process for the manufacture of high-density polyethene using catalysts of titanium(IV) chloride ($TiCl_4$) and aluminium alkyls (e.g. triethylaluminium, $Al(C_2H_5)_3$). The process was introduced in 1953 by the German chemist Karl Ziegler (1898-1973). It allowed the manufacture of polythene at lower temperatures (about 60 °C) and pressures (about 1 atm) than used in the original process.

Zig-zag projection A stereochemical projection for an acyclic molecule (or portion of a molecule) where the main chain is represented by a zig-zag line in the plane and the substituents are shown above or below the plane.

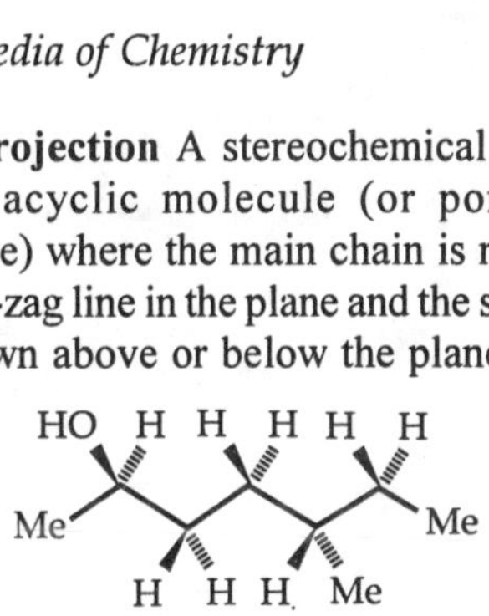

Zinc Zn. Element 30, atomic weight 65.37, a reactive gray metal that dissolves in acids, used to galvanize metals and in many alloys (e. g. brass and bronze).

Zinc finger A domain, found in certain DNA-binding proteins, comprising a helix-loop structure in which a zinc ion is coordinated to 2 - 4 cysteine sulfurs, the remaining ligands being histidines. In many proteins of this type the domain is repeated several times.

Zincography Process of etching unprotected parts of a zinc plate with strong acids to produce a printing surface.

Zirconium Zr. Element 40, atomic weight 91.22, a hard, grayish, highly flammable crystalline metal that dissolves in hot concentrated acids. Used in steel manufacture and in nuclear reactor chambers because of its transparency to neutrons. Its silicate is used to make zircon used as a gemstone.

Zone refining A method for purifying solids based on the fact that solutes tend to concentrate in the liquid when a solution is frozen. A solid bar is drawn slowly over a heat source and melted in a narrow band; impurities are carried along in the melted band until the end of the bar is reached.

Zucker-hammett hypothesis This hypothesis states that, if in an acid catalyzed reaction, lg k_1 (first-order rate constant of the reaction) is linear in H_o (Hammett acidity function), water is not involved in the transition state of the rate-controlling step. However, if lg k_1 is linear in lg(H^+), then water is involved. This has been shown to be incorrect by Hammett himself.

Z-value An index of the ionizing power of a solvent based on the frequency of the longest wavelength electronic absorption maximum of 1-ethyl-4-methoxycarbonylpyridinium iodide in the solvent. The Z-value is defined by

$$Z = 2.859 \times 10^4/$$

where Z is in kcal mol^{-1} and is in nm.

Zwitterion Zwitterion is an ion that has a positive and negative charge on the same group of atoms. Zwitterions can be formed from compounds that contain both acid groups and basic groups in their molecules. It is also called ampholyte ion.

Zwitterionic compound A neutral compound having electrical charges of opposite sign, delocalized or not on adjacent or nonadjacent atoms. Zwitterionic compounds have no uncharged canonical representations. Sometimes referred to as inner salts, ampholytes, dipolar ions (a misnomer).

For example: $H_3N^+CH_2C(=O)O^-$, glycine.

Zymase Enzymes present in yeast that catalyze fermentation of sugar into alcohol and carbon dioxide.

Zymogen Proenzyme. A protein that may be converted into an enzyme.